Noteables™

Interactive Study Notebook with FOLDABLES™

Geometry

Concepts and Applications

Contributing Author
Dinah Zike

FOLDABLES™

Consultant
Douglas Fisher, PhD
Director of Professional Development
San Diego State University
San Diego, CA

New York, New York Columbus, Ohio Chicago, Illinois Peoria, Illinois Woodland Hills, California

The McGraw·Hill Companies

Send all inquiries to:
The McGraw-Hill Companies
8787 Orion Place
Columbus, OH 43240-4027

ISBN: 0-07-872987-4

Geometry: Concepts and Applications (Student Edition)
Noteables™: Interactive Study Notebook with Foldables™

3 4 5 6 7 8 9 10 024 09 08 07 06

Contents

Organizing Your Foldables

Make this Foldable to help you organize and store your chapter Foldables. Begin with one sheet of 11" × 17" paper.

STEP 1 **Fold**
Fold the paper in half lengthwise. Then unfold.

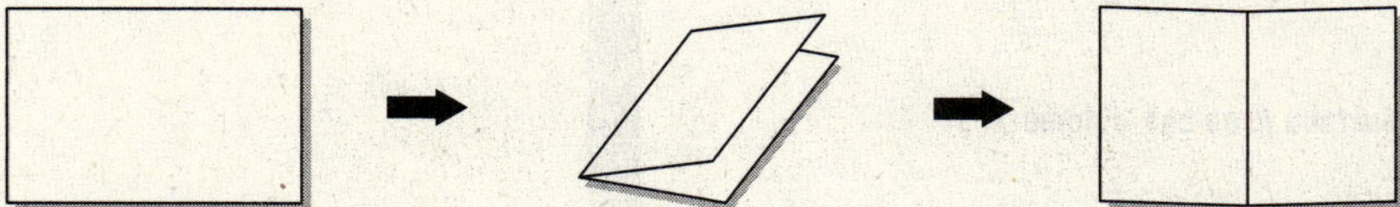

STEP 2 **Fold and Glue**
Fold the paper in half widthwise and glue all of the edges.

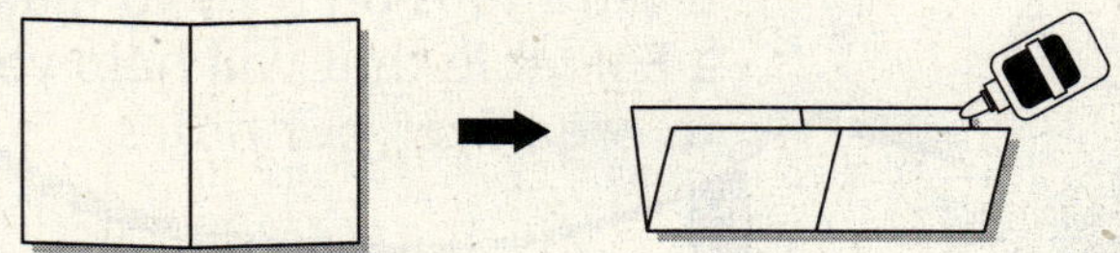

STEP 3 **Glue and Label**
Glue the left, right, and bottom edges of the Foldable to the inside back cover of your Noteables notebook.

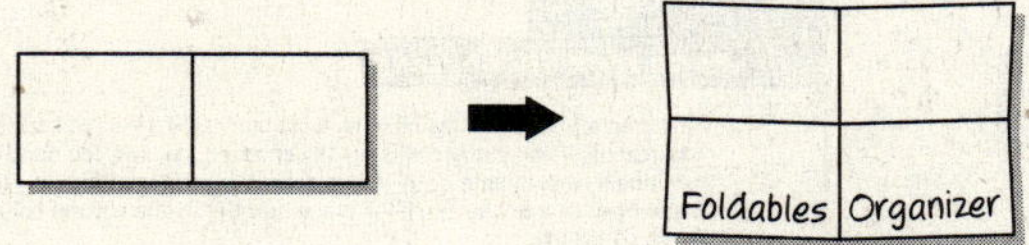

Reading and Taking Notes As you read and study each chapter, record notes in your chapter Foldable. Then store your chapter Foldables inside this Foldable organizer.

Using Your Noteables™ with Foldables™ Interactive Study Notebook

This note-taking guide is designed to help you succeed in *Geometry: Concepts and Applications*. Each chapter includes:

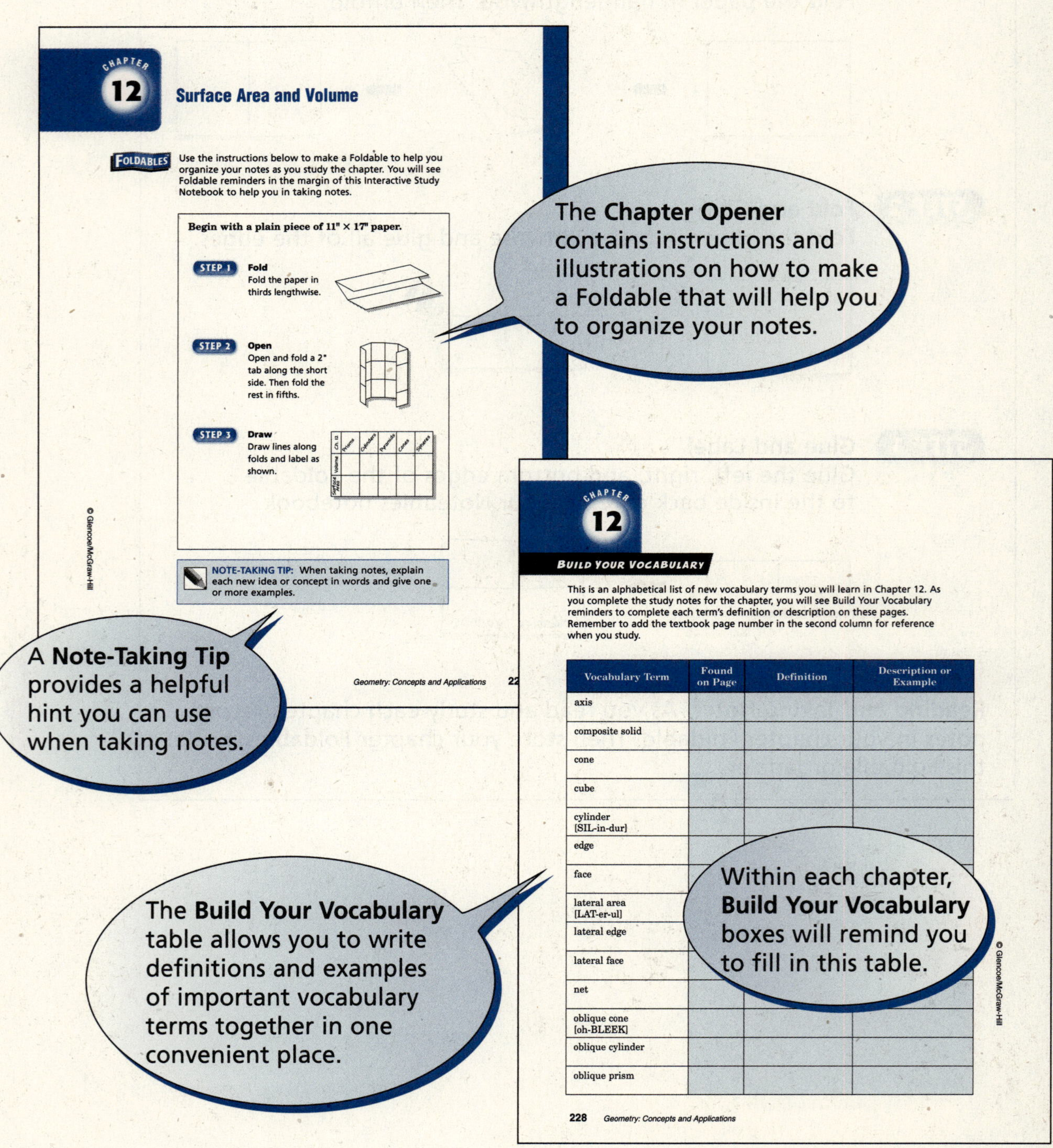
Chapter 12

Surface Area and Volume

FOLDABLES Use the instructions below to make a Foldable to help you organize your notes as you study the chapter. You will see Foldable reminders in the margin of this Interactive Study Notebook to help you in taking notes.

Begin with a plain piece of 11" × 17" paper.

STEP 1 **Fold** Fold the paper in thirds lengthwise.

STEP 2 **Open** Open and fold a 2" tab along the short side. Then fold the rest in fifths.

STEP 3 **Draw** Draw lines along folds and label as shown.

NOTE-TAKING TIP: When taking notes, explain each new idea or concept in words and give one or more examples.

Geometry: Concepts and Applications

© Glencoe/McGraw-Hill

Chapter 12

BUILD YOUR VOCABULARY

This is an alphabetical list of new vocabulary terms you will learn in Chapter 12. As you complete the study notes for the chapter, you will see Build Your Vocabulary reminders to complete each term's definition or description on these pages. Remember to add the textbook page number in the second column for reference when you study.

Vocabulary Term	Found on Page	Definition	Description or Example
axis			
composite solid			
cone			
cube			
cylinder [SIL-in-dur]			
edge			
face			
lateral area [LAT-er-ul]			
lateral edge			
lateral face			
net			
oblique cone [oh-BLEEK]			
oblique cylinder			
oblique prism			

228 Geometry: Concepts and Applications

© Glencoe/McGraw-Hill

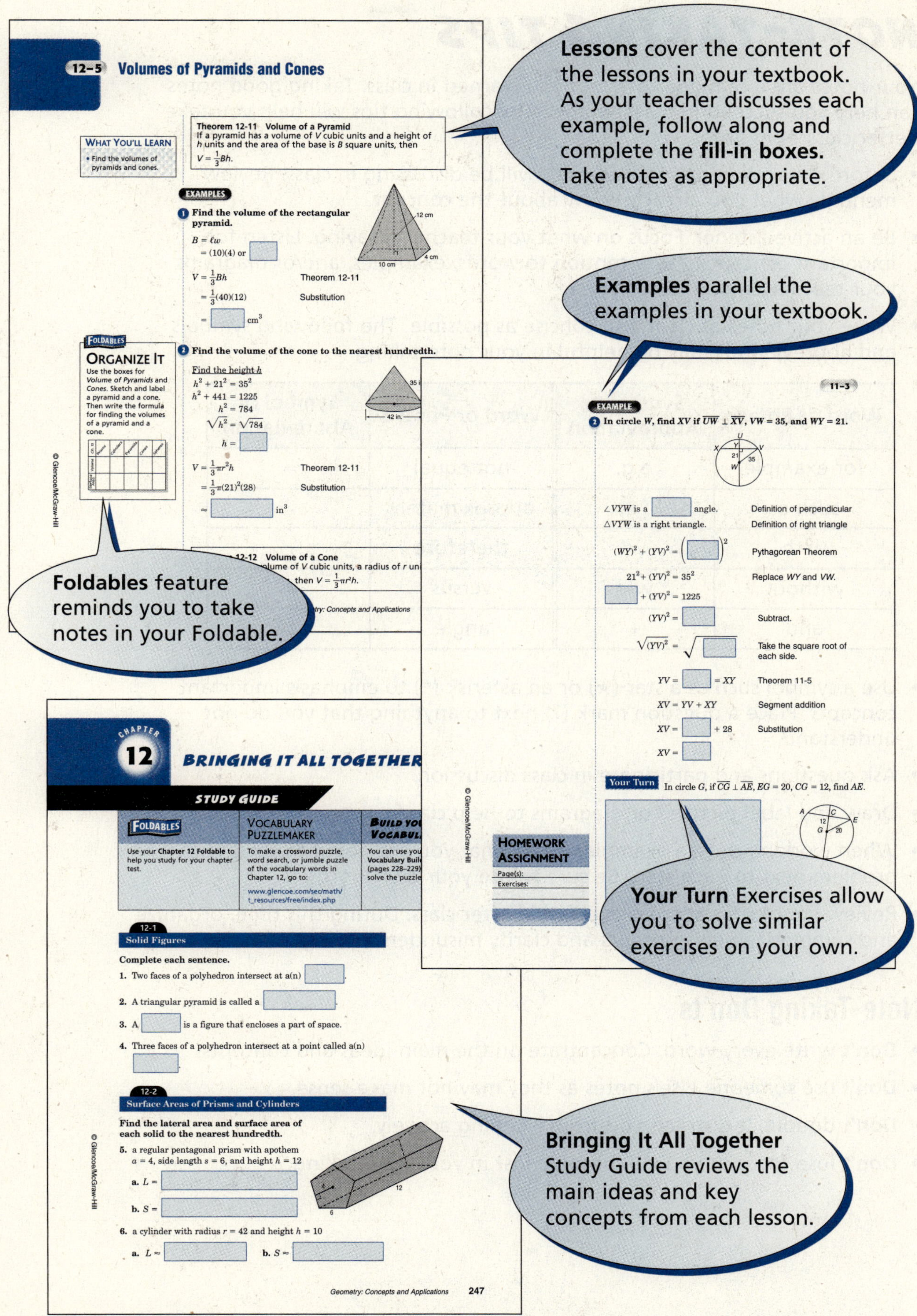
Lessons cover the content of the lessons in your textbook. As your teacher discusses each example, follow along and complete the fill-in boxes. Take notes as appropriate.
Examples parallel the examples in your textbook.
Foldables feature reminds you to take notes in your Foldable.
Your Turn Exercises allow you to solve similar exercises on your own.
Bringing It All Together Study Guide reviews the main ideas and key concepts from each lesson.

NOTE-TAKING TIPS

Your notes are a reminder of what you learned in class. Taking good notes can help you succeed in mathematics. The following tips will help you take better classroom notes.

- Before class, ask what your teacher will be discussing in class. Review mentally what you already know about the concept.
- Be an active listener. Focus on what your teacher is saying. Listen for important concepts. Pay attention to words, examples, and/or diagrams your teacher emphasizes.
- Write your notes as clear and concise as possible. The following symbols and abbreviations may be helpful in your note-taking.

Word or Phrase	Symbol or Abbreviation	Word or Phrase	Symbol or Abbreviation
for example	e.g.	not equal	$\neq$
such as	i.e.	approximately	$\approx$
with	w/	therefore	$\therefore$
without	w/o	versus	vs
and	+	angle	$\angle$

- Use a symbol such as a star (★) or an asterisk (*) to emphasis important concepts. Place a question mark (?) next to anything that you do not understand.
- Ask questions and participate in class discussion.
- Draw and label pictures or diagrams to help clarify a concept.
- When working out an example, write what you are doing to solve the problem next to each step. Be sure to use your own words.
- Review your notes as soon as possible after class. During this time, organize and summarize new concepts and clarify misunderstandings.

Note-Taking Don'ts

- **Don't** write every word. Concentrate on the main ideas and concepts.
- **Don't** use someone else's notes as they may not make sense.
- **Don't** doodle. It distracts you from listening actively.
- **Don't** lose focus or you will become lost in your note-taking.

Reasoning in Geometry

Use the instructions below to make a Foldable to help you organize your notes as you study the chapter. You will see Foldable reminders in the margin of this Interactive Study Notebook to help you in taking notes.

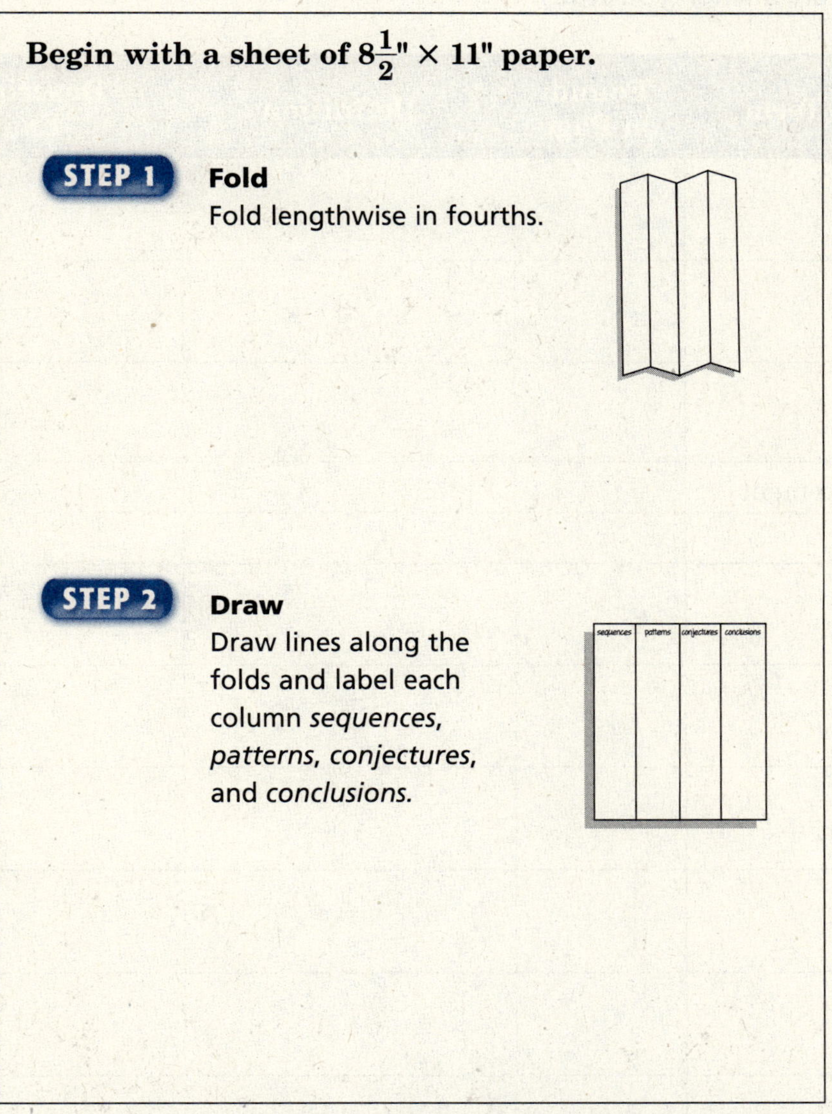

Begin with a sheet of $8\frac{1}{2}$" × 11" paper.

STEP 1 **Fold**
Fold lengthwise in fourths.

STEP 2 **Draw**
Draw lines along the folds and label each column *sequences*, *patterns*, *conjectures*, and *conclusions*.

NOTE-TAKING TIP: When you are taking notes, be sure to be an active listener by focusing on what your teacher is saying.

BUILD YOUR VOCABULARY

This is an alphabetical list of new vocabulary terms you will learn in Chapter 1. As you complete the study notes for the chapter, you will see Build Your Vocabulary reminders to complete each term's definition or description on these pages. Remember to add the textbook page number in the second column for reference when you study.

Vocabulary Term	Found on Page	Definition	Description or Example
collinear [co-LIN-ee-ur]			
compass			
conclusion			
conditional statement			
conjecture [con-JEK-shoor]			
construction			
contrapositive [con-tra-PAS-i-tiv]			
converse			
coplanar [co-PLAY-nur]			
counterexample			
endpoint			
formula			

Vocabulary Term	Found on Page	Definition	Description or Example
hypothesis [hi-PA-the-sis]			
if-then statement			
inductive reasoning [in-DUK-tiv]			
inverse [in-VURS]			
line			
line segment			
midpoint			
noncollinear			
noncoplanar			
plane			
point			
postulate [PAS-chew-let]			
ray			

1–1 Patterns and Inductive Reasoning

BUILD YOUR VOCABULARY (page 3)

When you make conclusions based on a 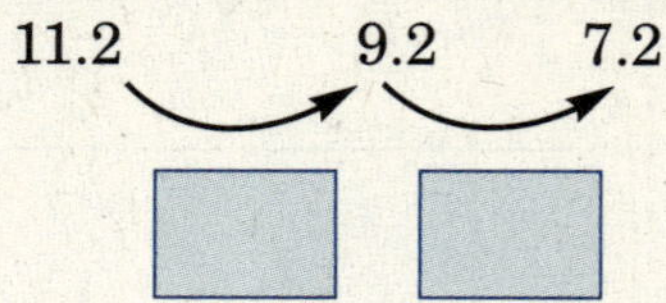 of examples or past events, you are using **inductive reasoning.**

WHAT YOU'LL LEARN

- Identify patterns and use inductive reasoning.

FOLDABLES

ORGANIZE IT

Write a sequence and a geometric pattern in your Foldable. Explain how to find the next 3 terms of each.

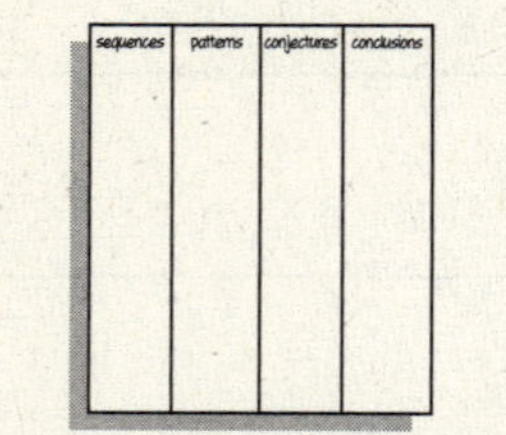

EXAMPLE

1 Find the next three terms of the sequence 11.2, 9.2, 7.2, ...

Study the pattern in the sequence.

11.2 9.2 7.2

Each term is less than the term before it. Assume this pattern continues.

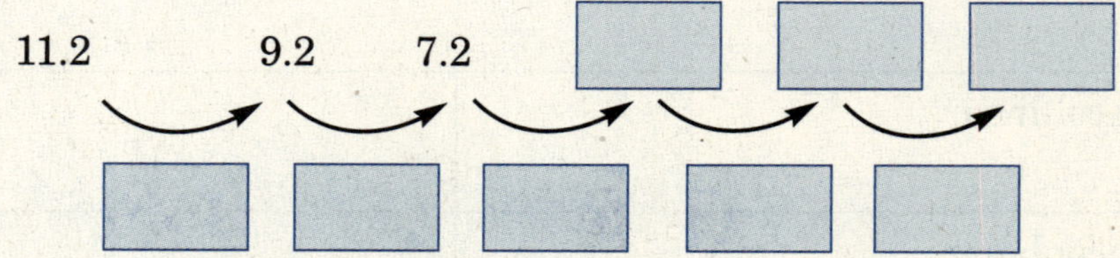

The next three items are .

Your Turn **Find the next three terms of each sequence.**

a. 3.7, 5.7, 7.7, . . .

b. 1, 3, 9, . . .

EXAMPLE

2 **Find the next three terms of the sequence 101, 102, 105, 110, 117, ...**

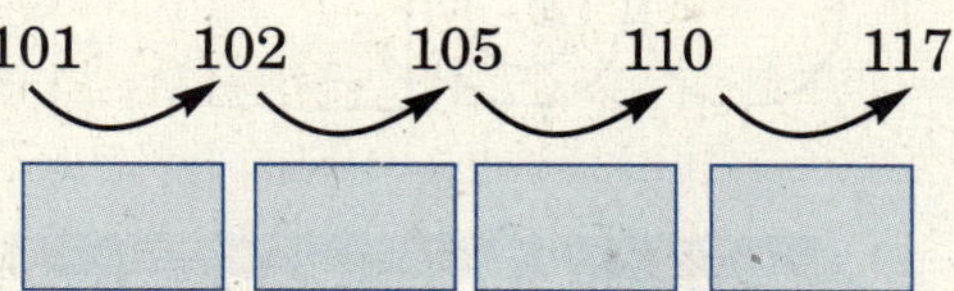

Notice the pattern. To find the next three terms in the sequence, add ______, ______, and ______.

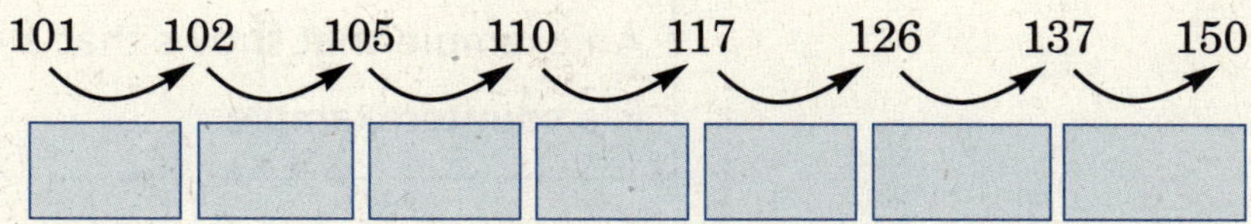

The next three terms are ______.

Your Turn Find the next four terms in the sequence 51, 53, 57, 63, 71, 81, 93, . . .

EXAMPLE

3 **Draw the next figure in the pattern.**
There are two patterns to study.

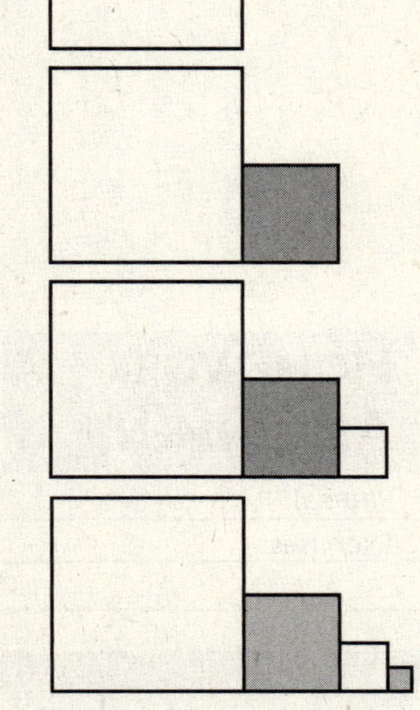

- The first pattern is size of the squares. The next square should be ______ the area of the previous square.
- The second pattern is shaded or unshaded. The next square should be ______.

Your Turn Draw the next figure in the pattern.

○ ◎ □ . . .

BUILD YOUR VOCABULARY (page 2)

A **conjecture** is a ______ based on inductive reasoning.

An example that shows that a conjecture is not ______ is a **counterexample**.

EXAMPLE

4 Minowa studied the data below and made the following conjecture. Find a counterexample for her conjecture.

Multiplying a number by −1 produces a product that is less than −1.

Number ×(−1)	Product
5(−1)	−5
15(−1)	−15
100(−1)	−100
300(−1)	−300

The product of −2 and −1 is 2 but 2 ______ −1. So, the conjecture is ______.

Your Turn Find a counterexample for this statement: *Division of a positive number by another positive number produces a quotient less than the dividend.*

HOMEWORK ASSIGNMENT

Page(s):

Exercises:

1–2 Points, Lines, and Planes

What You'll Learn

- Identify and draw models of points, lines, and planes, and determine their characteristics.

Build Your Vocabulary (pages 2–3)

A **point** is the basic unit of geometry.

A series of points that extends without end in ____ directions is a **line**.

Points that lie on the same ____ are said to be **collinear**.

Points that do not lie on the same line are said to be **noncollinear**.

A **ray** is part of a line that has a definite starting point and extends without end in ____ direction.

A **line segment** has a definite beginning and ____.

Examples

1 Name two points on the line.

Two points are point ____ and point ____.

2 Give three names for the line.

Any two points on the line or the script letter can be used to name it. Three names are ____.

Your Turn **Refer to the figure shown.**

a. Name two points on the line.

____

b. Give three names for the line.

____

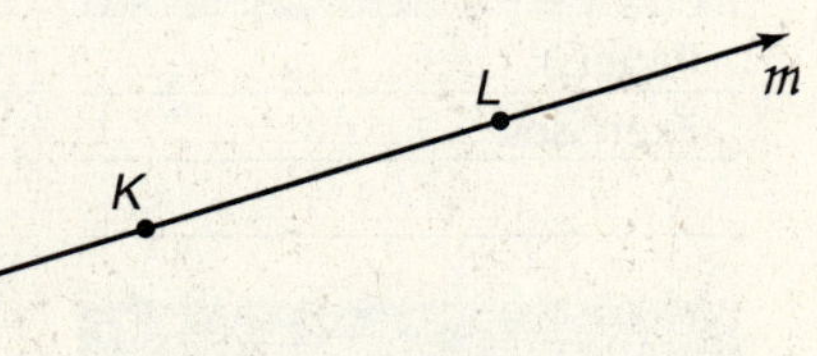

EXAMPLES

3 Name three points that are collinear and three points that are noncollinear.

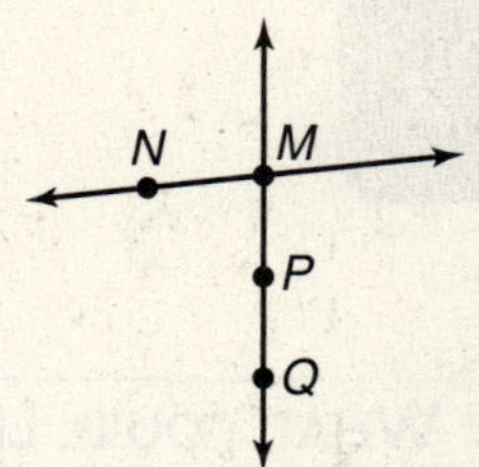

Points *M*, *P*, and *Q*, are ______.
Points *N*, *P*, and *Q* are ______.

REMEMBER IT

The order of the letters that identify a line can be switched but the order of the letters that identify a ray cannot.

4 Name three segments and one ray.

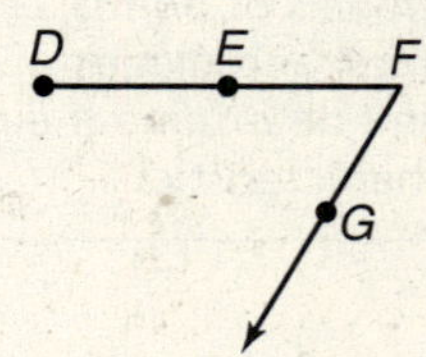

Three of the segments are ______.

One ray is ray ______.

Your Turn **Refer to the figure.**

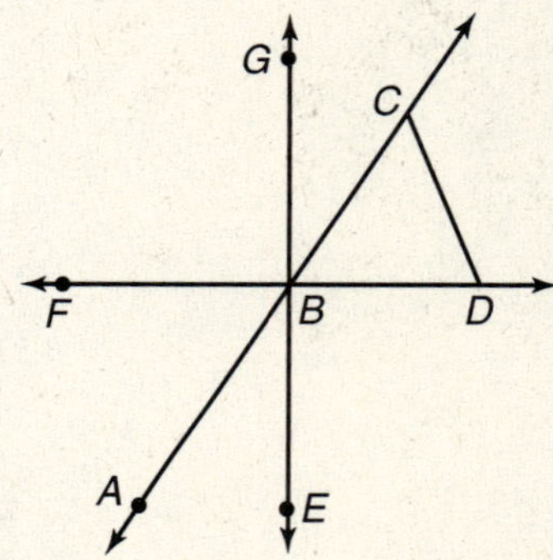

a. Name three collinear points and three noncollinear points.

b. Name three segments and one ray.

BUILD YOUR VOCABULARY (pages 2–3)

A **plane** is a ______ surface that extends without end in all directions.

Points that lie on the same ______ are **coplanar**.

Points that do not lie on the same ______ are **noncoplanar**.

HOMEWORK ASSIGNMENT

Page(s):

Exercises:

1–3 Postulates

What You'll Learn

- Identify and use basic postulates about points, lines, and planes.

Build Your Vocabulary (page 3)

Postulates are ______ in geometry that are accepted as ______.

Postulate 1–1	Two points determine a unique line.
Postulate 1–2	If two distinct lines intersect, then their intersection is a point.
Postulate 1–3	Three noncollinear points determine a unique plane.

Examples

In the figure, points *K*, *L*, and *M* are noncollinear.

•*K* •*M*
•*L*

1 Name all of the different lines that can be drawn through these points.

There is only one line through each pair of points. Therefore, the lines that contain points *K*, *L*, and *M*, taken two at a time, are ______.

2 Name the intersection of $\overleftrightarrow{KL}$ and $\overleftrightarrow{KM}$.

The intersection of $\overleftrightarrow{KL}$ and $\overleftrightarrow{KM}$ is ______.

Your Turn **Refer to the figure.**

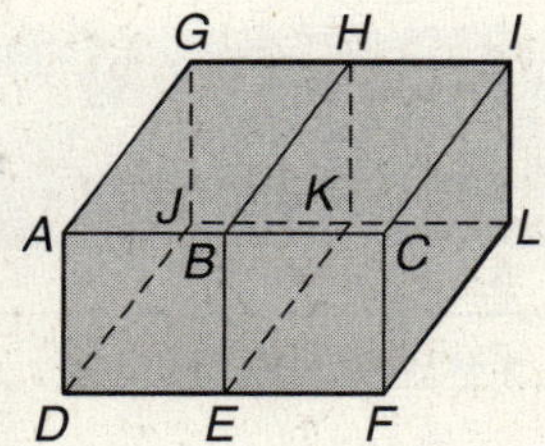

a. Name three different lines.

b. Name the intersection of $\overleftrightarrow{AC}$ and $\overleftrightarrow{BH}$.

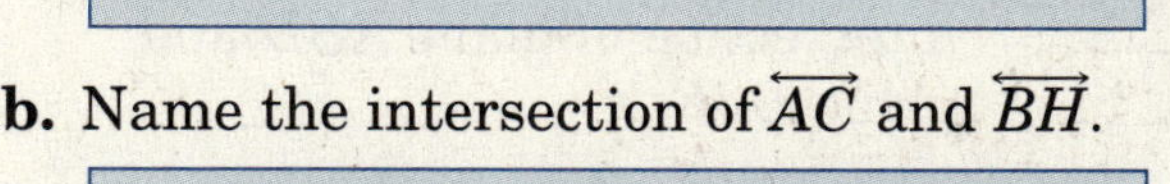

REMEMBER IT

Three noncollinear points determine a unique plane.

EXAMPLE

3 Name all of the planes that are represented in the prism.

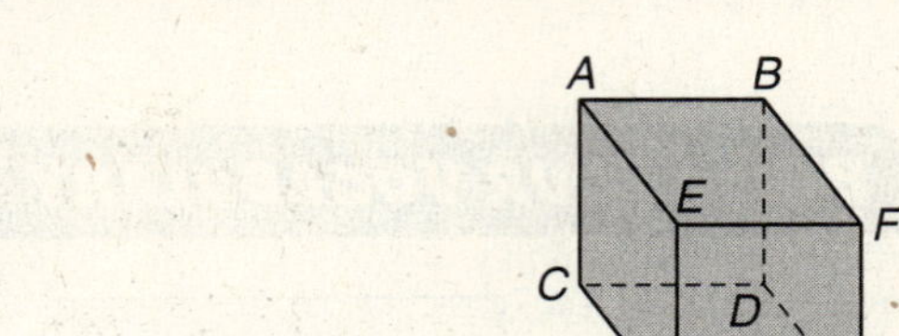

There are eight points, *A*, *B*, *C*, *D*, *E*, *F*, *G*, and *H*.

There is only ______ plane that contains three noncollinear points. The different planes are planes ______.

Your Turn Name four different planes in the figure.

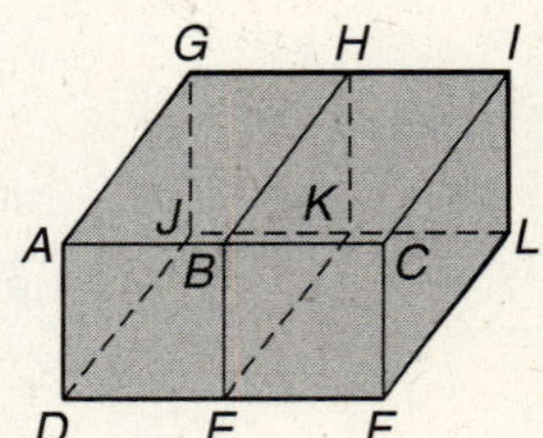

Postulate 1–4 If two distinct planes intersect, then their intersection is a line.

EXAMPLE

4 Name the intersection of plane *ABC* and plane *DEF*.

The intersection is ______.

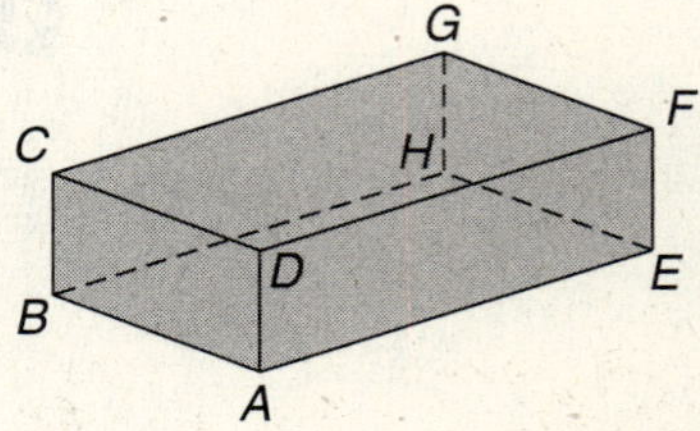

Your Turn Name the intersection of plane *ABD* and plane *DJK*.

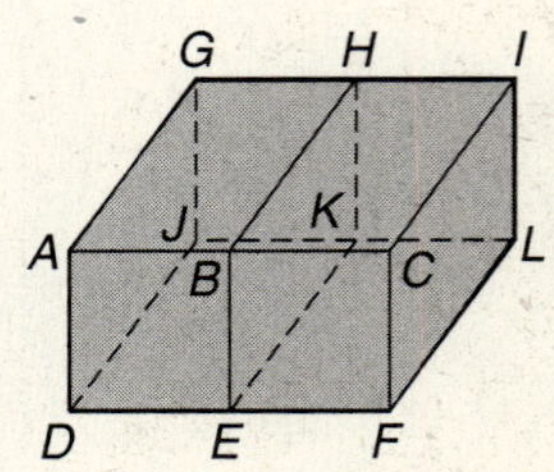

HOMEWORK ASSIGNMENT

Page(s):

Exercises:

1–4 Conditional Statements and Their Converses

WHAT YOU'LL LEARN

- Write statements in if-then form and write the converses of the statements.

BUILD YOUR VOCABULARY (pages 2–3)

If-then statements join two statements based on a condition.

If-then statements are also known as **conditional statements**.

In a conditional statement the part following *if* is the **hypothesis**. The part following then is the **conclusion**.

EXAMPLES

1 **Identify the hypothesis and conclusion in this statement.**

If it is raining, then we will read a book.

Hypothesis: ________

Conclusion: ________

2 **Write two other forms of this statement.**

If two lines are parallel, then they never intersect.

All ________ never intersect.

Lines never ________ if they are ________.

Your Turn

a. Identify the hypothesis and conclusion in this statement. *If you ski, then you like snow.*

____

b. Write two other forms of this statement. *If a figure is a rectangle, then it has four angles.*

____

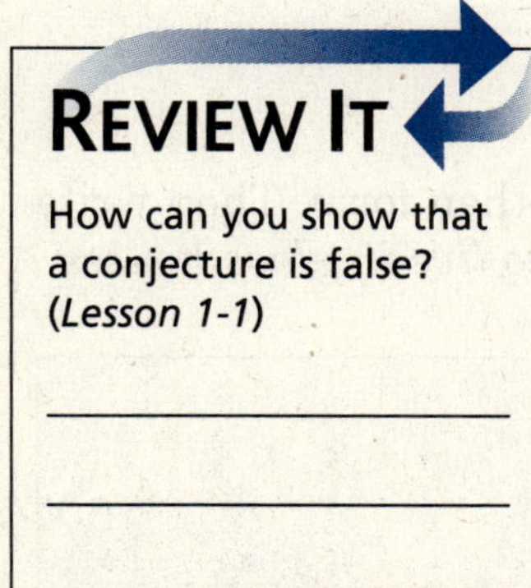

How can you show that a conjecture is false? (*Lesson 1-1*)

____

____

____

BUILD YOUR VOCABULARY (page 2)

The **converse** of a conditional statement is formed by exchanging the ______ and the conclusion.

EXAMPLE

3 Write the converse of this statement.

If today is Saturday, then there is no school.

If there is ______, then ______.

REMEMBER IT

The converse of a true statement is not necessarily true.

Your Turn Write the converse of this statement.

If it is −30° F, then it is cold.

______

EXAMPLE

4 Write the statement in if-then form. Then write the converse of the statement.

Every member of the jazz band must attend the rehearsal on Saturday.

If-then form: If a ______ is a member of the jazz band, then he or she must attend ______.

Converse: If a student ______ on Saturday, then he or she is a ______ member.

Your Turn Write the statement in if-then form. Then write the converse of the statement. *People who live in glass houses should not throw stones.*

______

HOMEWORK ASSIGNMENT

Page(s):

Exercises:

1–5 Tools of the Trade

WHAT YOU'LL LEARN

- Use geometry tools.

BUILD YOUR VOCABULARY (pages 2–3)

A **straightedge** is an object used to draw a ________ line.

A **compass** is commonly used for drawing arcs and ________.

In geometry, figures drawn using only a ________ and a ________ are **constructions**.

The **midpoint** is the ________ in the ________ of a line segment.

EXAMPLE

1 Find two lines or segments in a classroom that appear to be parallel. Use a ruler to determine whether they are parallel.

The opposite sides of a textbook represent two segments that appear to be parallel.

- Choose two points on one side of the textbook.
- Place the 0 mark of the ruler on each point. Make sure the ruler is perpendicular to the side at each chosen point.
- Measure the distance to the second side. If the distances are ________, then the sides are ________.

Your Turn Find another pair of lines or segments in a classroom that appear to be parallel. Use a ruler or a yardstick to determine if they are parallel.

____

EXAMPLES

2 On the figure shown, mark a point C on line ℓ that you judge will create $\overline{BC}$ that is the same length as $\overline{AB}$. Then measure to determine how accurate your guess was.

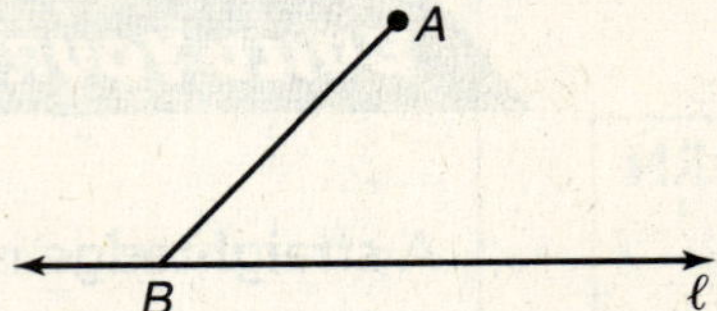

To draw an exact recreation of the length, place the point of a compass on point B. Place the point of the pencil on point

_____. Then draw a small arc on line ℓ without changing the

setting of the compass. This duplicates the measure of _____. 

REMEMBER IT 

An arc is part of a circle.

3 Use a compass and a straightedge to construct a six-pointed star.

Use the compass to draw a circle. Then using the same compass setting, put the compass point on the circle and draw a small arc on the circle.

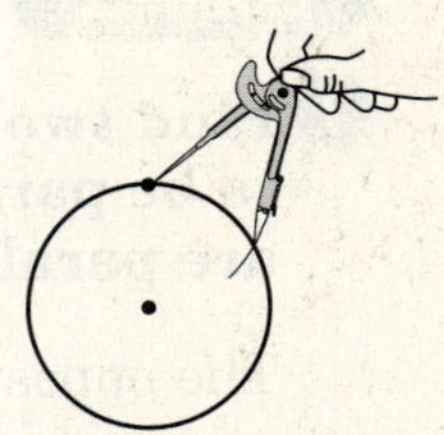

Move the compass point to the arc and, without changing the compass setting, draw another arc along the circle. Continue until there are six arcs.

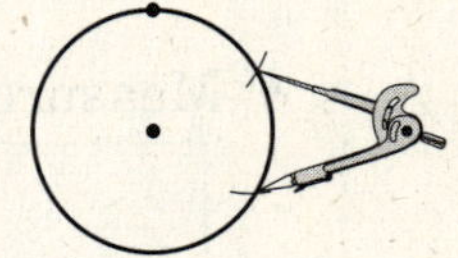

Draw two triangles by connecting alternating marks, resulting in a six-pointed star.

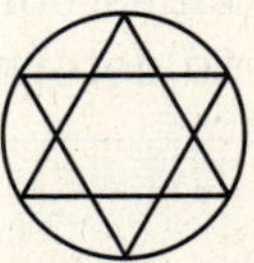

Your Turn

a. On the figure given, mark point Z on line m that you judge will create $\overline{WZ}$ that is the same length as $\overline{XY}$. Then measure to determine the accuracy of your guess.

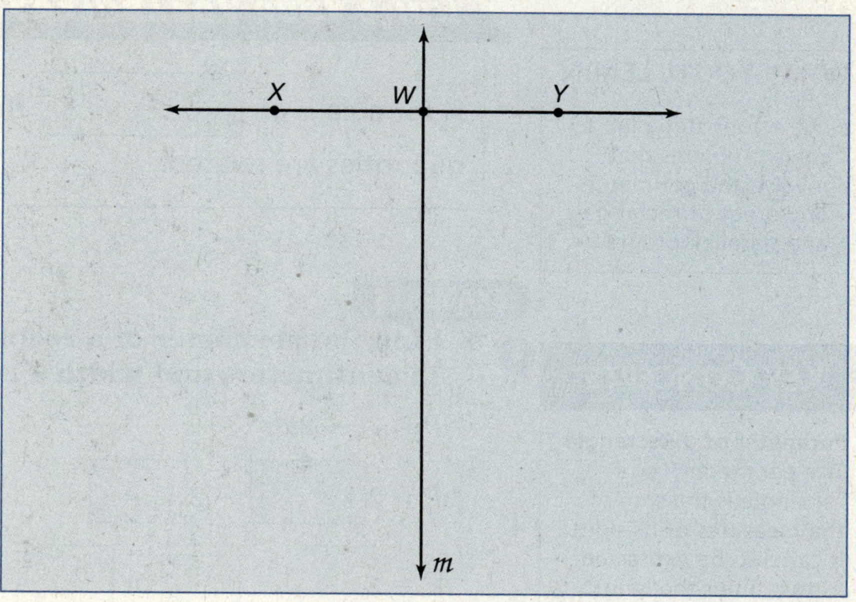

b. Use a compass and a straightedge to construct a triangle with sides of equal length.

Page(s):

Exercises:

1–6 A Plan for Problem Solving

BUILD YOUR VOCABULARY (page 2)

A **formula** is an ______ that shows how certain quantities are related.

WHAT YOU'LL LEARN

- Use a four-step plan to solve problems that involve the perimeters and areas of rectangles and parallelograms.

KEY CONCEPTS

Perimeter of a Rectangle
The perimeter *P* of a rectangle is the sum of the measures of its sides. It can also be expressed as two times the length ℓ plus two times the width *w*.

Area of a Rectangle
The area *A* of a rectangle is the product of the length ℓ and the width *w*.

EXAMPLES

1 a. Find the perimeter of a rectangle with length 12 centimeters and width 3 centimeters.

$P = 2\ell + 2w$

$P = 2$ ______ $+ 2$ ______

$P =$ ______ $+$ ______ or ______ centimeters

b. Find the perimeter of a square with side 10 feet long.

$P = 2\ell + 2w$

$P = 2(10) + 2(10)$

$P =$ ______ $+$ ______ or ______ feet

2 a. Find the area of a rectangle with length 12 kilometers and width 3 kilometers.

$A = \ell w$

$A = ($ ______ $)($ ______ $)$

$A =$ ______ square kilometers

b. Find the area of a square with sides 10 yards long.

$A = \ell w$

$A = ($ ______ $)($ ______ $)$

$A =$ ______ square yards

Write It

What is the difference between perimeter and area?

____

____

____

____

Your Turn

a. Find the perimeter of a rectangle with length 11 meters and width 4 meters.

b. Find the perimeter of a square with sides 7 centimeters long.

c. Find the area of a rectangle with length 14 inches and width 4 inches.

d. Find the area of a square with sides 11 feet long.

EXAMPLE

3 Find the area of a parallelogram with a height of 4 meters and a base of 5.5 meters.

$A = bh$

$A = (\square)(\square)$

$A = \square$ square meters

Key Concept

Area of a Parallelogram The area of a parallelogram is the product of the base *b* and the height *h*.

Your Turn Find the area of a parallelogram with a height of 6.4 inches and a base length of 10 inches.

KEY CONCEPT

Problem-Solving Plan

1. **Explore** the problem.
2. **Plan** the solution.
3. **Solve** the problem.
4. **Examine** the solution.

EXAMPLE

4 **A door is 3-feet wide and 6.5-feet tall. Chad wants to paint the front and back of the door. A one-pint can of paint will cover about 15 ft^2. Will two one-pint cans of paint be enough?**

EXPLORE You know the dimensions of the door and that one-pint can of paint covers about ______.

PLAN Use the formula for the area of a ______ to find the total area of the two sides of the door to be covered with paint.

SOLVE Area of both sides of the door

$A = 2\ell w$

$A = 2(\_\_\_\_)(\_\_\_\_) = \_\_\_\_$

One pint covers 15 ft^2. Two one-pint cans cover 2(15) or ______ ft^2. So, two one-pint cans will ______ be enough.

EXAMINE Since the area of one side of the door is (3)(6.5) or 19.5 ft^2 the answer is reasonable.

Chad will need ______ one-pint cans of paint.

REMEMBER IT

Abbreviations for units of area have exponent 2.

Square foot = ft^2

Square meter = m^2

Your Turn A building contractor needs to build a rectangular deck with an area of 484 ft^2. The side lengths must be whole numbers. The perimeter must be less than 260 ft. What are the possible dimensions for the deck?

HOMEWORK ASSIGNMENT

Page(s):

Exercises:

CHAPTER 1 BRINGING IT ALL TOGETHER

STUDY GUIDE

FOLDABLES™	VOCABULARY PUZZLEMAKER	BUILD YOUR VOCABULARY
Use your **Chapter 1 Foldable** to help you study for your chapter test.	To make a crossword puzzle, word search, or jumble puzzle of the vocabulary words in Chapter 1, go to: www.glencoe.com/sec/math/t_resources/free/index.php	You can use your completed **Vocabulary Builder** (pages 2–3) to help you solve the puzzle.

1-1 Patterns and Inductive Reasoning

Find the next three terms in the sequence.

1. 1, 1, 2, 3, 5, . . .

2. −1, 2, −4, 8, −16, . . .

3. Draw the next figure in the pattern.

1-2 Points, Lines, and Planes

Use the figure to match the example to the correct term.

4. collinear points 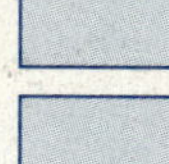

5. segment 

6. plane 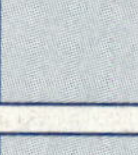 

7. ray 

a. G, F, C

b. $\overline{PB}$ 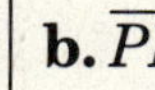

c. $\overrightarrow{AD}$

d. $\overleftrightarrow{PE}$ 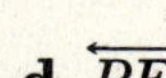

e. GBE

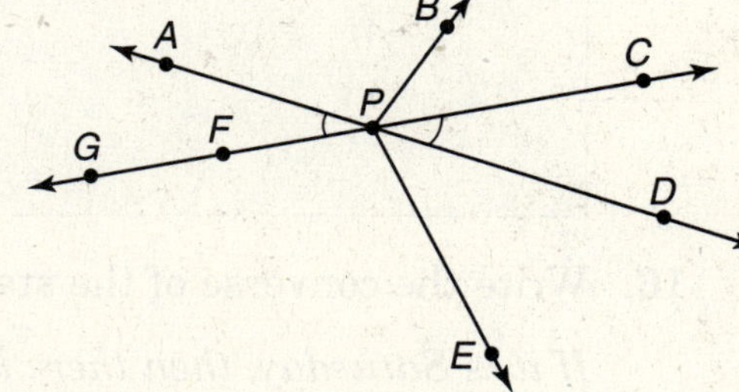

1-3 Postulates

Complete the sentence.

8. A(n)  is a statement in geometry that is accepted as true without proof.

Identify three planes in the figure shown.

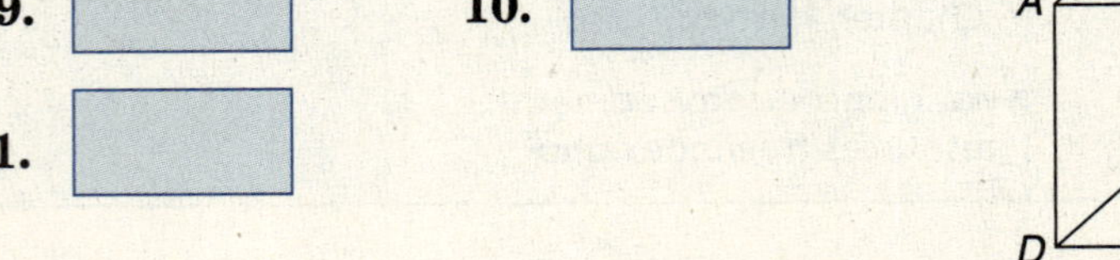

9.

10.

11.

12. Refer to the above figure. Where do planes ACF and DEF intersect? 

a. point F	**b.** $\overleftrightarrow{DF}$	**c.** plane DEF	**d.** point D

1-4 Conditional Statements and Their Converses

Underline the correct term that completes each sentence.

13. The "if" part of the *if-then statement* is the hypothesis/conclusion.

14. The "then" part of the *if-then statement* is the hypothesis/conclusion.

15. Rewrite the statement in *if-then* form.

Students who complete all assignments score higher on tests.

16. Write the converse of the statement.

If it is Saturday, then there is no school.

1-5 Tools of the Trade

Match the geometry tool to its function.

17. compass 

18. straightedge 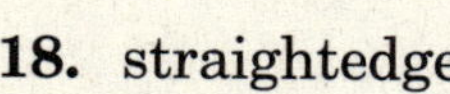 

19. protractor

20. patty paper

a. to plot points
b. to draw arcs and circles
c. to measure angles
d. to draw lines in constructions
e. to find the midpoint in constructions

21. Indicate whether the statement is *true* or *false*.
A conjecture is a special drawing that is created using only a straightedge and compass.

1-6 A Plan for Problem Solving

Complete each sentence.

22. The ______ is the distance around the edges of a figure.

23. The formula for the area of a rectangle is ______.

24. ______ is the formula to find the area of a parallelogram.

25. Find the area of a rectangle with length 8 feet and width 9 feet.

26. A framer must frame a piece of art. The frame is $1\frac{1}{2}$ inches wide, and its outer edge measures 24 inches by 36 inches. What is the area of the piece of art displayed in the center of the frame?

Checklist

ARE YOU READY FOR THE CHAPTER TEST?

Visit **geomconcepts.com** to access your textbook, more examples, self-check quizzes, and practice tests to help you study the concepts in Chapter 1.

Check the one that applies. Suggestions to help you study are given with each item.

☐ **I completed the review of all or most lessons without using my notes or asking for help.**

- You are probably ready for the Chapter Test.
- You may want to take the Chapter 1 Practice Test on page 45 of your textbook as a final check.

☐ **I used my Foldable or Study Notebook to complete the review of all or most lessons.**

- You should complete the Chapter 1 Study Guide and Review on pages 42–44 of your textbook.
- If you are unsure of any concepts or skills, refer back to the specific lesson(s).
- You may also want to take the Chapter 1 Practice Test on page 45 of your textbook.

☐ **I asked for help from someone else to complete the review of all or most lessons.**

- You should review the examples and concepts in your Study Notebook and Chapter 1 Foldable.
- Then complete the Chapter 1 Study Guide and Review on pages 42–44 of your textbook.
- If you are unsure of any concepts or skills, refer back to the specific lesson(s).
- You may also want to take the Chapter 1 Practice Test on page 45 of your textbook.

Student Signature

Parent/Guardian Signature

Teacher Signature

Segment Measure and Coordinate Graphing

Use the instructions below to make a Foldable to help you organize your notes as you study the chapter. You will see Foldable reminders in the margin of this Interactive Study Notebook to help you in taking notes.

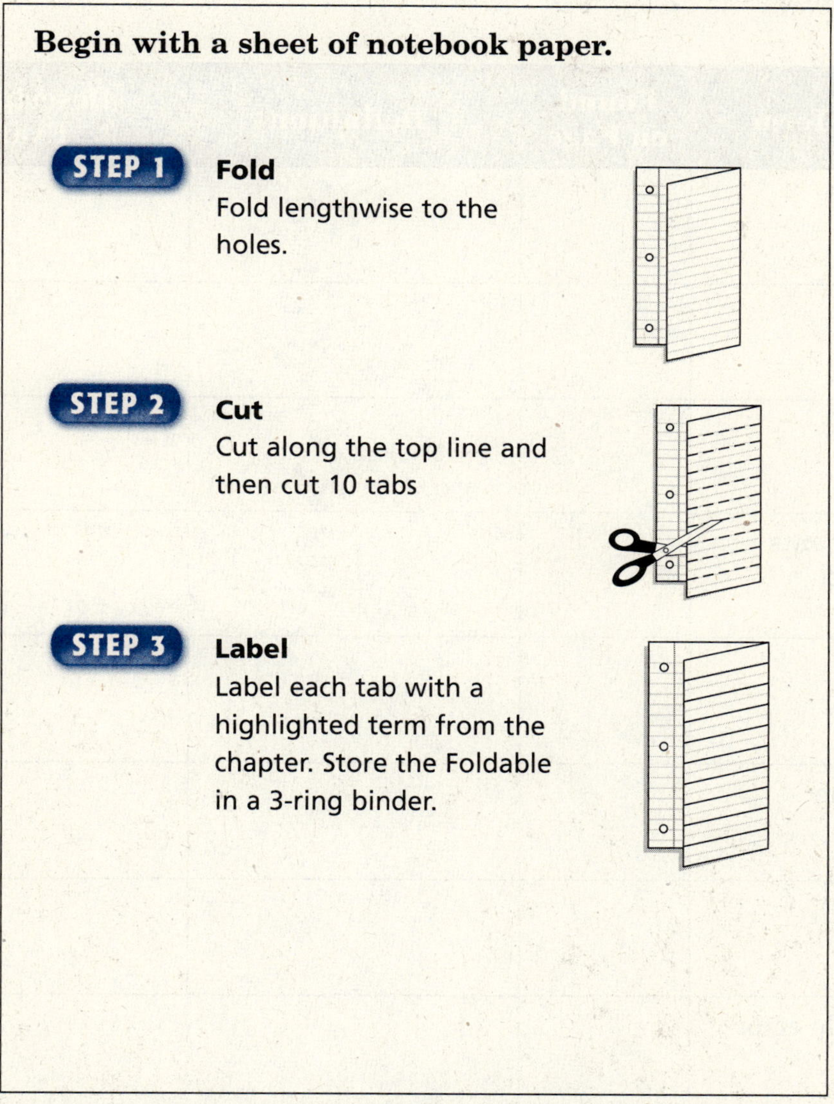

NOTE-TAKING TIP: When taking notes, it is helpful to record the main ideas as you listen to your teacher, or read through a lesson.

BUILD YOUR VOCABULARY

This is an alphabetical list of new vocabulary terms you will learn in Chapter 2. As you complete the study notes for the chapter, you will see Build Your Vocabulary reminders to complete each term's definition or description on these pages. Remember to add the textbook page number in the second column for reference when you study.

Vocabulary Term	Found on Page	Definition	Description or Example
absolute value			
betweenness			
bisect			
congruent segments [con-GROO-unt]			
coordinate [co-OR-duh-net]			
coordinate plane			
coordinates			
greatest possible error			
measure			
measurements			

Vocabulary Term	Found on Page	Definition	Description or Example
midpoint			
ordered pair			
origin [OR-a-jin]			
percent of error			
precision [pree-SI-zhun]			
quadrants [KWAH-druntz]			
theorem [THEE-uh-rem]			
unit of measure			
vector			
x-axis			
x-coordinate			
y-axis			
y-coordinate			

2–1 Real Numbers and Number Lines

What You'll Learn

- Find the distance between two points on a number line.

Postulate 2-1 Number Line Postulate
Each real number corresponds to exactly one point on a number line. Each point on a number line corresponds to exactly one real number.

EXAMPLES

For each situation, write a real number with ten digits to the right of the decimal point.

1 a rational number between 6 and 8 with a 2-digit repeating pattern

Sample answer: 7.3232323232 . . .

2 an irrational number greater than 5

Sample answer: 5.4344334443 . . .

FOLDABLES™

ORGANIZE IT

On the first tab of your Foldable, write *Rational Numbers* and on the second tab, write *Irrational Numbers*. Under each tab, describe the sets of rational and irrational numbers and give several examples of each.

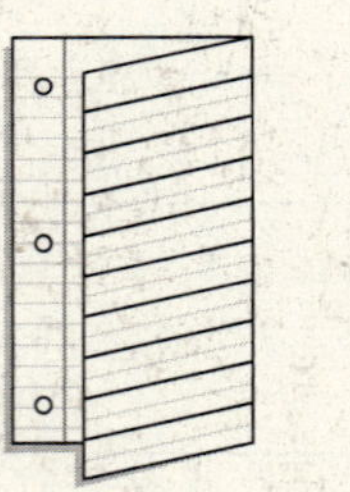

Your Turn **For each situation, write a real number with ten digits to the right of the decimal point.**

a. a rational number between −4 and −1 with a 3-digit repeating pattern

b. an irrational number less than −7

Postulate 2-2 Distance Postulate
For any two points on a line and a given unit of measure, there is a unique positive real number called the **measure** of the distance between the points.

Postulate 2-3 Ruler Postulate
The points on a line can be paired with the real numbers so that the measure of the distance between corresponding points is the positive difference of the numbers.

BUILD YOUR VOCABULARY (pages 24–25)

The number that corresponds to a point on a number line is called the **coordinate** of the point.

A point with coordinate ☐ is known as the **origin**.

The **absolute value** of a number is the number of units a number is from ☐ on the number line.

EXAMPLES

REMEMBER IT

XY represents the measure of the distance between points *X* and *Y*.

3 Use the number line below to find *CE*.

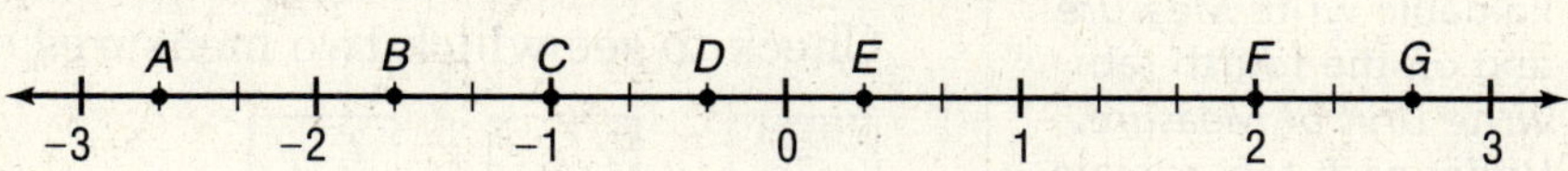

The coordinate of *C* is ☐, and the coordinate of *E* is ☐.

$CE = \left|-1 - \frac{1}{3}\right| = \left|-1\frac{1}{3}\right|$

$= \left|-1\frac{1}{3}\right|$ or ☐

4 Erin traveled on I-85 from Durham, North Carolina, to Charlotte. The Durham entrance to I-85 that she used is at the 173-mile marker, and the Charlotte exit she used is at the 39-mile marker. How far did Erin travel on I-85?

$|173 - 39| = |134| =$ ☐

She traveled ☐ miles on I-85.

Your Turn

a. Refer to Example 3. Find *AE*.

☐

b. Rahmi's drive starts at the 263-mile marker of I-35 and finishes at the 287-mile marker. How far did Rahmi drive on I-35?

☐

HOMEWORK ASSIGNMENT

Page(s):

Exercises:

2–2 Segments and Properties of Real Numbers

What You'll Learn

- Apply properties of real numbers to the measure of segments.

Build Your Vocabulary (page 24)

Point R is **between** points P and Q *if and only if* R, P and Q are ______ and $PR + RQ = PQ$.

Foldables™

Organize It

On the third tab of your Foldable write *Measure* and on the fourth tab write *Unit of Measure*. Under each tab, explain the differences between the terms and give examples of each.

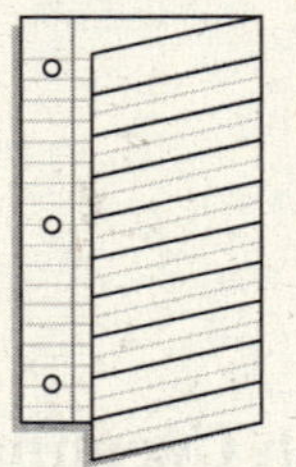

EXAMPLE

1 Points K, L, and J are collinear. If $KL = 31$, $JL = 16$, and $JK = 47$, determine which point is between the other two.

Check to see which two measures add to equal the third.

____ + ____ = ____

$KL + JL = JK$

Therefore, ____ is between ____ and ____.

Your Turn Points A, B, and C are collinear. If $AB = 54$, $BC = 33$, and $AC = 21$, determine which point is between the other two.

______

EXAMPLE

2 If $FG = 12$ and $FJ = 47$, find GJ.

$FG + GJ = FJ$	Definition of betweenness
____ $+ GJ =$ ____	Substitution Property
$12 + GJ$ ____ $= 47$ ____	Subtraction Property
$GJ =$ ____	Substitution Property

Key Concepts

Properties of Equality for Real Numbers

- **Reflexive Property** For any number a, $a = a$.
- **Symmetric Property** For any numbers a and b, if $a = b$, then $b = a$.
- **Transitive Property** For any numbers a, b, and c, if $a = b$ and $b = c$, then $a = c$.
- **Addition and Subtraction Properties** For any numbers a, b, and c, if $a = b$, then $a + c = b + c$, and $a - c = b - c$.
- **Multiplication and Division Properties** For any numbers a, b, and c, if $a = b$, then $a \cdot c = b \cdot c$, and if $c \neq 0$, then $\frac{a}{c} = \frac{b}{c}$.
- **Substitution Property** For any numbers a and b, if $a = b$, then a may be replaced by b in any equation.

Your Turn If $BE = 17$ and $AE = 25$, find AB.

A B C D E

Build Your Vocabulary (pages 24–25)

Measurements are composed of ______ parts; a number called the **measure** and the **unit of measure**.

The **precision** of a measurement depends on the ______ unit used to make the measurement.

The **greatest possible error** is ______ the smaller unit used to make the measurement.

The **percent of error** is the ______ of the greatest possible error with the measurement itself, multiplied by ______.

EXAMPLE

3 Use a ruler to draw a segment 8 centimeters long. Then find the length of the segment in inches.

Use a metric ruler to draw the segment. Mark a point and call it X. Then put the 0 point at point X and draw a line segment extending to the 8 centimeter mark. Mark the endpoint Y.

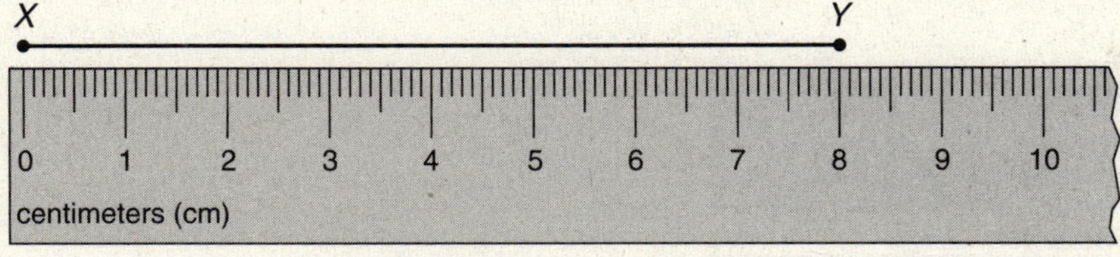

The length of $\overline{XY}$ is ______ centimeters.

Use a customary ruler to measure $\overline{XY}$ in inches. Put the 0 point at X and measure the distance to Y.

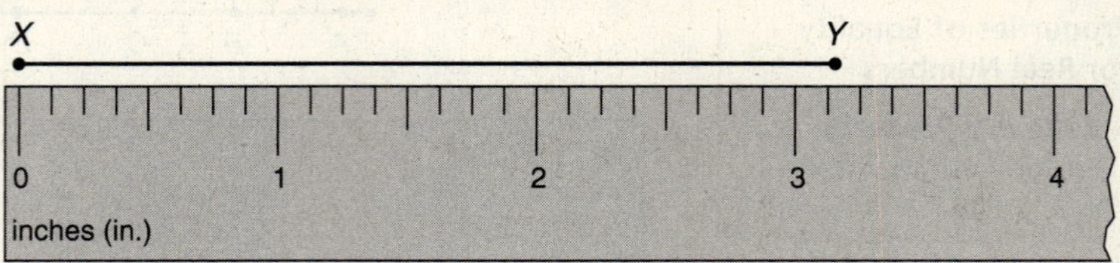

The length of $\overline{XY}$ is about ______ inches.

Your Turn Use a ruler to draw a segment 3 centimeters long. Then find the length of the segment in inches.

HOMEWORK ASSIGNMENT

Page(s):

Exercises:

2–3 Congruent Segments

What You'll Learn

- Identify congruent segments.
- Find midpoints of segments.

EXAMPLE

1 **Use the figure below to determine whether each statement is *true* or *false*. Explain your reasoning.**

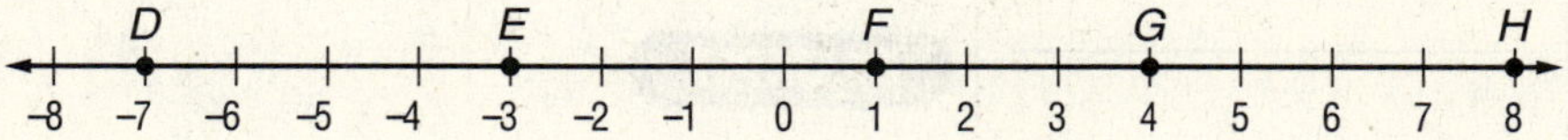

a. $\overline{DE} \cong \overline{GH}$

Because $DE = 4$ and $GH =$ ____, ____ $=$ ____.

So, ____ $\cong$ ____ is a true statement.

b. $\overline{EF} \cong \overline{FG}$

Because $EF =$ ____ and $FG =$ ____, $EF \neq FG$. So, $\overline{EF}$ is not congruent to $\overline{FG}$, and the statement is false.

Key Concept

Definition of Congruent Segments Two segments are congruent if and only if they have the same length.

FOLDABLES On the fifth tab of you Foldable, write *Congruent Segments*. Under the tab, write the definition and draw examples of congruent segments.

Your Turn **Use the figure below to determine whether each statement is *true* or *false*. Explain your reasoning.**

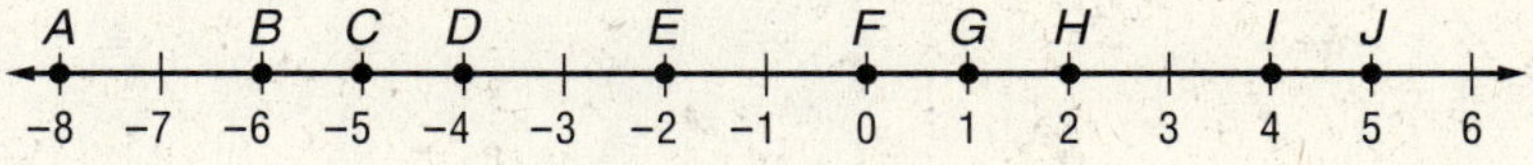

a. $\overline{AE} \cong \overline{BG}$

____

b. $\overline{DG} \cong \overline{FJ}$

____

BUILD YOUR VOCABULARY (pages 24–25)

Theorems are statements that can be justified by using ____ reasoning.

REVIEW IT

Write the converse of Theorem 2-2. Is the converse true? *(Lesson 1-4)*

____

____

____

____

Theorem 2-1
Congruence of segments is reflexive.

Theorem 2-2
Congruence of segments is symmetric.

Theorem 2-3
Congruence of segments is transitive.

EXAMPLE

2 **Determine whether the statement is *true* or *false*. Explain your reasoning.**

$\overline{CD}$ is congruent to $\overline{CD}$.

Congruence of segments is ______, so ______ $\cong$ ______.

Therefore, the statement is ______.

Your Turn Determine whether the statement is *true* or *false*. Explain your reasoning.

$\overline{MN}$ is congruent to $\overline{NM}$.

____

BUILD YOUR VOCABULARY (pages 24–25)

A unique point on every segment that separates the segment into ______ segments of ______ length is known as the **midpoint**.

To **bisect** something means to separate it into two ______ parts.

KEY CONCEPT

Definition of Midpoint A point M is the midpoint of a segment $\overline{ST}$ if and only if M is betweeen S and T and $SM = MT$.

FOLDABLES™ On the sixth tab of your Foldable, write *Midpoint*. Under the tab, write the definition and draw an example showing the midpoint of a line segment.

EXAMPLE

3 In the figure, K is the midpoint of $\overline{JL}$. Find the value of d.

J —— $d + 5$ —— K —— $2d$ —— L

You need to find the value of d. Since K is the midpoint of $\overline{JL}$, $JK = KL$. Write and solve an equation involving d, and solve for d.

$JK = KL$ Definition of ______

______ = ______ Substitution

______ = ______ Subtraction Property of Equality

______ $= d$

Your Turn In the figure, D is the midpoint of $\overline{XY}$. Find the value of a.

X —— $7a - 8$ —— D —— $5a$ —— Y

HOMEWORK ASSIGNMENT

Page(s):

Exercises:

2–4 The Coordinate Plane

What You'll Learn

- Name and graph ordered pairs on a coordinate plane.

Build Your Vocabulary (pages 24–25)

The [ ] of the grid used to locate points is known as the **coordinate plane**.

The [ ] number line is the ***y*-axis**.

The ***x*-axis** is the [ ] number line.

The two axes separate the coordinate plane into [ ] regions known as **quadrants**.

The two axes [ ] at a [ ] called the **origin**.

An **ordered pair** of real numbers, called the **coordinates** of a point, locates a [ ] on the coordinate plane.

The [ ] number of the ordered pair is called the ***x*-coordinate**.

The ***y*-coordinate** is the [ ] number of the ordered pair.

Foldables

Organize It

On the seventh tab of your Foldable, write *Coordinate Plane*. Under the tab, draw a coordinate plane, labeling the four quadrants and the two axes.

On the eighth tab of your Foldable, write *Ordered Pair and Coordinates*. Under the tab, give an example of an ordered pair. Label the *x*-coordinate and the *y*-coordinate for the pair.

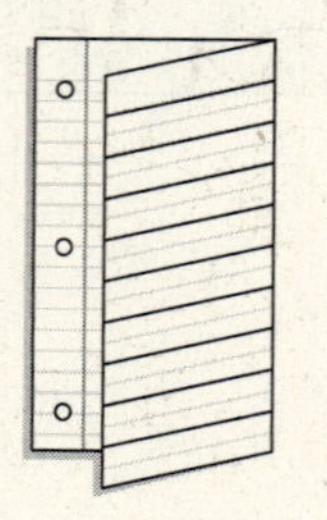

EXAMPLES

1 Graph point K at $(-4, 1)$.

Start at the origin. Move [ ] units to the left. Then, move [ ] unit up. Label this point K.

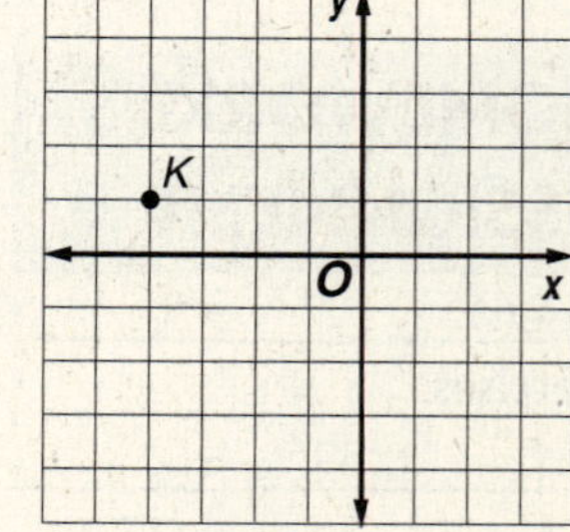

Postulate 2-4
Completeness Property for Points in the Plane
Each point in a coordinate plane corresponds to exactly one ordered pair of real numbers. Each ordered pair of real numbers corresponds to exactly one point in a coordinate plane.

Your Turn Graph point L at $(1, -4)$.

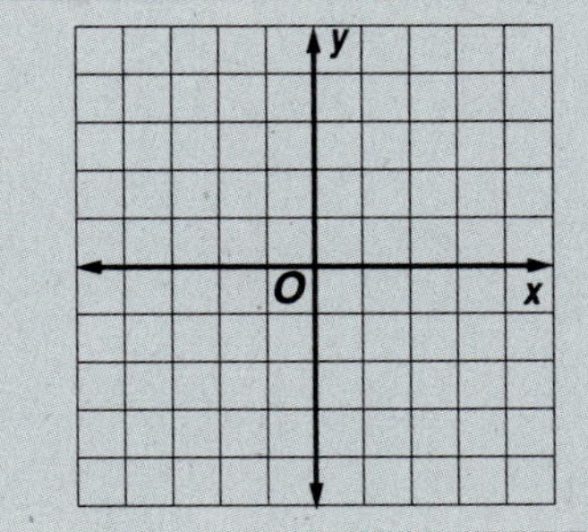

2 **Name the coordinates of points L and M.**

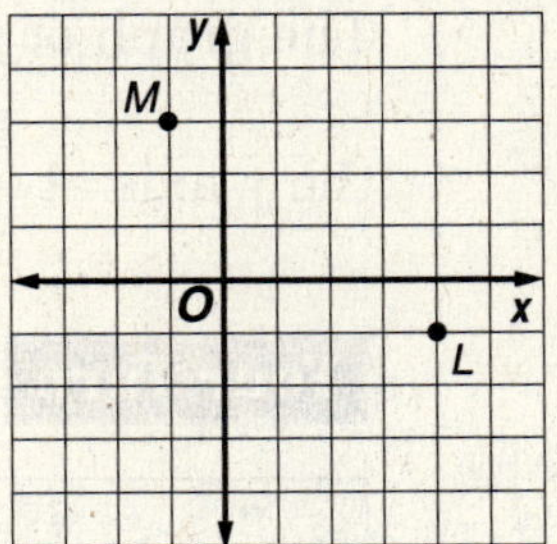

Point L is ☐ units to the right of the origin and ☐ unit below the origin. Its coordinates are ☐.

Point M is ☐ to the left of the origin and ☐ units above the origin. Its coordinates are .

Your Turn Name the coordinates of points P and Q.

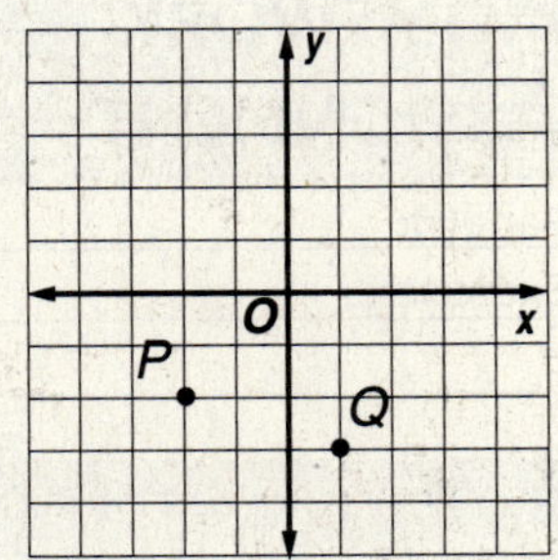

WRITE IT

Explain how to graph any ordered pair (x, y). Describe which direction you move when x or y are either positive or negative.

Theorem 2-4
If a and b are real numbers, a vertical line contains all points (x, y) such that $x = a$, and a horizontal line contains all points (x, y) such that $y = b$.

EXAMPLE

3 **Graph $y = -2$.**

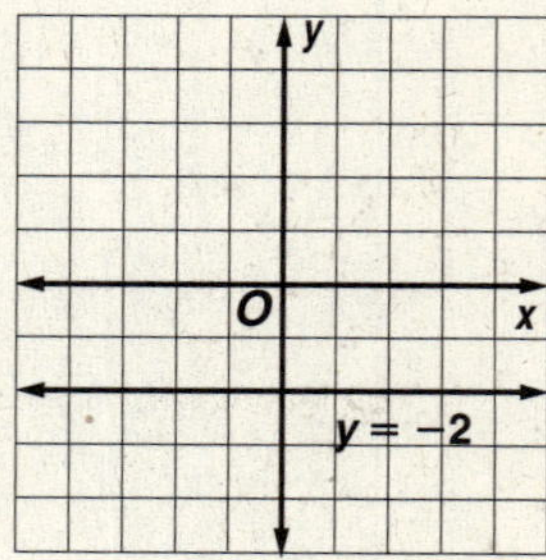

The graph of $y = -2$ is a ______ line that intersects the y-axis at ______.

Your Turn Graph $x = -1$.

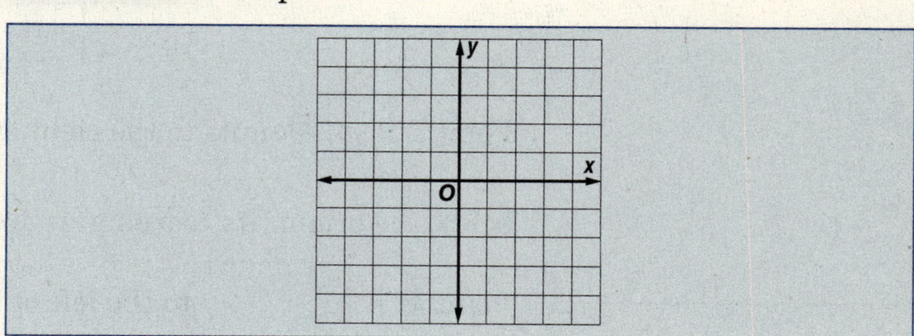

HOMEWORK ASSIGNMENT

Page(s):

Exercises:

2–5 Midpoints

WHAT YOU'LL LEARN

- Find the coordinates of the midpoint of a segment.

Theorem 2-5 Midpoint Formula for a Number Line
On a number line, the coordinate of the midpoint of a segment whose endpoints have coordinates *a* and *b* is $\frac{a+b}{2}$.

Theorem 2-6 Midpoint Formula for a Coordinate Plane
On a coordinate plane, the coordinates of the midpoint of a segment whose endpoints have coordinates (x_1, y_1) and (x_2, y_2) are $\left(\frac{x_1 + x_2}{2}, \frac{y_1 + y_2}{2}\right)$.

EXAMPLE

1 Find the coordinate of the midpoint of $\overline{AB}$.

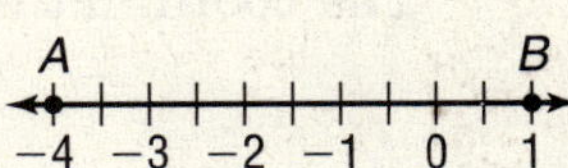

Use the Midpoint Formula to find the coordinate of the midpoint of $\overline{AB}$.

$$\frac{a+b}{2} = \frac{-4+1}{2}$$

$= \square$ or $\square$

The coordinate of the midpoint is $\square$.

FOLDABLES

ORGANIZE IT

On the ninth tab of your Foldable, write *Midpoint for a Number Line*. Under the tab, explain how to find the midpoint of a segment on a number line.

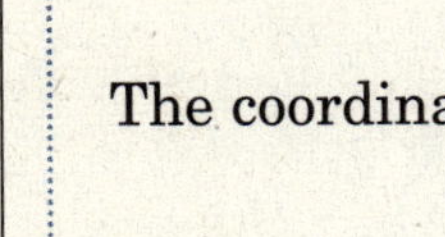

Your Turn Find the coordinate of the midpoint of $\overline{OK}$.

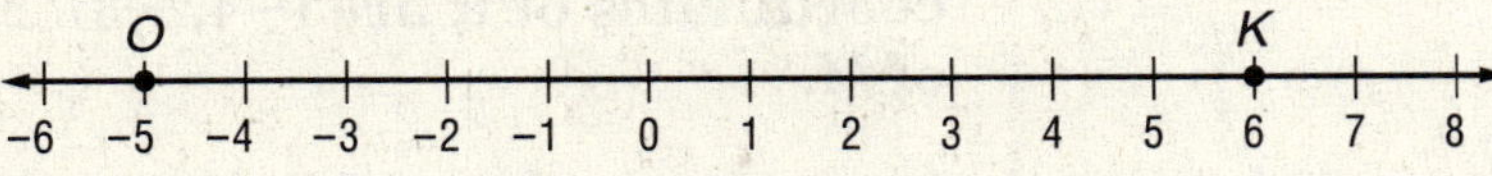

EXAMPLES

2 Find the coordinates of D, the midpoint of $\overline{CE}$, given endpoints $C(2, 1)$ and $E(16, 8)$.

Use the Midpoint Formula to find the coordinates of D.

$$\left(\frac{x_1 + x_2}{2}, \frac{y_1 + y_2}{2}\right) = \left(\frac{\square + \square}{2}, \frac{\square + \square}{2}\right)$$

$$= \left(\frac{\square}{2}, \frac{\square}{2}\right)$$

$$= \square$$

The coordinates of D are $\square$.

Your Turn Find the coordinates of Y, the midpoint of $\overline{XZ}$, given endpoints $X(-3, 5)$ and $Z(6, -1)$.

3 Suppose $L(2, -5)$ is the midpoint of $\overline{KM}$ and the coordinates of K are $(-4, -3)$. Find the coordinates of M.

Let (x_1, y_1) or $(-4, -3)$ be the coordinates of K and let (x_2, y_2) be the coordinates of M. So, $x_1 = \square$ and $y_1 = \square$. Use the Midpoint Formula.

$$\left(\frac{x_1 + x_2}{2}, \frac{y_1 + y_2}{2}\right) = \square$$

REMEMBER IT 

The x-coordinate of the midpoint is the *average* of the x-coordinates of the endpoints. The y-coordinate of the midpoint is the *average* of the y-coordinates of the endpoints.

x-coordinate of M

$\frac{-4 + x_2}{2} = \square$ Replace x_1 with -4.

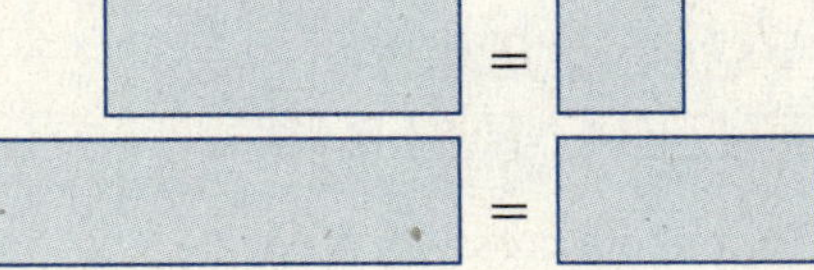

$\frac{-4 + x_2}{2}\square = 2\square$ Multiply each side by $\square$.

$\square = \square$

$\square = \square$ Add to isolate the variable.

$x_2 = \square$

y-coordinate of M

$\frac{-3 + y_2}{2} = \square$ Replace x_1 with -3.

$\frac{-3 + y_2}{2}\square = -5\square$ Multiply each side by $\square$.

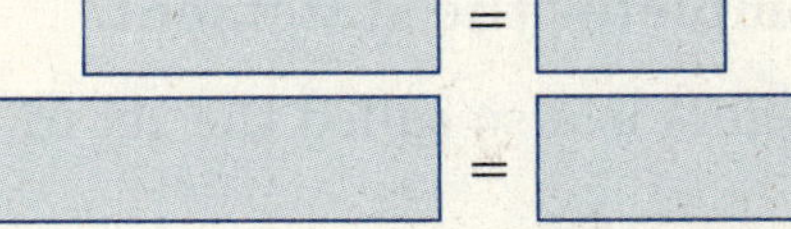

$\square = \square$

$\square = \square$ Add to isolate the variable.

$y_2 = \square$

The coordinates of M are $\square$.

Your Turn Suppose $S\left(-\frac{1}{2}, -\frac{3}{2}\right)$ is the midpoint of $\overline{RT}$ and the coordinates of R are $(-2, -5)$. Find the coordinates of T.

HOMEWORK ASSIGNMENT

Page(s):

Exercises:

CHAPTER 2

BRINGING IT ALL TOGETHER

STUDY GUIDE

FOLDABLES	VOCABULARY PUZZLEMAKER	BUILD YOUR VOCABULARY
Use your **Chapter 2 Foldable** to help you study for your chapter test.	To make a crossword puzzle, word search, or jumble puzzle of the vocabulary words in Chapter 2, go to: www.glencoe.com/sec/math/t_resources/free/index.php	You can use your completed **Vocabulary Builder** (pages 24–25) to help you solve the puzzle.

2-1 Real Numbers and Number Lines

Choose the term that best completes the statement.

1. The set of non-negative integers is also called the set of [natural/whole] numbers.

2. The quotient of two integers, where the denominator is not zero, is a(n) [rational/irrational] number.

3. Decimals that do not repeat or terminate are called [rational/irrational] numbers.

Find.

4. $|-4 - 1|$

5. $|-(-12)|$

6. $|11 + 2|$

2-2 Segments and Properties of Real Numbers

7. Points X, Y, and Z are collinear. If $XY = 10$ and $XZ = 3$, find YZ.

8. Points A, B, and C are collinear. If $AB = 6$, $BC = 8$, and $AC = 14$, which point is between the other two points?

9. Points M, N, and P are collinear. If P lies between M and N, $MP = 2$, and $PN = 1$, find MN.

2-3 Congruent Segments

Complete the statement.

10. Two segments are ______ if they are equal in length.

11. When a segment is separated into two congruent segments, the segment is ______.

12. Statements known as ______ can be justified using logical reasoning.

13. Points *A, B,* and *C* are collinear. If $\overline{AC} \cong \overline{CB}$, then the point *C* is the ______ of $\overline{AB}$.

2-4 The Coordinate Plane

Refer to the graph and name the ordered pair for each point.

14. point *P* 

15. point *L* 

16. point *A* 

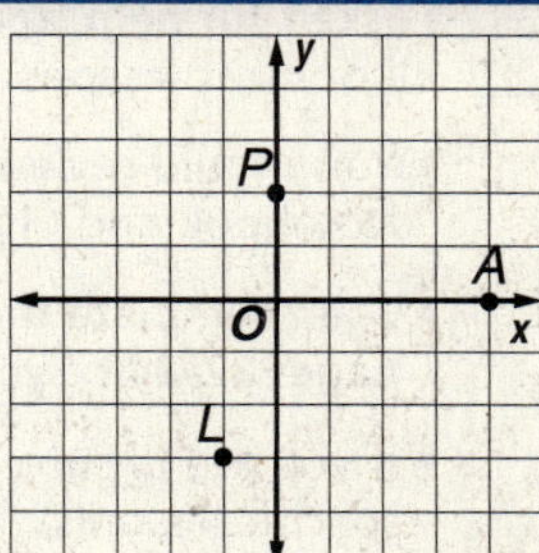

Graph and label the following points on the above coordinate plane.

17. point *N* $(-4, 2)$ **18.** point *E* $(3, 1)$ **19.** point *S* $(1, -5)$

2-5 Midpoints

20. On a number line, if $X = -2$ and $Y = 4$, what is the coordinate of midpoint *Z*? ______

21. Find the coordinates of the midpoint of a segment whose endpoints are $(-5, -1)$ and $(-3, 3)$. ______

22. Find the coordinates of the other endpoint of a segment whose midpoint has coordinates $(4, 5)$ and second endpoint at $(2, -1)$.

______

Checklist

ARE YOU READY FOR THE CHAPTER TEST?

Math Online

Visit **geomconcepts.com** to access your textbook, more examples, self-check quizzes, and practice tests to help you study the concepts in Chapter 2.

Check the one that applies. Suggestions to help you study are given with each item.

☐ **I completed the review of all or most lessons without using my notes or asking for help.**

- You are probably ready for the Chapter Test.
- You may want to take the Chapter 2 Practice Test on page 85 of your textbook as a final check.

☐ **I used my Foldable or Study Notebook to complete the review of all or most lessons.**

- You should complete the Chapter 2 Study Guide and Review on pages 82–84 of your textbook.
- If you are unsure of any concepts or skills, refer back to the specific lesson(s).
- You may also want to take the Chapter 2 Practice Test on page 85 of your textbook.

☐ **I asked for help from someone else to complete the review of all or most lessons.**

- You should review the examples and concepts in your Study Notebook and Chapter 2 Foldable.
- Then complete the Chapter 2 Study Guide and Review on pages 82–84 of your textbook.
- If you are unsure of any concepts or skills, refer back to the specific lesson(s).
- You may also want to take the Chapter 2 Practice Test on page 85 of your textbook.

Student Signature

Parent/Guardian Signature

Teacher Signature

Angles

Use the instructions below to make a Foldable to help you organize your notes as you study the chapter. You will see Foldable reminders in the margin of this Interactive Study Notebook to help you in taking notes.

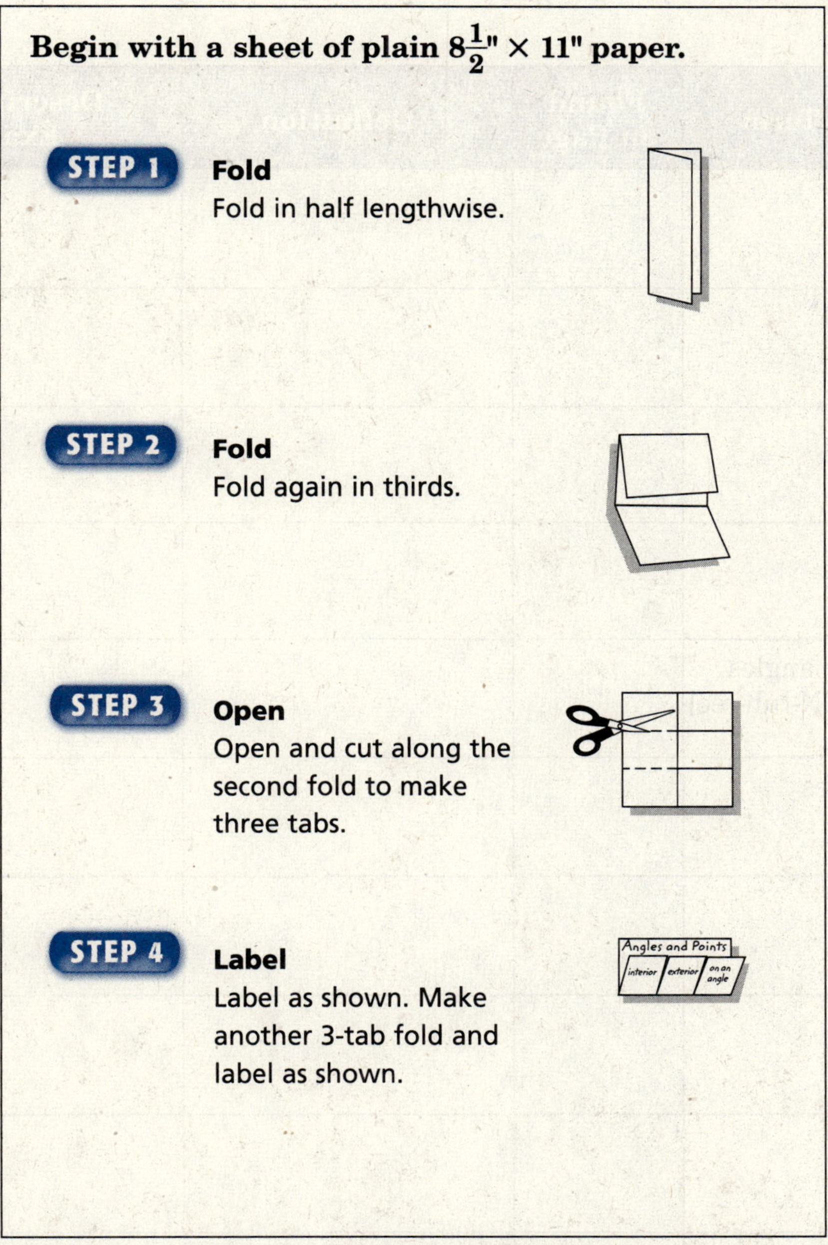

Begin with a sheet of plain $8\frac{1}{2}$" × 11" paper.

STEP 1 **Fold**
Fold in half lengthwise.

STEP 2 **Fold**
Fold again in thirds.

STEP 3 **Open**
Open and cut along the second fold to make three tabs.

STEP 4 **Label**
Label as shown. Make another 3-tab fold and label as shown.

NOTE-TAKING TIP: When you take notes, listen or read for main ideas. Then record those ideas in simplified form for future reference.

BUILD YOUR VOCABULARY

This is an alphabetical list of new vocabulary terms you will learn in Chapter 3. As you complete the study notes for the chapter, you will see Build Your Vocabulary reminders to complete each term's definition or description on these pages. Remember to add the textbook page number in the second column for reference when you study.

Vocabulary Term	Found on Page	Definition	Description or Example
acute angle [a-KYOOT]			
adjacent angles [uh-JAY-sent]			
angle			
angle bisector			
complementary angles [kahm-pluh-MEN-tuh-ree]			
congruent angles			
degrees			
exterior			
interior			
linear pair [LIN-ee-ur]			

Vocabulary Term	Found on Page	Definition	Description or Example
obtuse angle [ob-TOOS]			
opposite rays			
perpendicular [PER-pun-DI-kyoo-lur]			
protractor			
quadrilateral [KWAD-ruh-LAT-er-ul]			
right angle			
sides			
straight angle			
supplementary angles [SUP-luh-MEN-tuh-ree]			
triangle			
vertex [VER-teks]			
vertical angles			

3–1 Angles

What You'll Learn

- Name and identify parts of an angle.

Build Your Vocabulary (pages 44–45)

Opposite rays are two rays that are part of the same ________ and have only their ________ in common.

The figure formed by ________ is referred to as a **straight angle**.

Any case where two rays have a common ________ is known as **angle**.

The common ________ is called the **vertex**.

The two rays that make up the ________ are called the **sides** of the angle.

Remember It

Read the symbol ∠ as *angle*.

Example

1 Name the angle in four ways. Then identify its vertex and its sides.

The angle can be named in four ways: ________.

Its vertex is ________. Its sides are ________ and ________.

Your Turn Name the angle in four ways. Then identify its vertex and its sides.

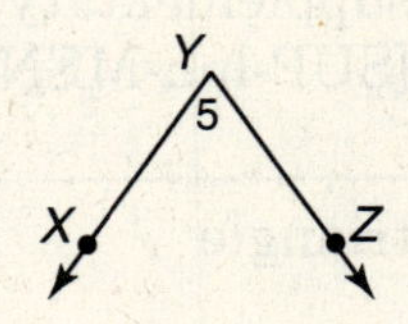

____

Review It

Name the sides of ∠*ABC*. (Lesson 1-2)

____

____

Example

2 Name all angles having *D* as their vertex.

There are ________ distinct angles with vertex *D*: ________.

Your Turn Name all angles having A as their vertex.

A, X, Y, Z, W, 3, 4, 5

BUILD YOUR VOCABULARY (page 44)

An angle separates a ______ into ______ parts: the **interior** of the angle, the **exterior** of the angle, and the **angle** itself.

EXAMPLES

Tell whether each point is in the *interior*, *exterior*, or *on the angle*.

C, A, B

FOLDABLES

ORGANIZE IT

In your first Foldable, explain and draw examples of *interior* points, *exterior* points, and points *on the angle*. Include under appropriate tab.

Angles and Points
interior | exterior | on an angle

3 **Point *A*:** Point A is on the ______ of the angle.

4 **Point *B*:** Point B is on the ______ of the angle.

5 **Point *C*:** Point C is ______.

Your Turn **Tell whether each point is in the *interior*, *exterior*, or *on the angle*.**

a. Point T

T

b. Point N

N

c. Point D

D, m, n

HOMEWORK ASSIGNMENT

Page(s):

Exercises:

3–2 Angle Measure

Build Your Vocabulary (pages 44–45)

Angles are measured in units called **degrees**.

A **protractor** is a tool used to measure angles and sketch angles of a given measure.

What You'll Learn

- Measure, draw, and classify angles.

Postulate 3-1 Angle Measurement Postulate
For every angle, there is a unique positive number between 0 and 180 called the *degree measure* of the angle.

EXAMPLE

1 Use a protractor to measure $\angle KLM$.

STEP 1 Place the center point of the protractor on vertex L. Align the straightedge with side $\overrightarrow{LM}$.

STEP 2 Use the scale that begins with 0 at $\overrightarrow{LM}$. Read where $\overrightarrow{LK}$ crosses this scale.

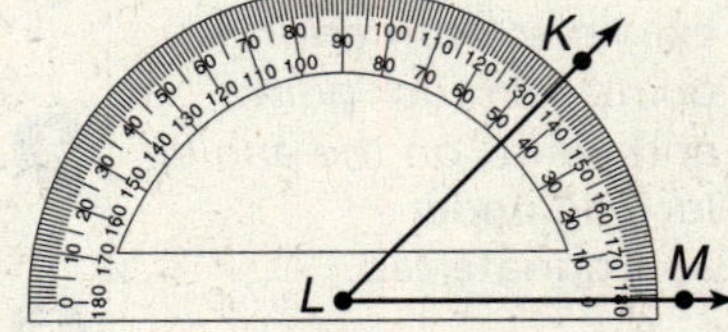

Angle KLM measures 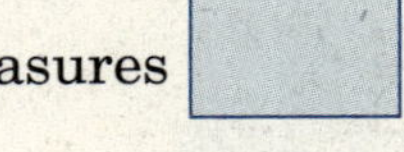.

Your Turn Use a protractor to measure $\angle XYZ$.

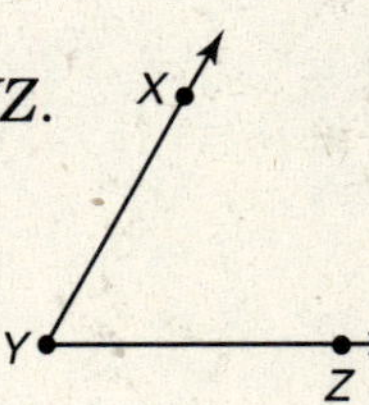

EXAMPLE

2 Find the measures of $\angle DHE$, $\angle EHG$, and $\angle FHG$.

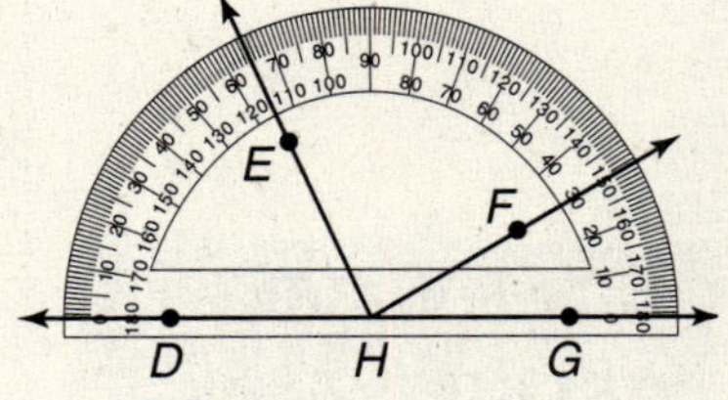

$m\angle DHE =$ $\overrightarrow{HD}$ is at 0° on the left.

$m\angle EHG =$ $\overrightarrow{HG}$ is at 0° on the right.

$m\angle FHG =$ $\overrightarrow{HG}$ is at 0° on the right.

Your Turn Find $m\angle PQR$, $m\angle RQS$, and $m\angle SQT$.

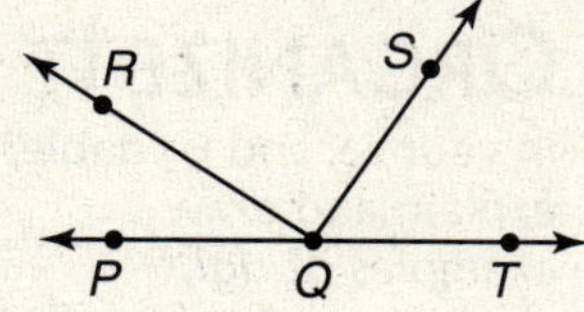

Postulate 3-2 Protractor Postulate
On a plane, given $\overrightarrow{AB}$ and a number r between 0 and 180, there is exactly one ray with endpoint A, extending on each side of $\overrightarrow{AB}$ such that the degree measure of the angle formed is r.

EXAMPLE

3 Use a protractor to draw an angle having a measure of 35°.

STEP 1 Draw $\overrightarrow{BC}$.

STEP 2 Place the center point of the protractor on B. Align the mark labeled ______ with the ray.

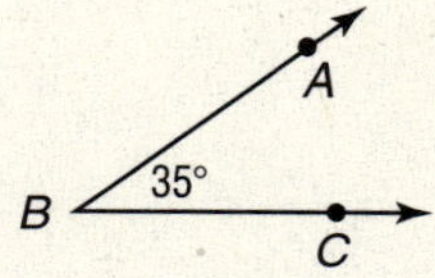

STEP 3 Locate and draw point A at the mark labeled ______. Draw $\overrightarrow{BA}$.

REMEMBER IT

Read $m\angle PQR = 75$ as *the degree measure of angle PQR is 75.*

Your Turn Use a protractor to draw an angle having a measure of 78°.

REMEMBER IT

The symbol ⌉ is used to indicate a right angle.

BUILD YOUR VOCABULARY (pages 44–45)

A **right angle** has a degree measure of 90.

The degree measure of an **acute angle** is greater than 0 and less than 90.

An **obtuse angle** has a degree measure greater than 90 and less than 180.

A three-sided closed figure with three interior angles is a **triangle**.

A four-sided closed figure with four interior angles is a **quadrilateral**.

EXAMPLES

Classify each angle as *acute*, *obtuse*, or *right*.

4

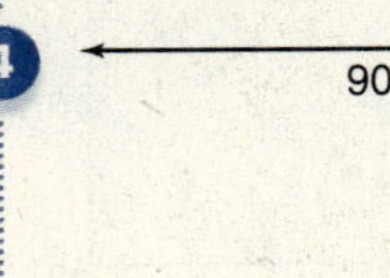

5

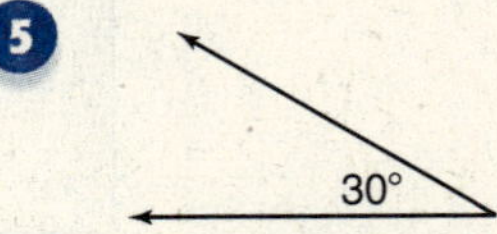

FOLDABLES™

ORGANIZE IT

In your second Foldable, explain and draw examples of *right* angles, *acute* angles, and *obtuse* angles. Include under appropriate tab.

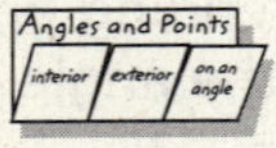

Your Turn **Classify each angle as *acute*, *obtuse*, or *right*.**

a.

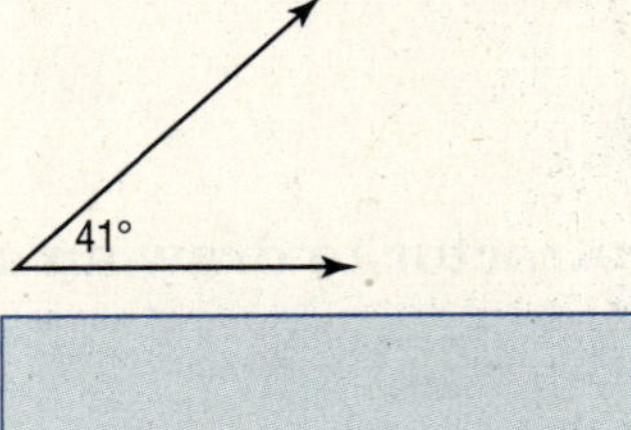

b.

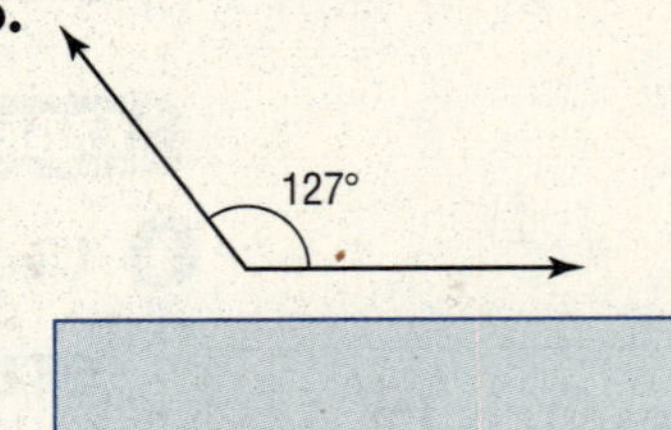

EXAMPLE

6 **The measure of $\angle A$ is 100. Solve for x.**

You know that $m\angle A = 100$

and $m\angle A =$ ______ $+ 10$.

Write and solve an equation.

______	$= 3x + 10$	Substitution
$100 -$ ______	$= 3x + 10 -$ ______	Subtract ______ from each side.
______	$= 3x$	
$\frac{90}{\_\_\_\_}$	$= \frac{3x}{\_\_\_\_}$	Divide each side by ______.
______	$= x$	

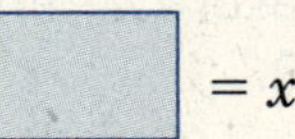

HOMEWORK ASSIGNMENT

Page(s):

Exercises:

Your Turn The measure of $\angle N$ is 135. Solve for x.

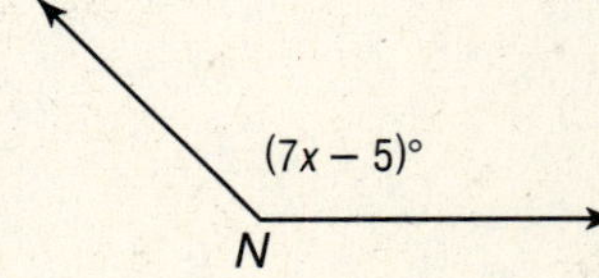

3–3 The Angle Addition Postulate

Postulate 3-3 Angle Addition Postulate
For any angle PQR, if A is in the interior of $\angle PQR$ then $m\angle PQA + m\angle AQR = m\angle PQR$.

What You'll Learn

- Find the measure and bisector of an angle.

EXAMPLES

1 **If $m\angle KNL = 110$ and $m\angle LNM = 25$, find $m\angle KNM$.**

$m\angle KNM = m\angle KNL + m\angle LNM$

$= \_\_\_\_ + 25$ Substitution

$= \_\_\_\_$ So, $m\angle KNM = \_\_\_\_$.

2 **Find $m\angle 2$ if $m\angle 1 = 75$ and $m\angle ABC = 140$.**

$m\angle 2 = m\angle ABC - m\angle 1$

$= \_\_\_\_ - \_\_\_\_$ Substitution

$= \_\_\_\_$ So, $m\angle 2 = \_\_\_\_$.

3 **Find $m\angle JKL$ and $m\angle LKM$ if $m\angle JKM = 140$.**

$m\angle JKL + m\angle LKM = m\angle JKM$

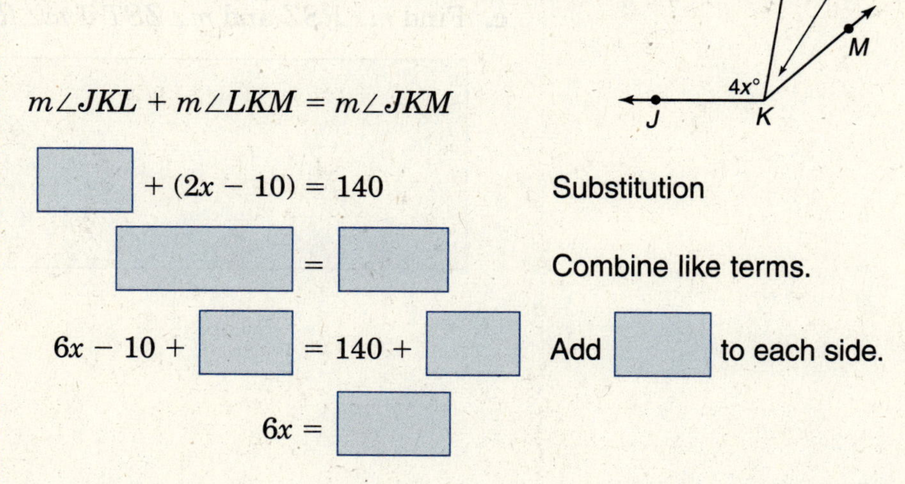

$\_\_\_\_ + (2x - 10) = 140$ Substitution

$\_\_\_\_ = \_\_\_\_$ Combine like terms.

$6x - 10 + \_\_\_\_ = 140 + \_\_\_\_$ Add ____ to each side.

$6x = \_\_\_\_$

$x = \_\_\_\_$ Divide each side by ____.

Replace x with   in each expression.

$m\angle JKL = 4x$ $\qquad$ $m\angle LKM = 2x - 10$

$= 4$ $\qquad$ $= 2$ $- 10$

$=$ $\qquad$ $=$ $- 10 =$

Therefore, $m\angle JKL =$ and $m\angle LKM =$.

Your Turn

a. If $m\angle ABC = 95$ and $m\angle CBD = 65$, find $m\angle ABD$.

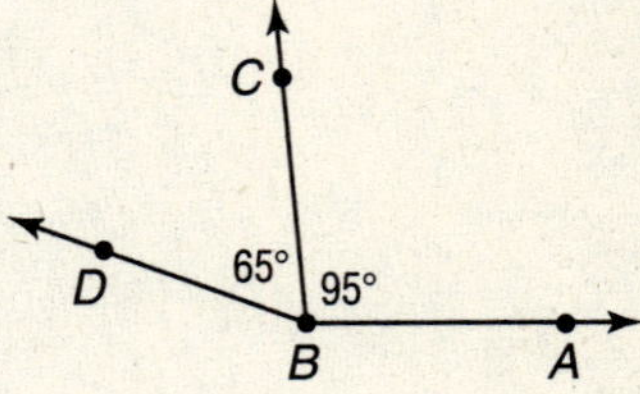

b. If $m\angle XYZ = 110$ and $m\angle XYW = 22$, find $m\angle WYZ$.

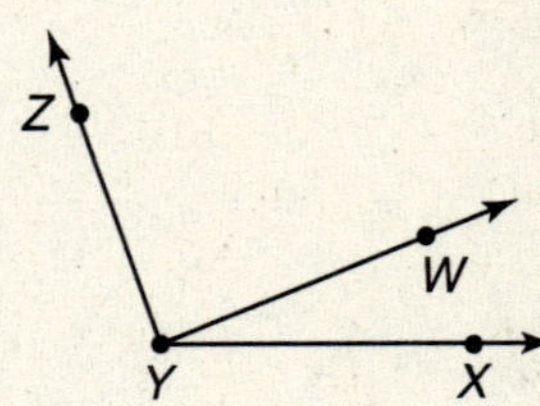

c. Find $m\angle RSZ$ and $m\angle ZST$ if $m\angle RST = 135$.

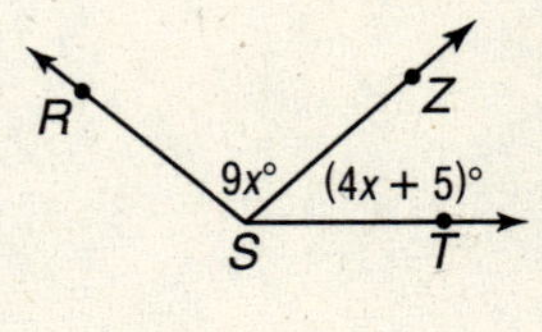

BUILD YOUR VOCABULARY (page 44)

A ______ that divides an angle into ______ angles of equal ______ is called the **angle bisector**.

REVIEW IT

$\overline{DF}$ is bisected at point E, and $DF = 8$. What do you know about the lengths DE and EF? (*Lesson 2-3*)

______

______

EXAMPLE

4 **If $\overrightarrow{FD}$ bisects $\angle CFE$ and $m\angle CFE = 70$, find $m\angle 1$ and $m\angle 2$.**

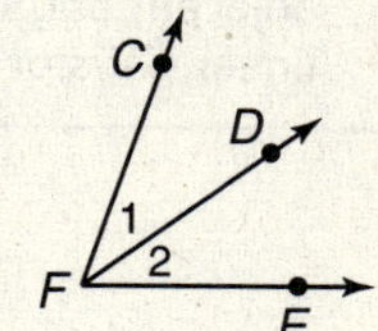

Since $\overrightarrow{FD}$ bisects $\angle CFE$, $m\angle 1 = m\angle 2$.

$m\angle 1 + m\angle$ ______ $= m\angle CFE$ Postulate 3–3

$m\angle 1 + m\angle 2 =$ ______ Replace $m\angle CFE$ with ______.

$m\angle 1 +$ ______ $= 70$ Replace $m\angle 2$ with ______.

$2(m\angle$ ______ $) = 70$ Combine like terms.

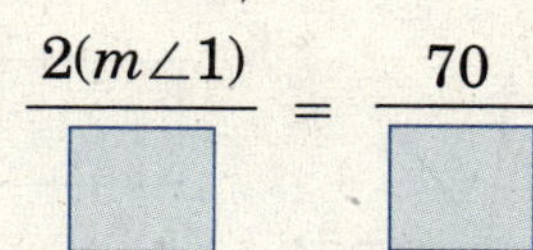

$\frac{2(m\angle 1)}{\square} = \frac{70}{\square}$ Divide each side by ______.

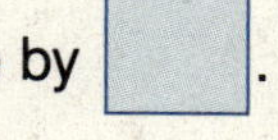

$m\angle 1 =$ ______

Since $m\angle 1 = m\angle 2$, $m\angle 2 =$ ______.

Your Turn If $\overrightarrow{EG}$ bisects $\angle FEH$ and $m\angle FEH = 98$, find $m\angle 1$ and $m\angle 2$.

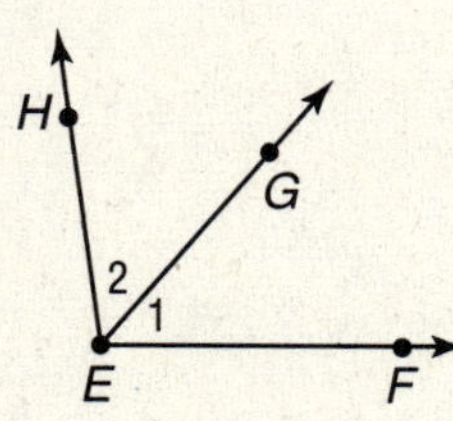

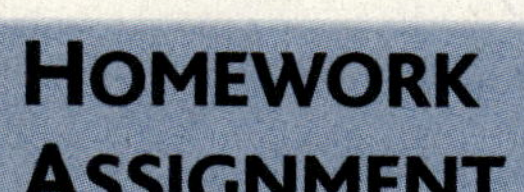

Page(s):

Exercises:

3–4 Adjacent Angles and Linear Pairs of Angles

WHAT YOU'LL LEARN

- Identify and use adjacent angles and linear pairs of angles.

BUILD YOUR VOCABULARY (page 44)

Adjacent angles share a common side and a vertex, but have no ______ points in common.

When the noncommon sides of adjacent angles form a ______, the angles are said to form a **linear pair.**

EXAMPLES

Determine whether ∠1 and ∠2 are adjacent angles.

1 ______. They have the same ______ but no ______.

2 ______. They have the same ______ and a common ______ with no interior points in common.

3 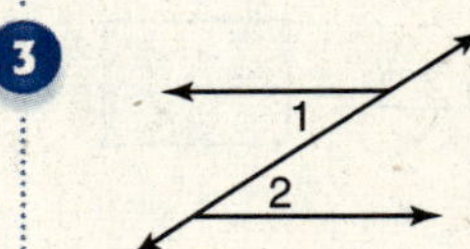

______. They have a ______ but no common ______.

Your Turn **Determine whether ∠1 and ∠2 are adjacent angles.**

a. 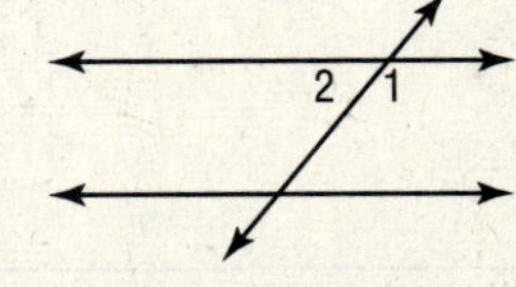

______

b. 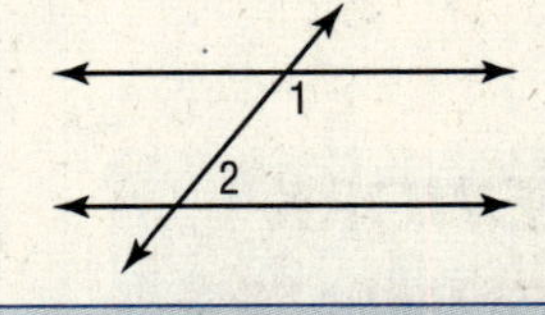

______

c. 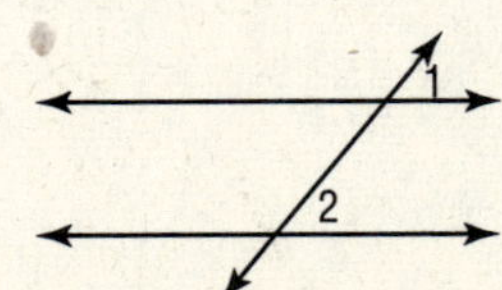

______

EXAMPLES

$\overrightarrow{CM}$ and $\overrightarrow{CE}$ are opposite rays.

4 Name the angle that forms a linear pair with ∠*TCM*.

∠*TCE* and ∠*TCM* have a common side ______, the same vertex ______, and opposite rays ______ and ______.

So, ∠*TCE* forms a linear pair with ∠*TCM*.

5 Do ∠1 and ∠*TCE* form a linear pair? Justify your answer.

______, they are not ______ angles.

WRITE IT

List the differences and similarities between linear pairs of angles and adjacent angles.

Your Turn **Refer to Examples 4 and 5.**

a. Name the angle that forms a linear pair with ∠*HCE*.

b. Determine if ∠*TCA* and ∠*TCH* form a linear pair. Justify your answer.

EXAMPLE

6 List at least two models of linear pairs in your classroom or home.

Your Turn List at least two models of adjacent angles on a school playground.

HOMEWORK ASSIGNMENT

Page(s):

Exercises:

3–5 Complementary and Supplementary Angles

What You'll Learn

- Identify and use complementary and supplementary angles.

BUILD YOUR VOCABULARY (pages 44–45)

Complementary angles are two angles whose degree measures total 90.

Supplementary angles are two angles whose degree measures total 180.

EXAMPLES

1 Use the figure to name a pair of nonadjacent supplementary angles.

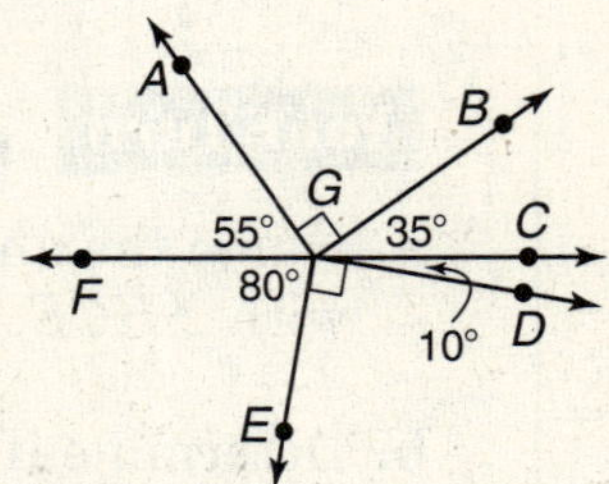

$m\angle AGB$ + ______ = ______, and they have the same vertex ______, but ______ sides. Therefore, $\angle AGB$ and ______ are nonadjacent supplementary angles.

2 Use the above figure to find the measure of an angle that is supplementary to $\angle BGC$.

Let x = measure of angle supplementary to $\angle BGC$.

$m\angle BGC + x = 180$	Defn. of Supplementary Angles
______ $+ x = 180$	$m\angle BGC =$ ______
$35 + x -$ ______ $= 180 -$ ______	Subtract ______ from each side.
$x =$ ______	

Your Turn

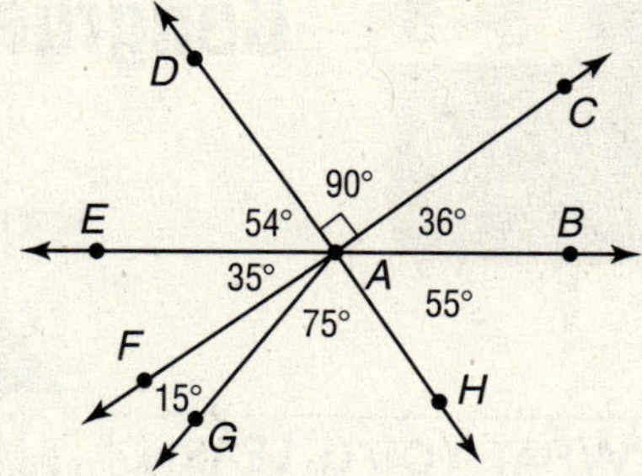

a. In the figure, name a pair of nonadjacent supplementary angles.

b. In the figure, find an angle with a measure supplementary to $\angle BAF$.

REMEMBER IT 

Supplementary angles must be adjacent angles to form a linear pair.

EXAMPLE

3 **Angles C and D are supplementary. If $m\angle C = 12x$ and $m\angle D = 4(x + 5)$, find x. Then find $m\angle C$ and $m\angle D$.**

$m\angle C + m\angle D = 180$	Defn. of Supplementary Angles
$\square + 4(x + 5) = 180$	Substitution
$12x + 4x + \square = 180$	Distributive Property
$\square = 160$	Combine like terms.
$\frac{16x}{16} = \frac{160}{16}$	Divide each side by 16.
$x = \square$	

Replace x with $\square$ in each expression.

$m\angle C = 12x$

$= 12\square$ or $\square$

$m\angle D = 4(x + 5)$

$= 4\left(\square + 5\right)$ or $\square$

HOMEWORK ASSIGNMENT

Page(s):

Exercises:

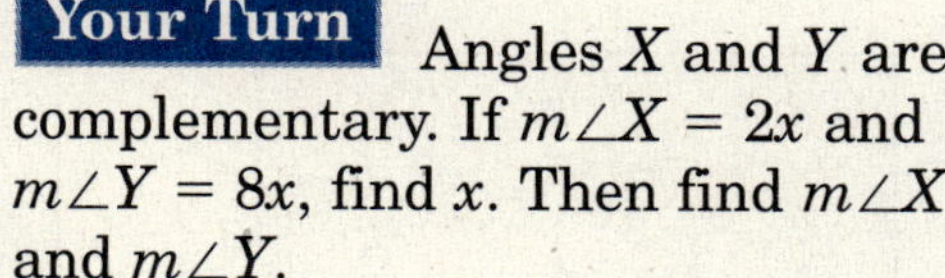

Your Turn Angles X and Y are complementary. If $m\angle X = 2x$ and $m\angle Y = 8x$, find x. Then find $m\angle X$ and $m\angle Y$.

3–6 Congruent Angles

BUILD YOUR VOCABULARY (pages 44–45)

Congruent angles have the same measure.

When two lines ________, ________ angles are formed. There are two pairs of nonadjacent angles. These pairs are **vertical angles**.

WHAT YOU'LL LEARN

- Identify and use congruent and vertical angles.

Theorem 3-1 Vertical Angle Theorem
Vertical angles are congruent.

EXAMPLES

Find the value of x in each figure.

1

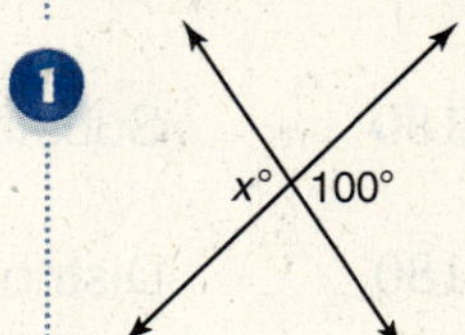

The angles are ________ angles.

So, $x =$ ________.

2

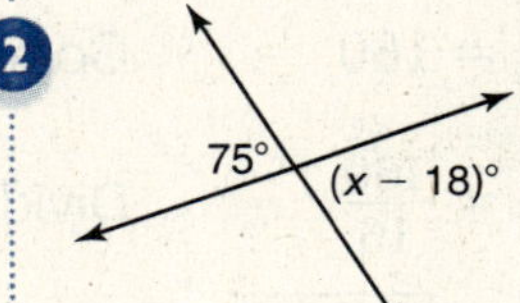

Since the angles are vertical angles, they are congruent.

$x - 18 =$ ________

$x - 18 +$ ________ $= 75 +$ ________ Add ________ to each side.

$x =$ ________

Your Turn **Find the value of x in each figure.**

a.

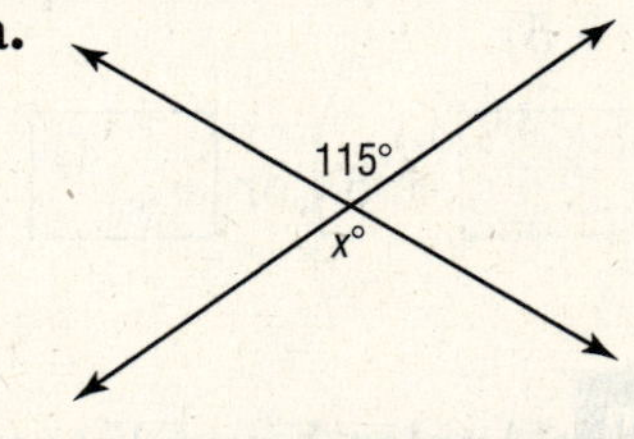

b.

38° (3x − 10)°

REMEMBER IT

The notation $\angle A \cong \angle B$ is read as *angle A is congruent to angle B.*

Theorem 3–2 If two angles are congruent, then their complements are congruent.

Theorem 3–3 If two angles are congruent, then their supplements are congruent.

Theorem 3–4 If two angles are complementary to the same angle, then they are congruent.

Theorem 3–5 If two angles are supplementary to the same angle, then they are congruent.

Theorem 3–6 If two angles are congruent and supplementary, then each is a right angle.

Theorem 3–7 All right angles are congruent.

EXAMPLES

3 **Suppose $\angle A \cong \angle B$ and $m\angle B = 47$. Find the measure of an angle that is supplementary to $\angle A$.**

Since $\angle A \cong \angle B$, their supplements are congruent.

The supplement of $\angle B$ is 180 − 47 or ______. So, the measure of an angle that is supplementary to $\angle A$ is ______.

4 **In the figure, $\angle 1$ is supplementary to $\angle 2$, $\angle 3$ is supplementary to $\angle 2$, and $m\angle 2$ is 105. Find $m\angle 1$ and $m\angle 3$.**

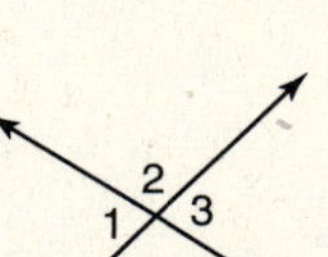

$\angle 1$ and $\angle 2$ are supplementary.

So, $m\angle 1$ = ______ − 105 or . $\angle 3$ and $\angle 2$ are supplementary. So, $m\angle 3$ = ______ − 105 or ______.

Your Turn

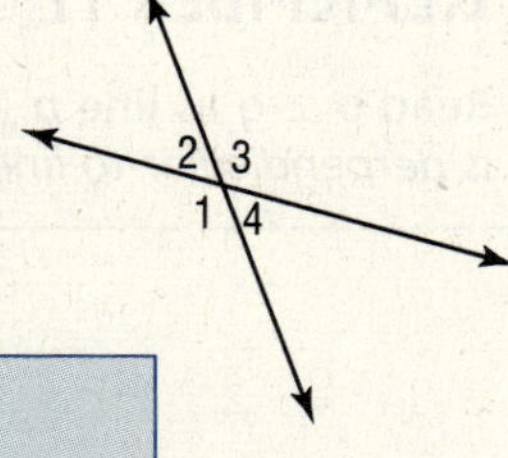

a. Suppose $\angle X \cong \angle Y$ and $m\angle Y = 82$. Find the measure of an angle that is supplementary to $\angle X$.

______

b. In the figure, $\angle 1$ is supplementary to $\angle 2$ and $\angle 4$. If $m\angle 4 = 54$, find $m\angle 1$, $m\angle 2$, and $m\angle 3$.

______

HOMEWORK ASSIGNMENT

Page(s):

Exercises:

3–7 Perpendicular Lines

WHAT YOU'LL LEARN

- Identify, use properties of, and construct perpendicular lines and segments.

BUILD YOUR VOCABULARY (page 45)

Lines that ______ at an angle of ______ degrees are said to be **perpendicular lines**.

Theorem 3–8
If two lines are perpendicular, then they form right angles.

EXAMPLES

Refer to the figure to determine whether each of the following is *true* or *false*.

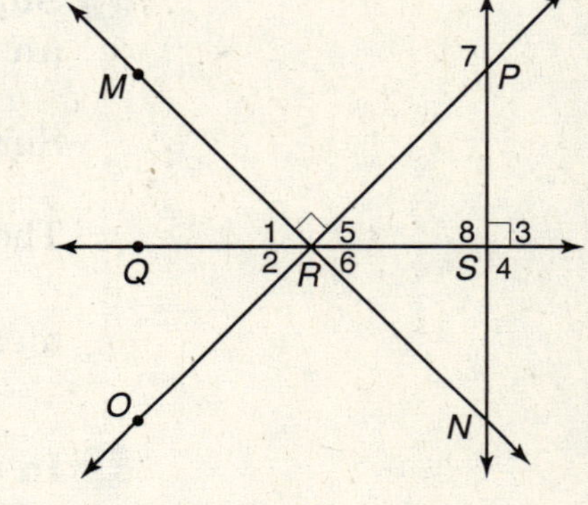

1 $\overline{QS} \perp \overline{OP}$

______. $\overline{QS}$ and $\overline{OP}$ do not form ______ angles.

Therefore, they ______ perpendicular.

2 **∠7 is an obtuse angle.**

______. ∠7 forms a ______ with an acute angle.

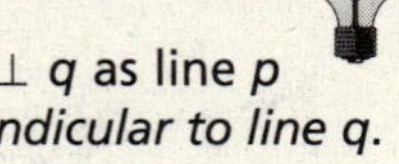

REMEMBER IT

Read $p \perp q$ as line p *is perpendicular to line q.*

Your Turn **In the figure $\overline{WY} \perp \overline{ZT}$. Determine whether each of the following is *true* or *false*.**

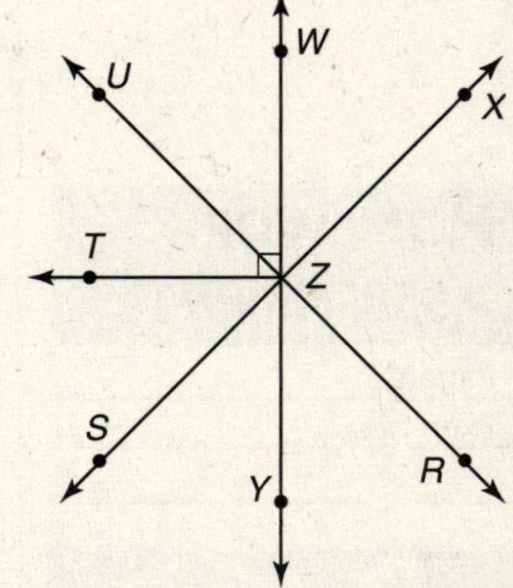

a. $m\angle WZU + m\angle UZT = 90$

______

b. $\angle SZY$ is obtuse.

______

EXAMPLE

3 **Find $m\angle 1$ and $m\angle 2$ if $\overline{AC} \perp \overline{BD}$, $m\angle 1 = 8x - 2$ and $m\angle 2 = 16x - 4$.**

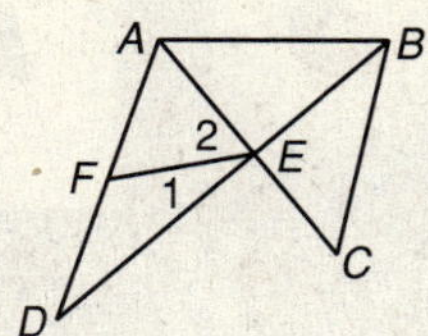

Since $\overline{AC} \perp \overline{BD}$, $\angle AED$ is a right angle.

$m\angle AED = 90$	Definition of perpendicular lines
$\angle 1 + \angle \square = \angle AED$	Angle Addition Postulate
$m\angle 1 + m\angle \square = m\angle AED$	
$m\angle 1 + m\angle 2 = \square$	Substitution
$(8x - 2) + (16x - 4) = 90$	Substitution
$24x - 6 = 90$	Combine like terms.
$24x - 6 + 6 = 90 + 6$	Add 6 to each side.
$24x = 96$	
$\frac{24x}{24} = \frac{96}{24}$	Divide each side by 24.
$x = \square$	

Replace x with $\square$ to find $m\angle 1$ and $m\angle 2$.

$m\angle 1 = 8x - 2$
$= 8(\square) - 2$
$= 32 - 2$ or 30

$m\angle 2 = 16x - 4$
$= 16(\square) - 4$
$= 64 - 4$ or 60

Therefore, $m\angle 1 = 30$ and $m\angle 2 = 60$.

HOMEWORK ASSIGNMENT

Page(s):

Exercises:

Your Turn Find $m\angle 3$ and $m\angle 4$ if $\overline{AC} \perp \overline{BF}$, $m\angle 3 = 7x + 6$ and $m\angle 4 = 12x + 27$.

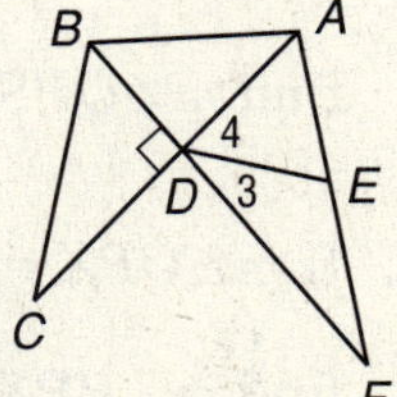

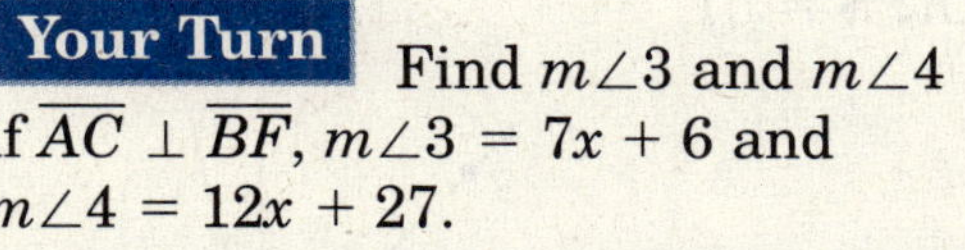

CHAPTER 3

BRINGING IT ALL TOGETHER

STUDY GUIDE

FOLDABLES™	VOCABULARY PUZZLEMAKER	BUILD YOUR VOCABULARY
Use your **Chapter 3 Foldable** to help you study for your chapter test.	To make a crossword puzzle, word search, or jumble puzzle of the vocabulary words in Chapter 3, go to: www.glencoe.com/sec/math/t.resources/free/index.php	You can use your completed **Vocabulary Builder** (pages 44–45) to help you solve the puzzle.

3-1 Angles

Indicate whether the statement is *true* or *false*.

1. $\overrightarrow{XY}$ and $\overrightarrow{YZ}$ are the sides of $\angle XYZ$.

2. The vertex of an angle is a point where two rays intersect.

3. A straight angle is also a line.

3-2 Angle Measure

Use a protractor to measure the specified angles. Then, classify them as *acute*, *right*, or *obtuse* angles.

4. $\angle BAC$

5. $\angle CAE$

6. $\angle DAE$

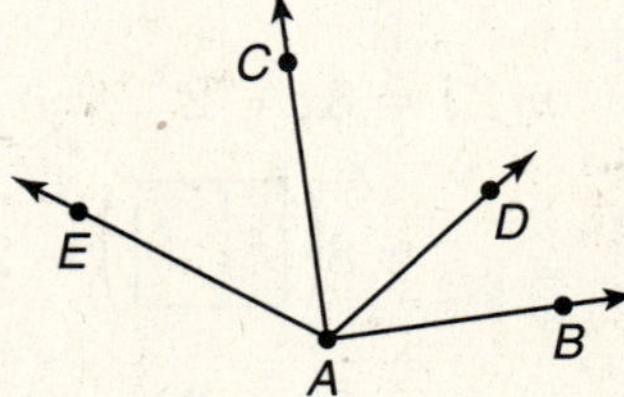

3-3 The Angle Addition Postulate

7. If $m\angle QPR = 30$ and $m\angle RPS = 51$, find $m\angle QPS$.

8. If $m\angle QPX = 137$ and $m\angle QPR = 30$, find $m\angle RPX$.

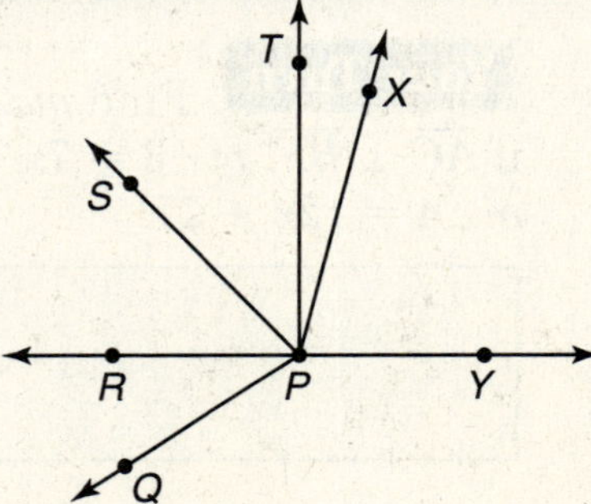

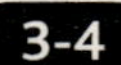

3-4 Adjacent Angles and Linear Pairs of Angles

9. In the figure $\overrightarrow{QN}$ and $\overrightarrow{QP}$ are opposite rays. Name the angles that form a linear pair.

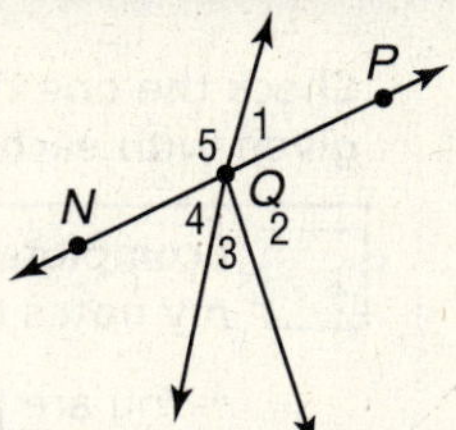

3-5 Complementary and Supplementary Angles

10. If $m\angle 1 = 36$, what is the measure of its complement?

11. What is the measure of an angle supplementary to $m\angle 1 = 36$?

3-6 Congruent Angles

Lines *m* and *n* intersect at point *P*. What is the measure of each of the four angles formed?

12. 　　**13.**

14. 　　**15.**

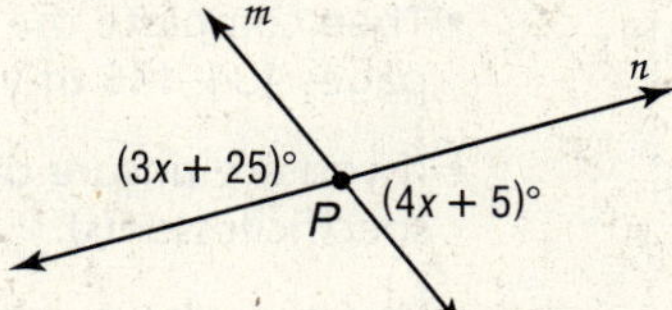

3-7 Perpendicular Lines

If $\overline{WZ}$ is constructed perpendicular to $\overline{XY}$, list six terms that describe $\angle XWZ$ and $\angle YWZ$.

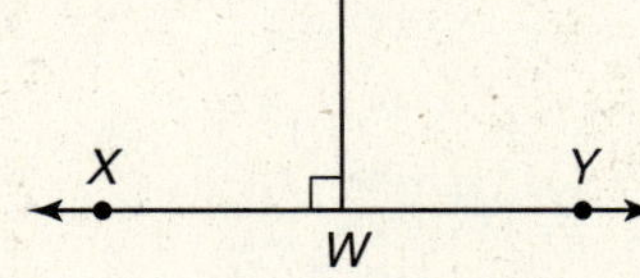

16.

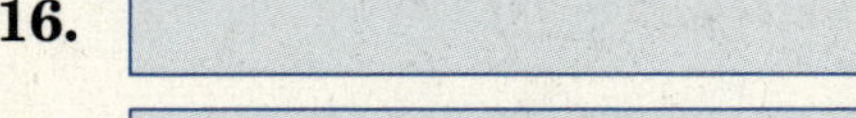

17.

18. 　　**20.**

19. 　　**21.**

ARE YOU READY FOR THE CHAPTER TEST?

Visit **geoconcepts.com** to access your textbook, more examples, self-check quizzes, and practice tests to help you study the concepts in Chapter 3.

Check the one that applies. Suggestions to help you study are given with each item.

☐ **I completed the review of all or most lessons without using my notes or asking for help.**

- You are probably ready for the Chapter Test.
- You may want to take the Chapter 3 Practice Test on page 137 of your textbook as a final check.

☐ **I used my Foldable or Study Notebook to complete the review of all or most lessons.**

- You should complete the Chapter 3 Study Guide and Review on pages 134–136 of your textbook.
- If you are unsure of any concepts or skills, refer back to the specific lesson(s).
- You may also want to take the Chapter 3 Practice Test on page 137.

☐ **I asked for help from someone else to complete the review of all or most lessons.**

- You should review the examples and concepts in your Study Notebook and Chapter 3 Foldable.
- Then complete the Chapter 3 Study Guide and Review on pages 134–136 of your textbook.
- If you are unsure of any concepts or skills, refer back to the specific lesson(s).
- You may also want to take the Chapter 3 Practice Test on page 137.

Student Signature

Parent/Guardian Signature

Teacher Signature

Parallels

Use the instructions below to make a Foldable to help you organize your notes as you study the chapter. You will see Foldable reminders in the margin of this Interactive Study Notebook to help you in taking notes.

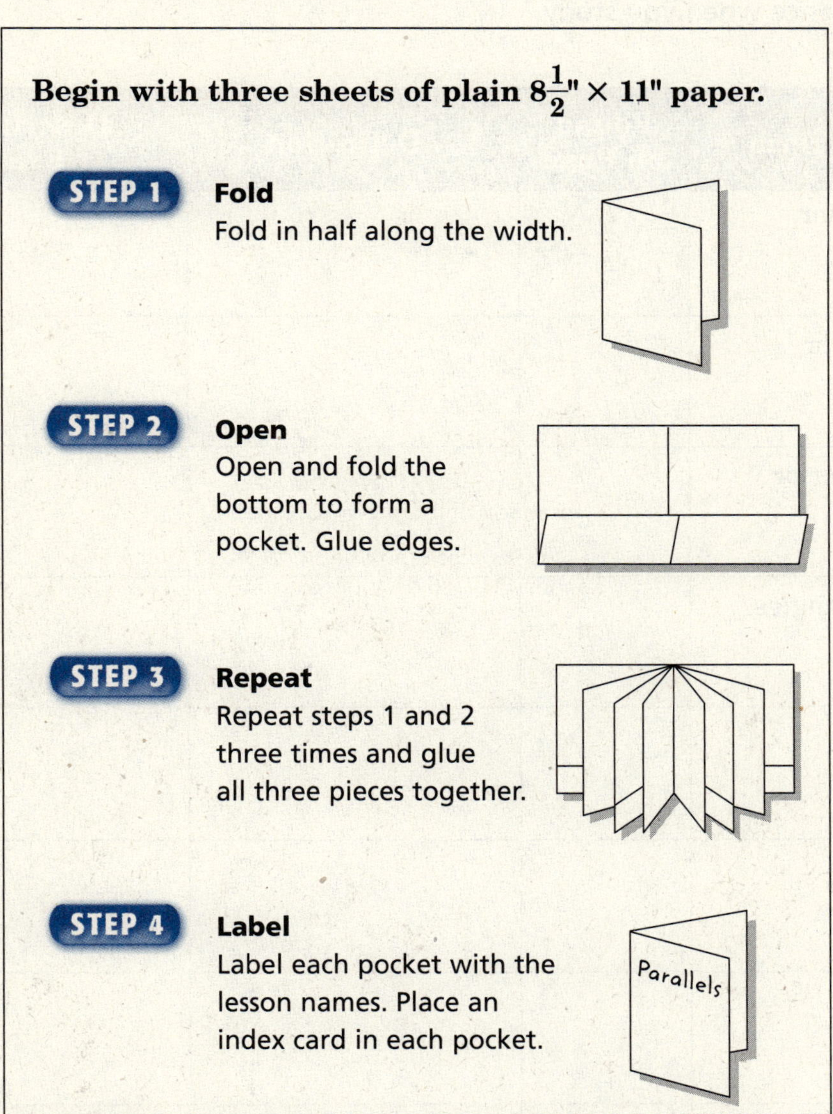

Begin with three sheets of plain $8\frac{1}{2}$" × 11" paper.

STEP 1 **Fold**
Fold in half along the width.

STEP 2 **Open**
Open and fold the bottom to form a pocket. Glue edges.

STEP 3 **Repeat**
Repeat steps 1 and 2 three times and glue all three pieces together.

STEP 4 **Label**
Label each pocket with the lesson names. Place an index card in each pocket.

NOTE-TAKING TIP: When taking notes, it is often a good idea to write in your own words a summary of the lesson. Be sure to paraphrase key points.

BUILD YOUR VOCABULARY

This is an alphabetical list of new vocabulary terms you will learn in Chapter 4. As you complete the study notes for the chapter, you will see Build Your Vocabulary reminders to complete each term's definition or description on these pages. Remember to add the textbook page number in the second column for reference when you study.

Vocabulary Term	Found on Page	Definition	Description or Example
alternate exterior angles			
alternate interior angles			
consecutive interior angles			
corresponding angles			
exterior angles			
finite			
great circle			
interior angles			
line			

Vocabulary Term	Found on Page	Definition	Description or Example
line of latitude			
line of longitude			
linear equation			
parallel lines [PARE-uh-lel]			
parallel planes			
skew lines [SKYOO]			
slope			
slope-intercept form			
transversal			
y-intercept			

4–1 Parallel Lines and Planes

What You'll Learn

- Describe relationships among lines, parts of lines, and planes.

BUILD YOUR VOCABULARY (pages 66–67)

Parallel lines are two lines in the same ______ that do not intersect.

Parallel planes are the same ______ apart at all points and ______ intersect.

Lines that do not ______ and are not in the ______ plane are said to be **skew lines**.

Foldables™

ORGANIZE IT

Use the index card labeled *Parallel Lines and Planes* to record the definitions in this lesson, along with examples to help you remember the main idea.

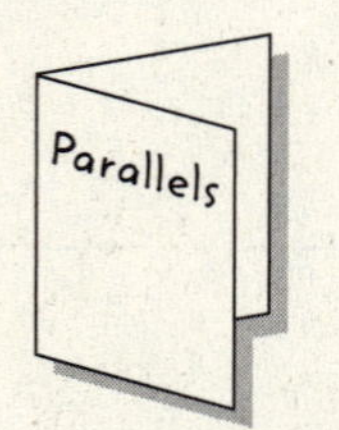

EXAMPLES

Name the parts of the prism shown below. Assume segments that look parallel are parallel.

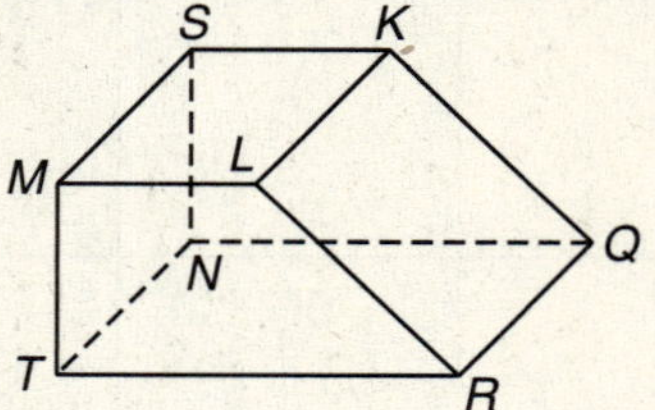

1 all planes parallel to plane *SKL*

Plane ______ is parallel to plane *SKL*.

2 all segments that intersect $\overline{MT}$

______ intersect $\overline{MT}$.

3 all segments parallel to $\overline{MT}$

______ is parallel to $\overline{MT}$.

4 all segments skew to $\overline{MT}$

______ are skew to $\overline{MT}$.

Your Turn **Name the parts of the prism shown below. Assume segments that look parallel are parallel.**

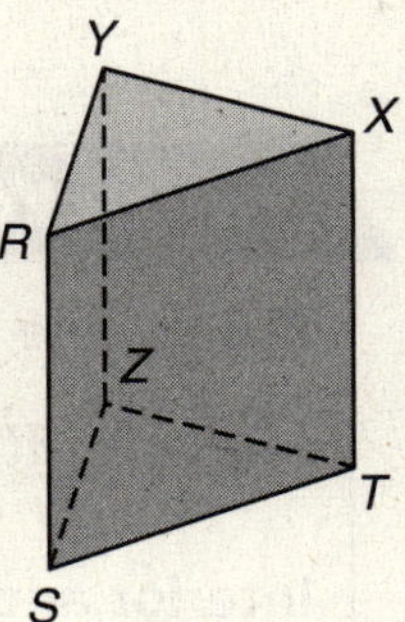

a. all segments parallel to $\overline{RS}$

b. all segments that intersect $\overline{RS}$

c. a pair of parallel planes

d. all segments skew to $\overline{XT}$

REMEMBER IT

A plane that passes through points *A*, *B*, *C* and *D* can be named using any three of the points.

HOMEWORK ASSIGNMENT

Page(s):

Exercises:

4–2 Parallel Lines and Transversals

WHAT YOU'LL LEARN

- Identify the relationships among pairs of interior and exterior angles formed by two parallel lines and a transversal.

BUILD YOUR VOCABULARY (pages 66–67)

A line, line segment, or ray that intersects two or more lines at different ______ is known as a **transversal**.

Interior angles lie in between the two lines.

Alternate interior angles are on ______ sides of the transversal.

Consecutive interior angles are on the ______ side of the transversal.

Exterior angles lie ______ the two lines.

Alternate exterior angles are on ______ sides of the transversal.

FOLDABLES™

ORGANIZE IT

Use the index card labeled *Parallel Lines and Transversals* to record the definitions and theorems in this lesson. Draw pictures and examples to help you remember them.

EXAMPLES

Identify each pair of angles as *alternate interior*, *alternate exterior*, *consecutive interior*, or *vertical*.

1 2
3 4
5 6
7 8

1 **∠3 and ∠5**

∠3 and ∠5 are interior angles on the same side as the transversal, so they are ______ angles.

2 **∠1 and ∠8**

∠1 and ∠8 are exterior angles on opposite sides of the transversal, so they are ______ angles.

Your Turn **Identify each pair of angles as *alternate interior, alternate exterior, consecutive interior*, or *vertical*.**

a. $\angle 3$ and $\angle 5$

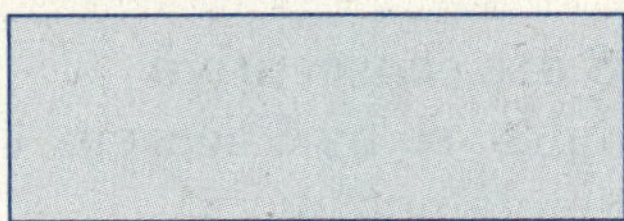

b. $\angle 3$ and $\angle 6$

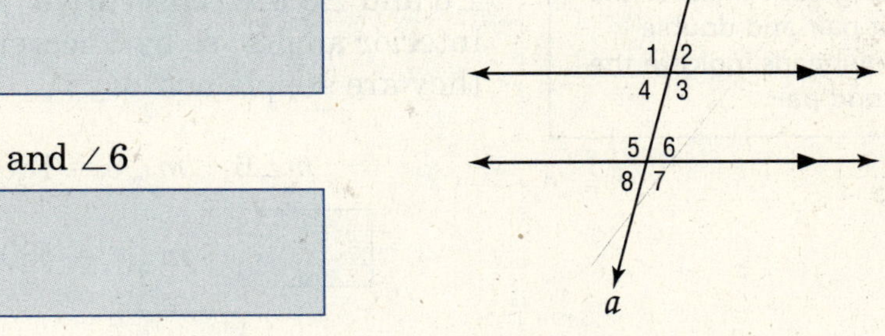

Theorem 4-1 Alternate Interior Angles
If two parallel lines are cut by a transversal, then each pair of alternate interior angles is congruent.

Theorem 4-2 Consecutive Interior Angles
If two parallel lines are cut by a transversal, then each pair of consecutive interior angles is supplementary.

Theorem 4-3 Alternate Exterior Angles
If two parallel lines are cut by a transversal, then each pair of alternate exterior angles is congruent.

REVIEW IT

The sum of the degree measures of three angles is 180. Are the three angles supplementary? Explain. (*Lesson 3-5*)

______

______

______

EXAMPLE

3 In the figure, $p \| q$, and r is a transversal. If $m\angle 6 = 115$, find $m\angle 7$.

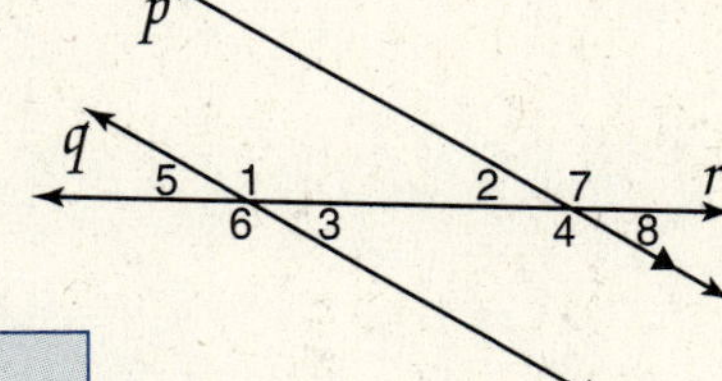

$\angle 6$ and $\angle 7$ are alternate ______

angles, so by Theorem 4-3, they are .

Therefore, $m\angle 7 =$ ______.

Your Turn If $m\angle 1 = 50$, find $m\angle 8$.

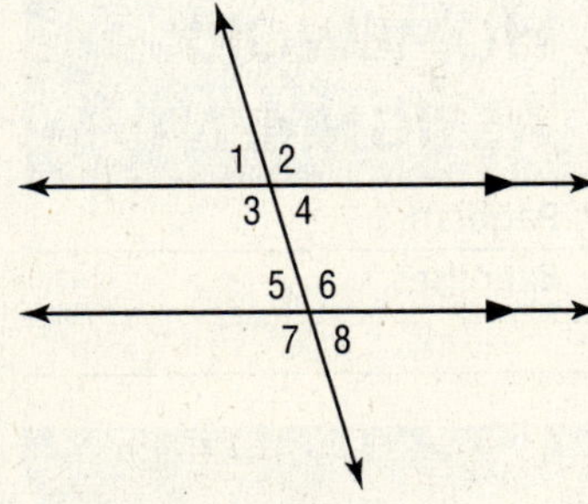

EXAMPLE

REMEMBER IT

In figures with two pairs of parallel lines, arrowheads indicate the first pair and double arrowheads indicate the second pair.

4 **In the figure, $\overleftrightarrow{AB} \parallel \overleftrightarrow{CD}$, and t is a transversal. If $m\angle 6 = 128$, find $m\angle 7$, $m\angle 8$, and $m\angle 9$.**

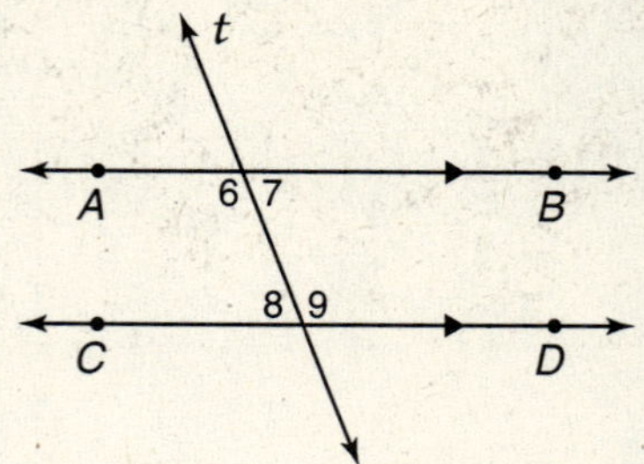

$\angle 6$ and $\angle 8$ are consecutive interior angles, so by Theorem 4-2 they are supplementary.

$$m\angle 6 + m\angle 8 = 180$$

$\_\_\_\_\_ + m\angle 8 = 180$ — Replace $m\angle 6$.

$128 + m\angle 8 - \_\_\_\_\_ = 180 - \_\_\_\_\_$ — Subtract 128 from each side.

$m\angle 8 = \_\_\_\_\_$

REVIEW IT

If angles P and Q are vertical angles and $m\angle P = 47$, what is $m\angle Q$? *(Lesson 3-6)*

____

$\angle 7$ and $\angle 8$ are alternate interior angles, so by Theorem 4-1 they are congruent. Therefore, $m\angle 7 = \_\_\_\_\_$.

$\angle 6$ and $\angle 9$ are __________ angles, so by Theorem 4-1 they are congruent. Therefore, $m\angle 9 = \_\_\_\_\_$.

Your Turn In the figure, $n \parallel m$, and a is a transversal. If $m\angle 6 = 73$, find $m\angle 1$, $m\angle 4$, and $m\angle 7$.

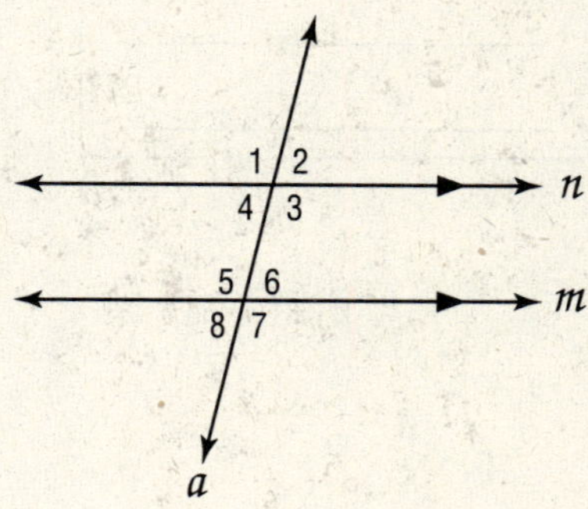

HOMEWORK ASSIGNMENT

Page(s):

Exercises:

4–3 Transversals and Corresponding Angles

What You'll Learn

- Identify the relationships among pairs of corresponding angles formed by two parallel lines and a transversal.

BUILD YOUR VOCABULARY (page 66)

When a ______ crosses two lines, an interior angle and an exterior angle that are on the ______ side of the transversal and have different verticies are called **corresponding angles**.

ORGANIZE IT

Use the index card labeled *Transversals and Corresponding Angles* to record the postulates, theorems, and other main ideas in this lesson. Draw pictures and examples as needed.

EXAMPLE

1 **Lines *a* and *b* are cut by transversal *c*. Name two pairs of corresponding angles.**

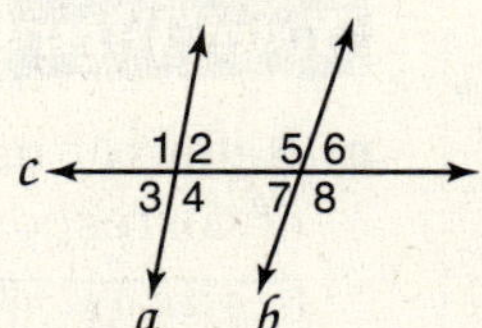

Corresponding angles lie on the same ______ of the transversal and have ______ vertices. Two pairs of corresponding angles are ______.

Postulate 4-1 Corresponding Angles
If two parallel lines are cut by a transversal, then each pair of corresponding angles is congruent.

EXAMPLES

In the figure, $a \parallel b$, and *k* is a transversal.

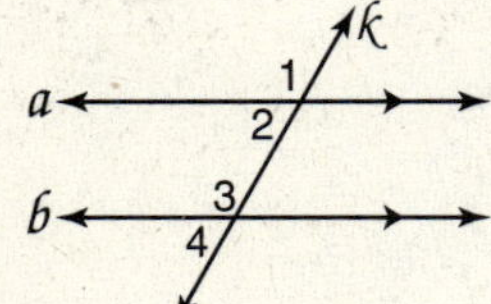

2 **Which angle is congruent to ∠1? Explain your answer.**

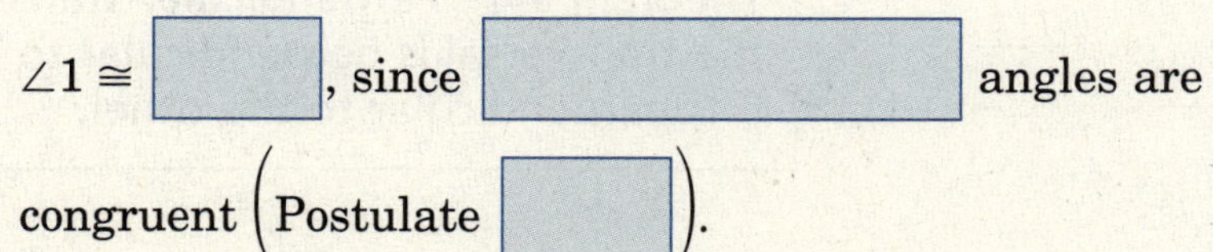

$\angle 1 \cong$ ______, since ______ angles are congruent (Postulate ______).

Key Concepts

Types of angle pairs formed when a transversal cuts two parallel lines.

1. Congruent
   a. alternate interior
   b. alternate exterior
   c. corresponding
2. Supplementary
   a. consecutive interior

3 Find the measure of $\angle 1$ if $m\angle 4 = 60$.

$m\angle 1 = m\angle 3$

$\angle 3$ and $\angle 4$ are a linear pair, so they are supplementary.

$m\angle 3 + m\angle 4 = 180$

$m\angle 3 + \square = 180$ Replace $m\angle 4$ with $\square$.

$m\angle 3 + 60 - \square = 180 - \square$ Subtract 60 from each side.

$m\angle 3 = \square$

$m\angle 1 = \square$ Substitution

Your Turn

a. Refer to the figure in Example 1. Name two different pairs of corresponding angles.

b. Refer to the figure in Example 2. Which angle is congruent to $\angle 2$? Explain your answer.

c. Refer to the figure in Example 2. Find the measure of $\angle 2$ if $m\angle 3 = 145$.

Theorem 4-4 Perpendicular Transversal
If a transversal is perpendicular to one of two parallel lines, it is perpendicular to the other.

EXAMPLE

REMEMBER IT There are always four pairs of corresponding angles when two lines are cut by a transversal.

4 **In the figure, $p \parallel q$, and transversal r is perpendicular to q. If $m\angle 2 = 3(x + 2)$, find x.**

$p \perp r$	Theorem 4-4
$\angle 2$ is a right angle.	Definition of perpendicular lines
$m\angle 2 =$	Definition of right angles
$m\angle 2 =$	Given
$= 3(x + 2)$	Replace $m\angle 2$ with .
$=$	Distributive Property
$90 - \quad = 3x + 6 -$	Subtract 6 from each side.
$84 = 3x$	
$\frac{84}{3} = \frac{3x}{3}$	Divide each side by .
$= x$	

Your Turn In the figure, $a \parallel b$ and r is a transversal. If $m\angle 1 = 3x - 5$ and $m\angle 2 = 2x + 35$, find x.

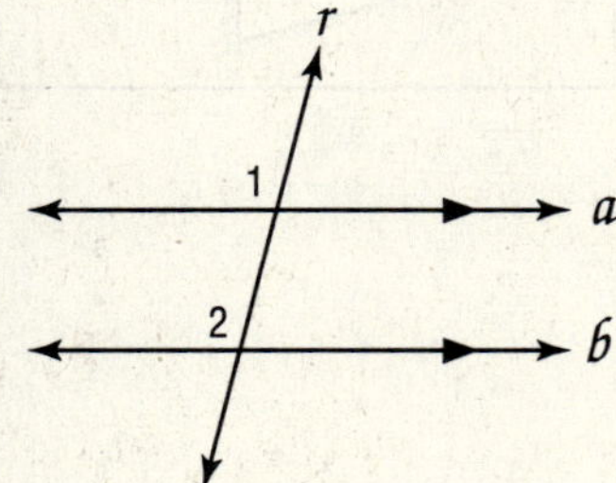

HOMEWORK ASSIGNMENT

Page(s):

Exercises:

4–4 Proving Lines Parallel

WHAT YOU'LL LEARN

- Identify conditions that produce parallel lines and construct parallel lines.

FOLDABLES™

ORGANIZE IT

Use the index card labeled *Proving Lines Parallel* to record the postulates, theorems, and important concepts in this lesson. Record examples to help you remember the main idea.

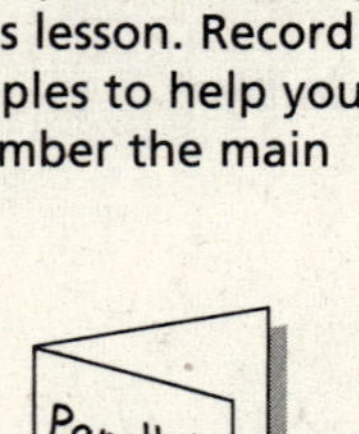

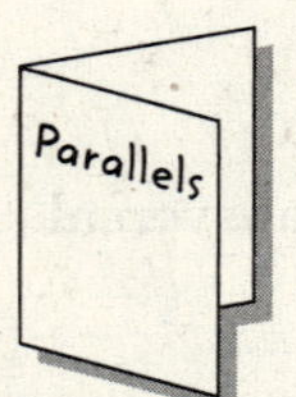

Postulate 4-2 In a plane, if two lines are cut by a transversal so that a pair of corresponding angles is congruent, then the lines are parallel.

EXAMPLE

1 If $m\angle 1 = 5x + 10$ and $m\angle 2 = 6x - 4$, find x so that $a \parallel b$.

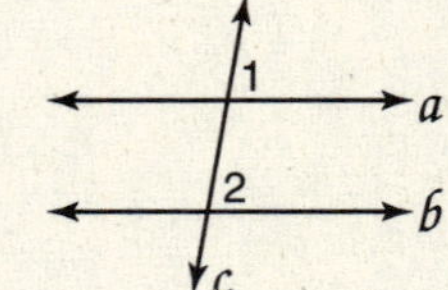

From the figure, you know that $\angle 1$ and $\angle 2$ are corresponding angles. According to Postulate 4-2, if $m\angle 1 = m\angle 2$, then $a \parallel b$.

$m\angle 1 = m\angle 2$	
______ = ______	Substitution
$5x - 5x + 10 = 6x - 5x - 4$	Subtract $5x$ from each side.
$10 = x - 4$	
$10 + 4 = x - 4 + 4$	Add 4 to each side.
______ $= x$	

Your Turn Find c so that $r \parallel s$.

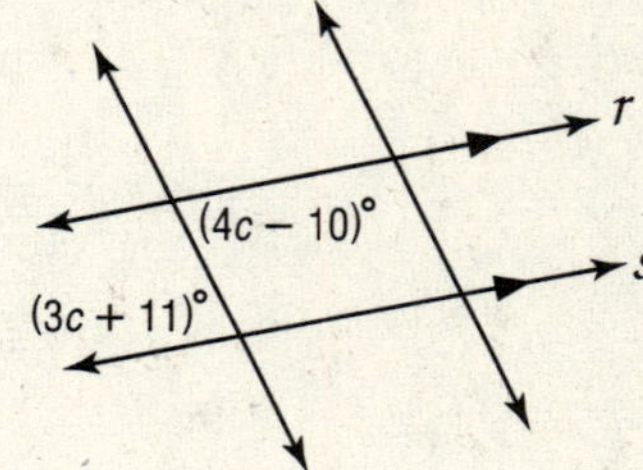

Theorem 4-5 In a plane, if two lines are cut by a transversal so that a pair of alternate interior angles is congruent, then the two lines are parallel.

Theorem 4-6 In a plane, if two lines are cut by a transversal so that a pair of alternate exterior angles is congruent, then the two lines are parallel.

Theorem 4-7 In a plane, if two lines are cut by a transversal so that a pair of consecutive interior angles is supplementary, then the two lines are parallel.

Theorem 4-8 In a plane, if two lines are perpendicular to the same line, then the two lines are parallel.

EXAMPLE

2 Identify the parallel segments in the letter *E*.

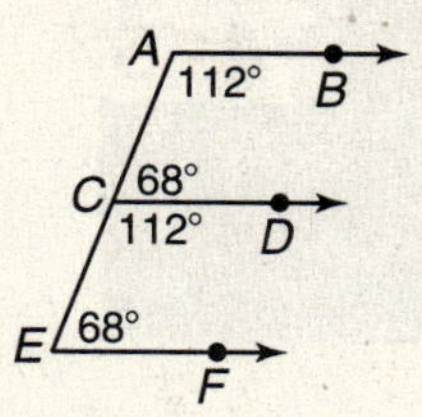

$\angle FEC$ and $\angle DCA$ are corresponding angles.

$m\angle FEC = m\angle DCA$	Both angles measure 68°.
$\overline{EF} \parallel \overline{CD}$	Postulate 4-2

$\angle BAC$ and $\angle DCE$ are corresponding angles.

$m\angle BAC = m\angle DCE$	Both angles measure 112°.
$\overline{AB} \parallel \overline{CD}$	Postulate 4-2
$\overline{AB} \parallel \overline{CD} \parallel \overline{EF}$	Transitive Property

Your Turn Identify the parallel lines in the figure.

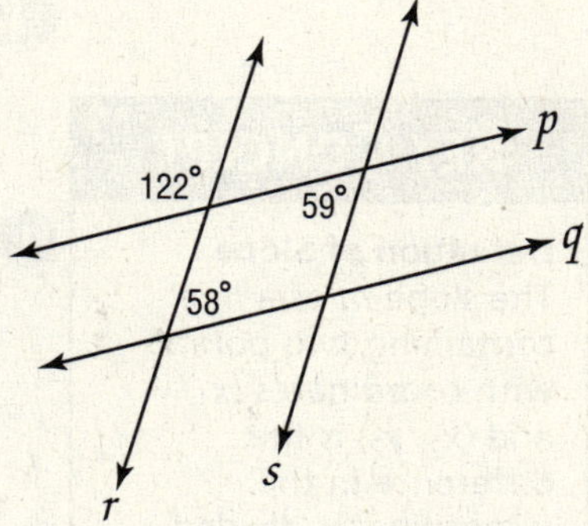

EXAMPLE

3 Find the value of x so that $\overleftrightarrow{KL} \parallel \overleftrightarrow{MN}$.

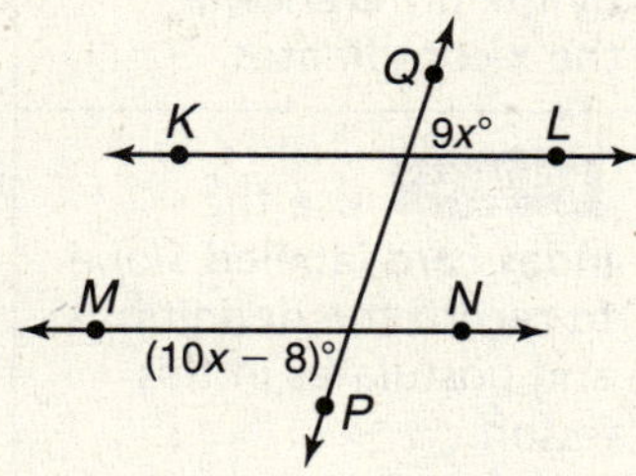

$\overleftrightarrow{PQ}$ is a transversal for $\overleftrightarrow{KL}$ and $\overleftrightarrow{MN}$. If $(9x)° = (10x - 8)°$, then $\overleftrightarrow{KL} \parallel \overleftrightarrow{MN}$ by Theorem 4-6.

$9x = 10x - 8$

$9x - 9x = 10x - 9x - 8$ Subtract $9x$ from each side.

$0 = x - 8$

$0 + 8 = x - 8 + 8$ Add ______ to each side.

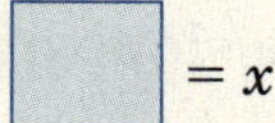

______ $= x$

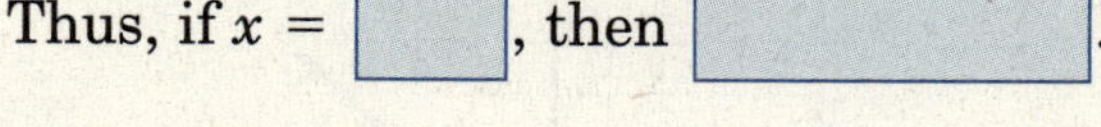

Thus, if $x =$ ______, then ______.

Your Turn Find c so that $r \parallel s$.

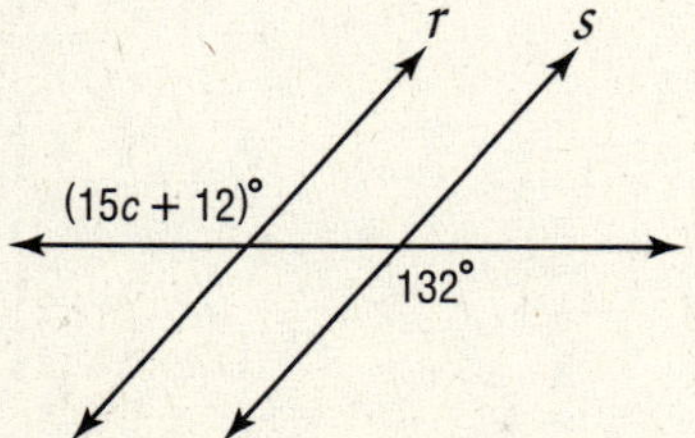

REVIEW IT

What is the relationship between Theorem 4-1 and Theorem 4-5? (*Lesson 4-2*)

HOMEWORK ASSIGNMENT

Page(s):

Exercises:

4–5 Slope

What You'll Learn

- Find the slopes of lines and use slope to identify parallel and perpendicular lines.

Build Your Vocabulary (page 67)

Slope is the ratio of the vertical change to the horizontal change, or the ________ to the ________, as you move from one point on the line to another.

Key Concept

Definition of Slope
The slope m of a line containing two points with coordinates (x_1, y_1) and (x_2, y_2) is the difference in the y-coordinates divided by the difference in the x-coordinates.

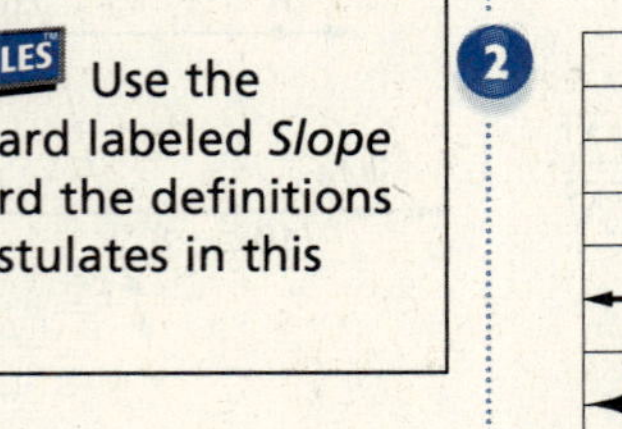

Foldables Use the index card labeled *Slope* to record the definitions and postulates in this lesson.

Examples

Find the slope of each line.

1

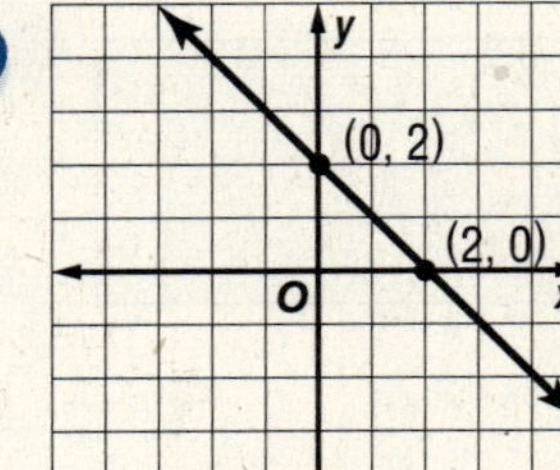

$$m = \frac{0 - 2}{2 - 0} = \frac{-2}{2} = \_\_\_\_$$

2

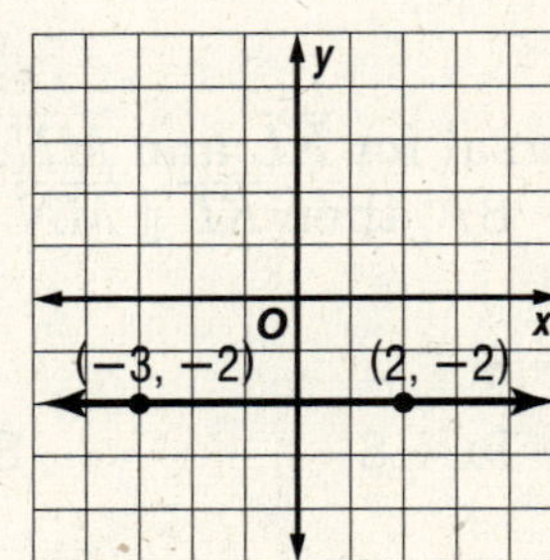

$$m = \frac{-2 - (-2)}{2 - (-3)} = \frac{-2 + 2}{5} = \frac{0}{5} = \_\_\_\_$$

Your Turn **Find the slope of each line.**

a.

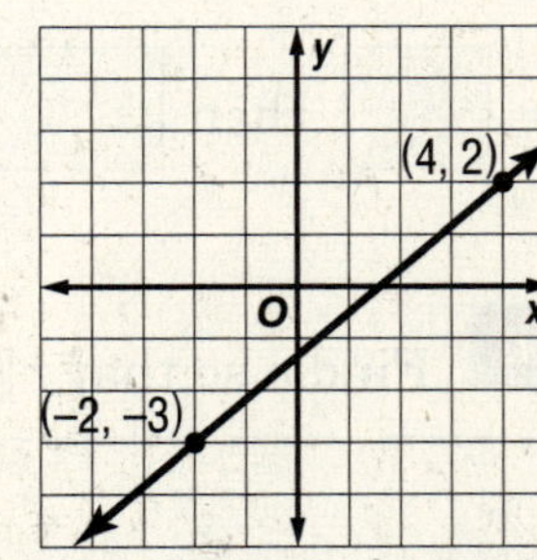

b.

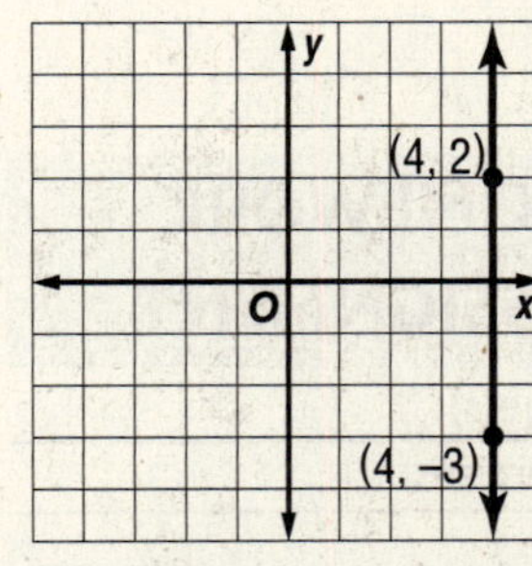

WRITE IT

Explain how you can determine whether a line has a positive or negative slope by observing its graph.

Postulate 4-3
Two distinct nonvertical lines are parallel if and only if they have the same slope.

Postulate 4-4
Two nonvertical lines are perpendicular if and only if the product of their slopes is -1.

EXAMPLE

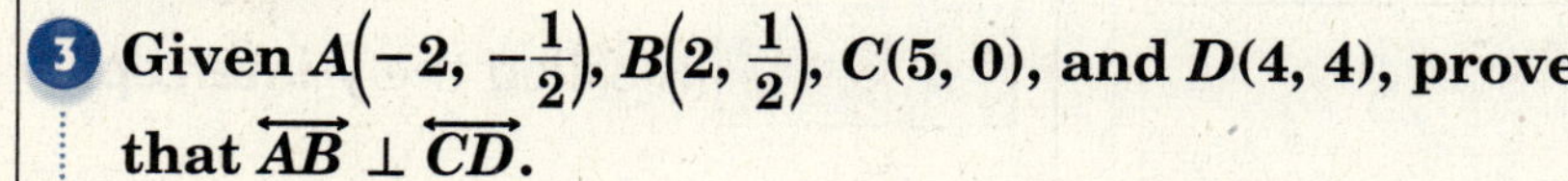

3 Given $A\left(-2, -\frac{1}{2}\right)$, $B\left(2, \frac{1}{2}\right)$, $C(5, 0)$, and $D(4, 4)$, prove that $\overleftrightarrow{AB} \perp \overleftrightarrow{CD}$.

First, find the slopes of $\overleftrightarrow{AB}$ and $\overleftrightarrow{CD}$.

$$\text{slope of } \overleftrightarrow{AB} = \frac{\frac{1}{2} - \left(-\frac{1}{2}\right)}{2 - (-2)} = \frac{\frac{1}{2} + \frac{1}{2}}{2 + 2} = \square$$

$$\text{slope of } \overleftrightarrow{CD} = \frac{4 - 0}{4 - 5} = \frac{4}{-1} = \square$$

The product of the slopes for $\overleftrightarrow{AB}$ and $\overleftrightarrow{CD}$ is $\square(-4)$

or $\square$. Therefore, $\square \perp \square$.

Your Turn Given $A(-3, -4)$, $B(-1, 7)$, $C(2, -5)$, and $D(4, 6)$, prove that $\overleftrightarrow{AB} \parallel \overleftrightarrow{CD}$.

HOMEWORK ASSIGNMENT

Page(s):

Exercises:

4–6 Equations of Lines

What You'll Learn

- Write and graph equations of lines.

Build Your Vocabulary (pages 66–67)

The graph of a **linear equation** is a straight line.

The y-value of the point where the line crosses the

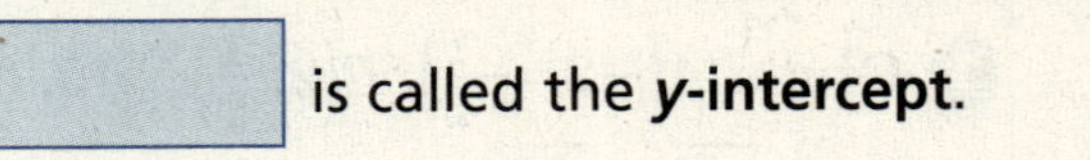 is called the **y-intercept**.

The **slope-intercept form** of a linear equation is written as

, where m is the slope and b is the y-intercept.

Key Concept

Slope-Intercept Form
An equation of the line having slope m and y-intercept b is $y = mx + b$.

Examples

Name the slope and y-intercept of the graph of each equation.

1 $y = \frac{2}{3}x + 6$ The slope is 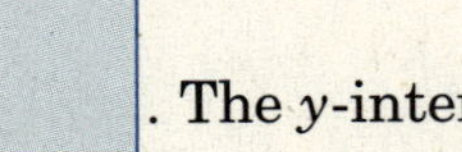. The y-intercept .

2 $y = 0$ The slope is . The y-intercept .

3 $x = 7$ The graph is a line.The slope is undefined. There is no y-intercept.

4 $3y + 12 = 6x$

Rewrite the equation in slope-intercept form by solving for y.

$$3y + 12 = 6x$$

$$3y + 12 - 12 = 6x - 12$$ Subtract 12 from each side.

$$3y = 6x - 12$$

$$\frac{3y}{3} = \frac{6x - 12}{3}$$ Divide each side by 3.

$$y = 2x - 4$$ Simplify. This is written in slope-intercept form.

The slope m = . The y-intercept is .

FOLDABLES™

ORGANIZE IT

Use the index card labeled *Equations of Lines* to record important formulas and ideas in this lesson. Give examples that show the most important ideas in the lesson.

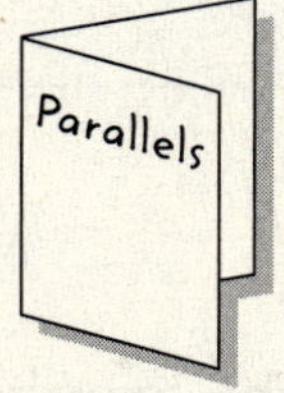

Your Turn **Name the slope and y-intercept of the graph of each equation.**

a. $y = -6x + 13$

b. $y = 8$

c. $x = 7$

d. $4x + 3y = 5$

EXAMPLE

5 Graph $2x - y = 4$ using the slope and y-intercept.

First, rewrite the equation in slope-intercept form.

$$2x - y = 4$$

$2x - y -$ ____ $= 4 -$ ____ Subtract $2x$ from each side.

$-y =$ ____

$\frac{-y}{-1} = \frac{4 - 2x}{-1}$ Divide each side by -1.

$y =$ ____ Slope-intercept form

The y-intercept is -4. So, the point $(0, -4)$ is on the line. Since the slope is 2, or $\frac{2}{1}$, plot a point by using a *rise* of ____ units (up) and a *run* of  unit (right). Draw a line through the two points.

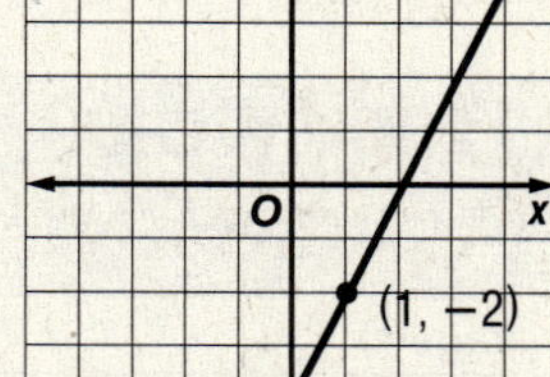

Your Turn Graph $3x - 4y = -8$ using the slope and y-intercept.

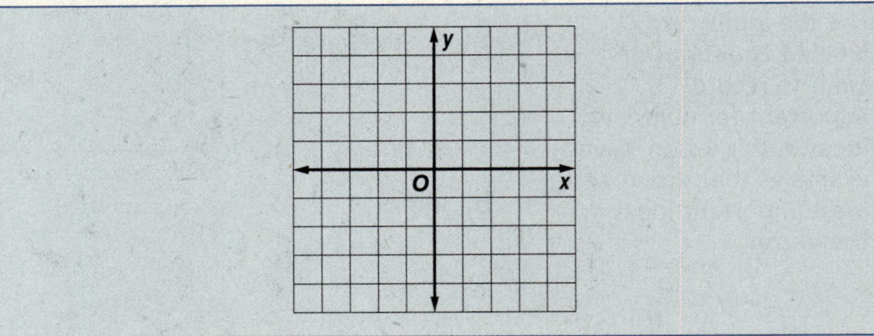

WRITE IT

Explain how you can find the slope of a line perpendicular to a given line.

______

______

______

______

EXAMPLE

6 Write an equation of the line parallel to the graph of $y = -2x + 3$ that passes through the point at (0, 1).

Because the lines are parallel, they must have the same slope. So, $m =$ ______.

To find b, use the ordered pair (0, 1) and substitute for m, x, and y in the slope-intercept form.

$y = mx + b$

$1 =$ ______ $(0) + b$ $\quad m =$ ______, $(x, y) =$ ______

$1 = 0 + b$

______ $= b$

The value of b is ______. So, the equation of the line is ______.

Your Turn

a. Write an equation of the line parallel to the graph of $-5x + y = 6$ that passes through the point $(-1, 3)$.

b. Write an equation of the line perpendicular to the graph of $y = -2x + 1$ that passes through the point $(4, -5)$.

HOMEWORK ASSIGNMENT

Page(s):

Exercises:

CHAPTER 4

BRINGING IT ALL TOGETHER

STUDY GUIDE

FOLDABLES™	VOCABULARY PUZZLEMAKER	BUILD YOUR VOCABULARY
Use your **Chapter 4 Foldable** to help you study for your chapter test.	To make a crossword puzzle, word search, or jumble puzzle of the vocabulary words in Chapter 4, go to: www.glencoe.com/sec/math/t_resources/free/index.php.	You can use your completed **Vocabulary Builder** (pages 66–67) to help you solve the puzzle.

4-1

Parallel Lines and Planes

Choose the term that best completes each sentence.

1. (Skew/Parallel) lines always lie on the same plane.

2. (Perpendicular/Skew) lines never have any points in common.

3. (Parallel/Perpendicular) lines never intersect.

4-2

Parallel Lines and Transversals

Refer to the figure and match the term with the best representative angle pair. Angle pairs cannot be matched more than once.

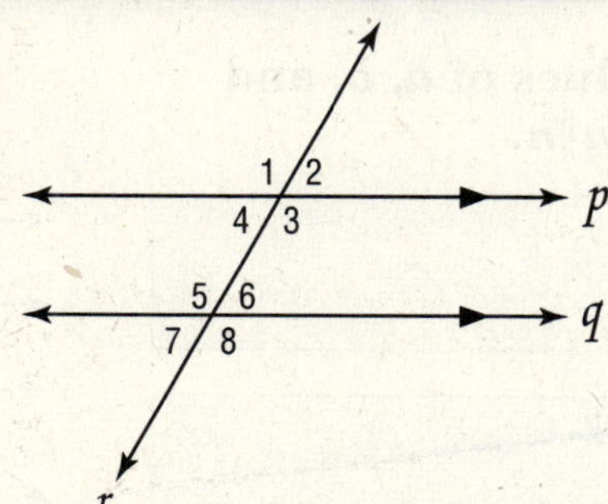

4. consecutive interior angles

5. exterior angles

6. alternate interior angles

7. alternate exterior angles

a. ∠2 and ∠7

b. ∠3 and ∠6

c. ∠4 and ∠6

d. ∠1 and ∠7

e. ∠3 and ∠4

4-3

Transversals and Corresponding Angles

In the figure, $\ell \parallel m$, and transversal r is perpendicular to m. Name all angles congruent to the given angle.

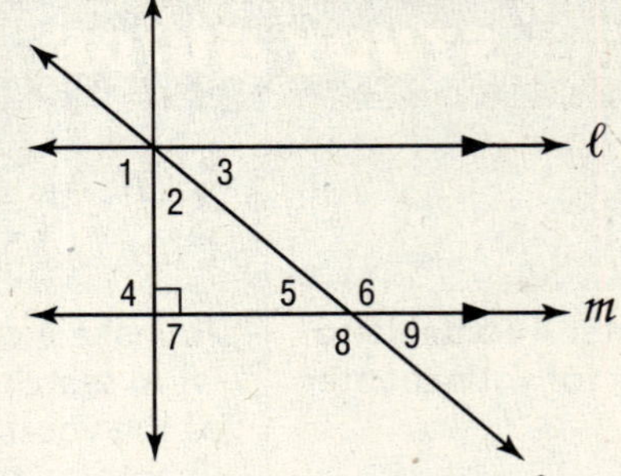

8. ∠4

9. ∠3

10. ∠9

Refer to the above figure to find the measure of the specified angle if $m\angle 3 = 40$.

11. ∠4

12. ∠5

13. ∠8

14. ∠2

4-4

Proving Lines Parallel

Find the values of a, b, and c so that $\ell \parallel m \parallel n$.

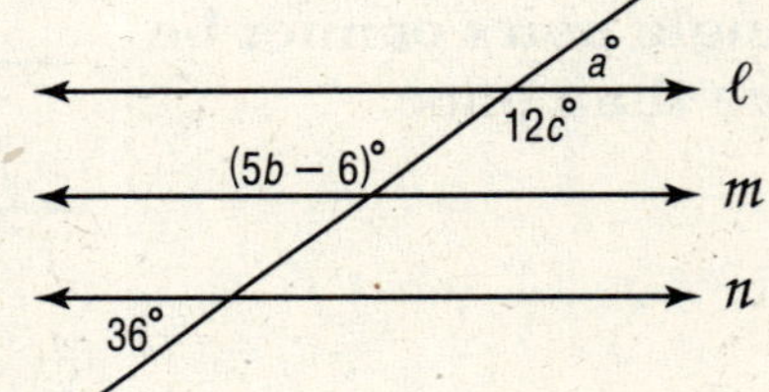

15. $a =$

16. $b =$

17. $c =$

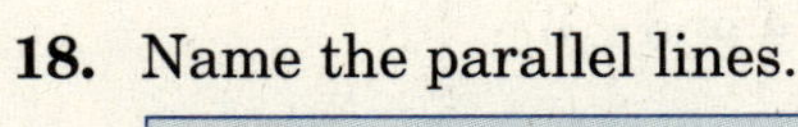

18. Name the parallel lines.

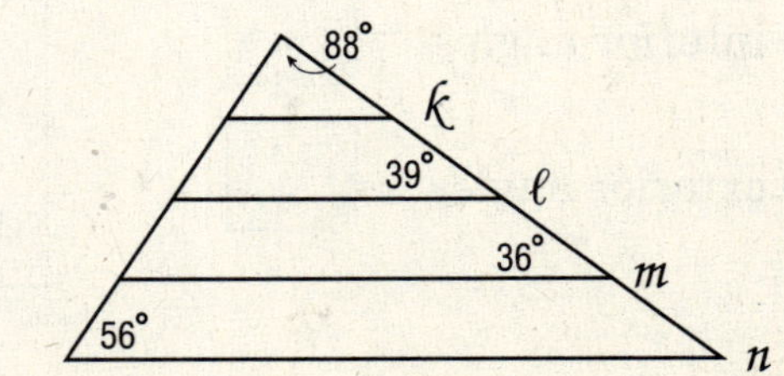

4-5

Slope

A wheelchair access ramp must be added to a home. One plan showed a ramp that started 30 feet away from the entrance. The entrance was 3 feet higher than ground level. The second plan started the ramp 15 feet from the same 3-foot high entrance.

19. What is the slope of each ramp?

20. Which slope is steeper?

21. Given $A(0, 4)$, $B(3, 6)$, $C(1, 2)$, and $D(3, -1)$, determine whether $\overleftrightarrow{AB}$ and $\overleftrightarrow{CD}$ are *parallel, perpendicular*, or *neither*.

4-6

Equations of Lines

Identify the slope and *y*-intercept of each equation.

22. $y = -6x + \frac{1}{2}$

23. $5x - 4y = 7$

24. $y = -2$

25. $x = 5$

26. Write an equation of a line parallel to $y = 3x + 2$ that passes through the point $(-1, -4)$.

Checklist

ARE YOU READY FOR THE CHAPTER TEST?

Visit **geomconcepts.com** to access your textbook, more examples, self-check quizzes, and practice tests to help you study the concepts in Chapter 4.

Check the one that applies. Suggestions to help you study are given with each item.

☐ **I completed the review of all or most lessons without using my notes or asking for help.**

- You are probably ready for the Chapter Test.
- You may want to take the Chapter 4 Practice Test on page 183 of your textbook as a final check.

☐ **I used my Foldable or Study Notebook to complete the review of all or most lessons.**

- You should complete the Chapter 4 Study Guide and Review on pages 180–182 of your textbook.
- If you are unsure of any concepts or skills, refer back to the specific lesson(s).
- You may also want to take the Chapter 4 Practice Test on page 183.

☐ **I asked for help from someone else to complete the review of all or most lessons.**

- You should review the examples and concepts in your Study Notebook and Chapter 4 Foldable.
- Then complete the Chapter 4 Study Guide and Review on pages 180–182 of your textbook.
- If you are unsure of any concepts or skills, refer back to the specific lesson(s).
- You may also want to take the Chapter 4 Practice Test on page 183.

Student Signature

Parent/Guardian Signature

Teacher Signature

Triangles and Congruence

Use the instructions below to make a Foldable to help you organize your notes as you study the chapter. You will see Foldable reminders in the margin of this Interactive Study Notebook to help you in taking notes.

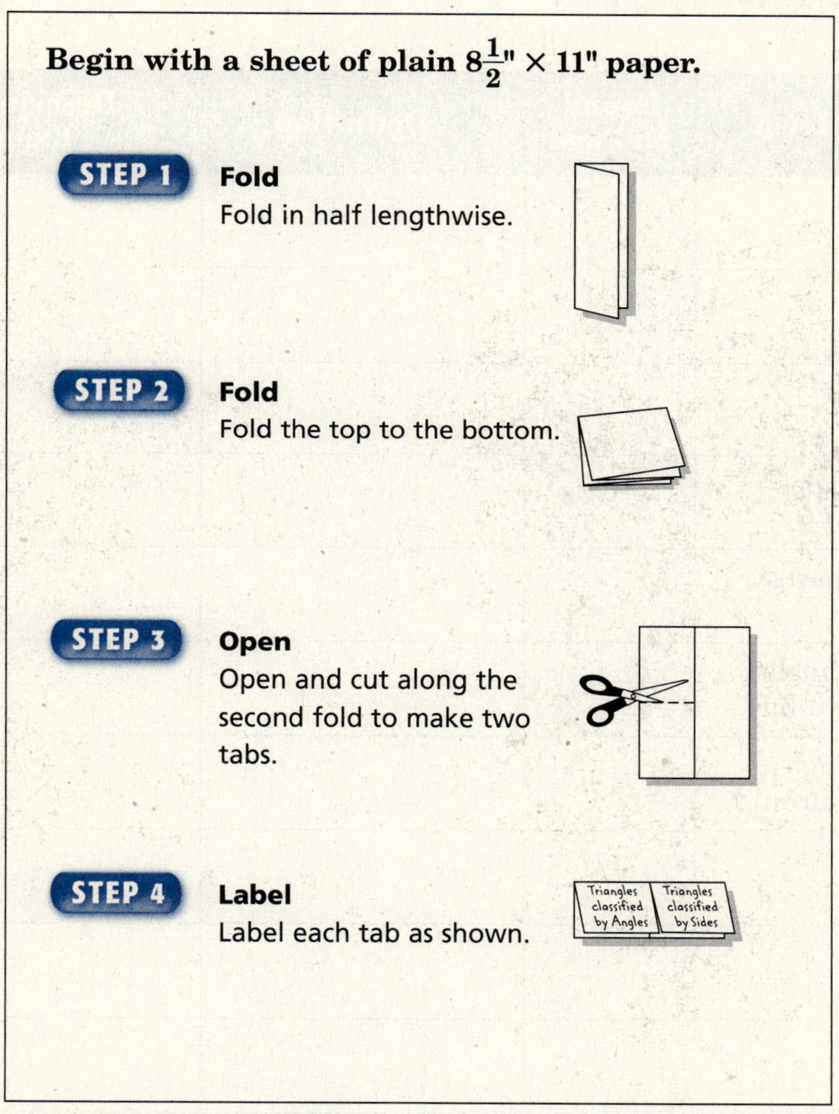

Begin with a sheet of plain $8\frac{1}{2}$" × 11" paper.

STEP 1 **Fold**
Fold in half lengthwise.

STEP 2 **Fold**
Fold the top to the bottom.

STEP 3 **Open**
Open and cut along the second fold to make two tabs.

STEP 4 **Label**
Label each tab as shown.

NOTE-TAKING TIP: When you take notes, define new terms and write about the new concepts you are learning in your own words. Then, write your own examples that use the new terms and concepts.

CHAPTER 5

BUILD YOUR VOCABULARY

This is an alphabetical list of new vocabulary terms you will learn in Chapter 5. As you complete the study notes for the chapter, you will see Build Your Vocabulary reminders to complete each term's definition or description on these pages. Remember to add the textbook page number in the second column for reference when you study.

Vocabulary Term	Found on Page	Definition	Description or Example
acute triangle			
base			
base angles			
congruent triangles			
corresponding parts			
equiangular triangle [eh-kwee-AN-gyu-lur]			
equilateral triangle [EE-kwuh-LAT-ur-ul]			
image			
included angle			
included side			
isometry [eye-SAH-muh-tree]			
isosceles triangle [eye-SAHS-uh-LEEZ]			

Vocabulary Term	Found on Page	Definition	Description or Example
legs			
mapping			
obtuse triangle			
preimage			
reflection			
right triangle			
rotation			
scalene triangle [SKAY-leen]			
transformation			
translation			
vertex			
vertex angle			

5–1 Classifying Triangles

WHAT YOU'LL LEARN

- Identify the parts of triangles and classify triangles by their parts.

BUILD YOUR VOCABULARY (pages 88–89)

The side that is opposite the vertex angle in an ______ triangle is called the **base**.

In an isosceles triangle, the two angles formed by the ______ and one of the congruent ______ are called **base angles**.

The congruent sides in an isosceles triangle are the **legs**.

The vertex of each angle of a ______ is a **vertex** of the triangle.

The angle formed by the ______ sides in an ______ triangle is called the **vertex angle**.

FOLDABLES ORGANIZE IT

Draw examples of acute, obtuse, right, scalene, isosceles, and equilateral triangles in your notes.

Triangles classified by Angles | Triangles classified by Sides

EXAMPLES

Classify each triangle by its angles and by its sides.

1

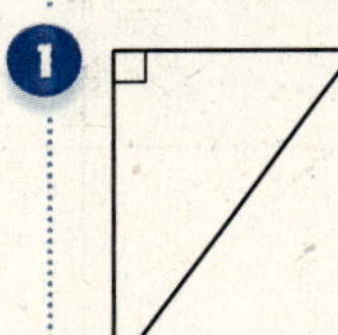

The triangle is a ______ triangle.

2

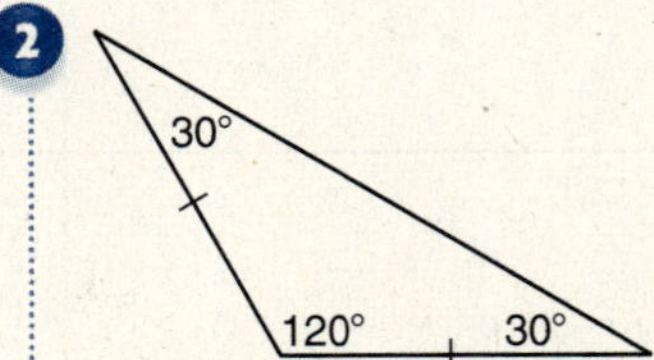

The triangle is an ______ triangle.

Your Turn **Classify each triangle by its angles and by its sides.**

a.

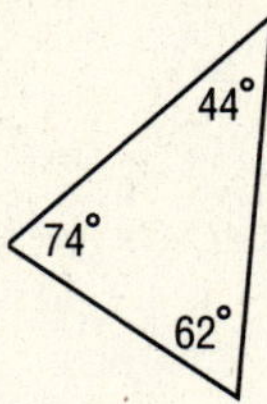

b.

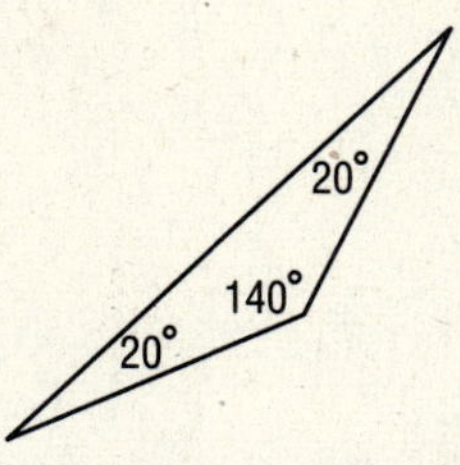

EXAMPLE

> **REMEMBER IT**
>
> The vertex of each angle is a vertex of the triangle.

3 **Find the measures of $\overline{XY}$ and $\overline{YZ}$ of isosceles triangle *XYZ* if $\angle X$ is the vertex angle.**

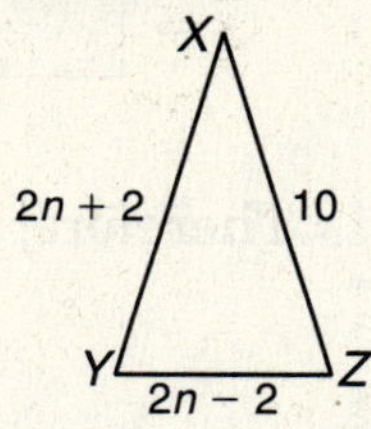

Since $\angle X$ is the vertex angle,  $\cong$ .

So, $XY =$ ______. Write and solve an equation.

$XY =$ ______

______ $=$ ______ Substitution

$2n + 2 -$ ______ $= 10 -$ ______ Subtract ______ from each side.

______ $=$ ______

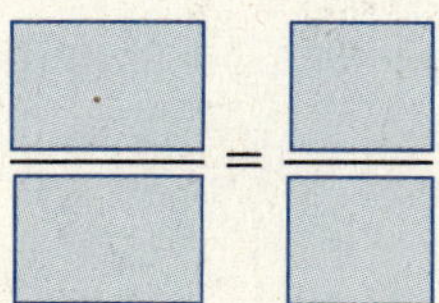 $=$ 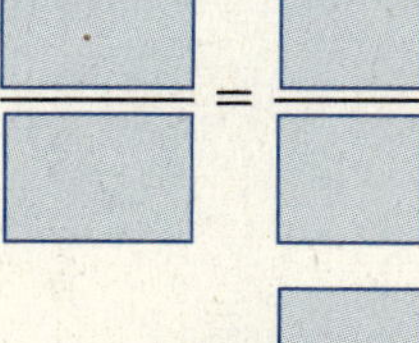 Divide each side by 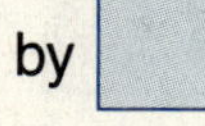.

$n =$ ______

The value of n is ______.

To find the measures of $\overline{XY}$ and $\overline{YZ}$, replace n with ☐ in the expression for each measure.

$XY = 2n + 2$

$= 2(\square) + 2$

$= \square + 2$

$= \square$

$YZ = 2n - 2$

$= 2(\square) - 2$

$= \square - 2$

$= \square$

Therefore, $XY = \square$ and $YZ = \square$.

Your Turn Triangle DEF is an isosceles triangle with base $\overline{EF}$. Find DE and EF.

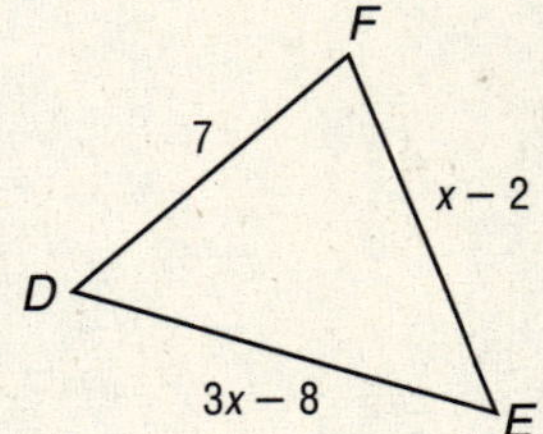

HOMEWORK ASSIGNMENT

Page(s):

Exercises:

5–2 Angles of a Triangle

What You'll Learn

- Use the Angle Sum Theorem

Theorem 5-1 Angle Sum Theorem
The sum of the measures of the angles of a triangle is 180.

EXAMPLES

1 Find $m\angle P$ in $\triangle MNP$ if $m\angle M = 80$ and $m\angle N = 45$.

$m\angle P + m\angle M + m\angle N = 180$ Angle Sum Theorem

$m\angle P + \square + \square = 180$ Substitution

$m\angle P + 125 = 180$

$m\angle P + 125 - \square = 180 - \square$ Subtract.

$m\angle P = \square$

2 Find the value of each variable in $\triangle ABC$.

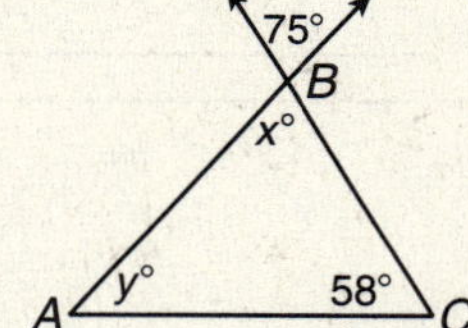

$\angle ABC$ is a vertical angle to the given angle measure of 75. Since vertical angles are congruent, $m\angle ABC = 75 = x$.

$m\angle ABC + m\angle BCA + m\angle CAB = 180$ Angle Sum Theorem

$\square + 58 + \square = 180$ Substitution

$133 + y = 180$

$133 + y - 133 = 180 - 133$ Subtract.

$y = \square$

Therefore, $x = \square$ and $y = \square$.

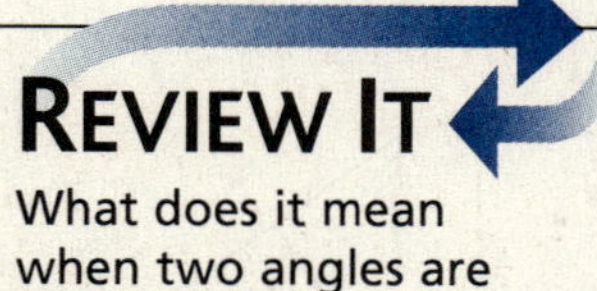

Review It

What does it mean when two angles are complementary? *(Lesson 3-5)*

____

____

____

____

Your Turn Find the value of each variable.

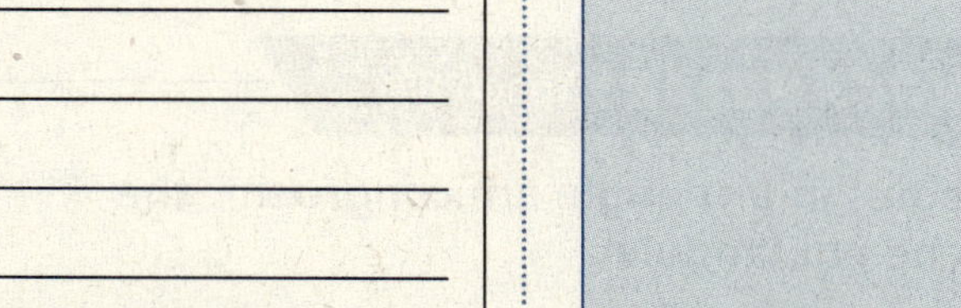

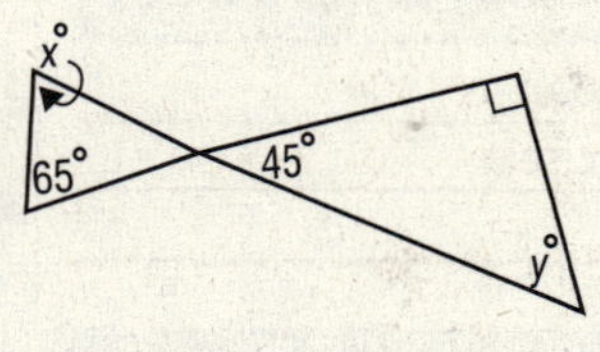

Theorem 5-2
The acute angles of a right triangle are complementary.

Theorem 5-3
The measure of each angle of an equiangular triangle is 60.

EXAMPLE

3 **Find $m\angle J$ and $m\angle K$ in right triangle *JKL*.**

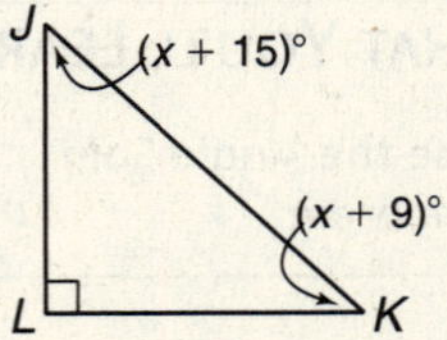

$m\angle J + m\angle K = 90$	Theorem 5-2
$(x + 15) + (x + 9) = 90$	Substitution
$\square + 24 = 90$	Combine like terms.
$2x + 24 - \square = 90 - \square$	Subtract.
$2x = 66$	
$\frac{2x}{\square} = \frac{66}{\square}$	Divide.
$x = \square$	

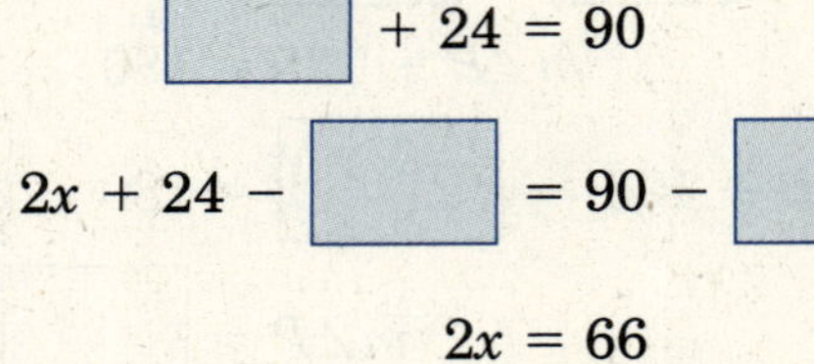

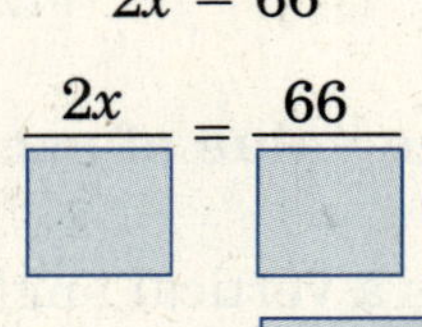

Replace x with $\square$ in each angle expression.

$m\angle J = \square + 15$ or $\square$

$m\angle K = \square + 9$ or $\square$

Therefore, $m\angle J = \square$ and $m\angle K = \square$.

Your Turn Find the value of a, b, and c.

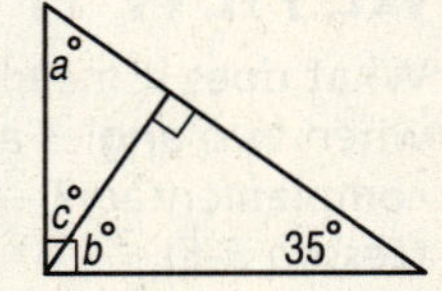

WRITE IT

Is it possible to have two right angles in a triangle? Justify your answer.

______

______

______

______

HOMEWORK ASSIGNMENT

Page(s):

Exercises:

BUILD YOUR VOCABULARY (page 88)

When all three angles in a triangle are congruent, the triangle is said to be **equiangular**.

5–3 Geometry in Motion

What You'll Learn

- Identify translations, reflections, and rotations and their corresponding parts.

Build Your Vocabulary (page 89)

________ a figure from one position to another without turning it is called a **translation**.

________ a figure over a line creates the mirror image of the figure, or a **reflection**.

________ a figure around a fixed point creates a **rotation**.

EXAMPLES

Identify each motion as a *translation*, *reflection*, or *rotation*.

1 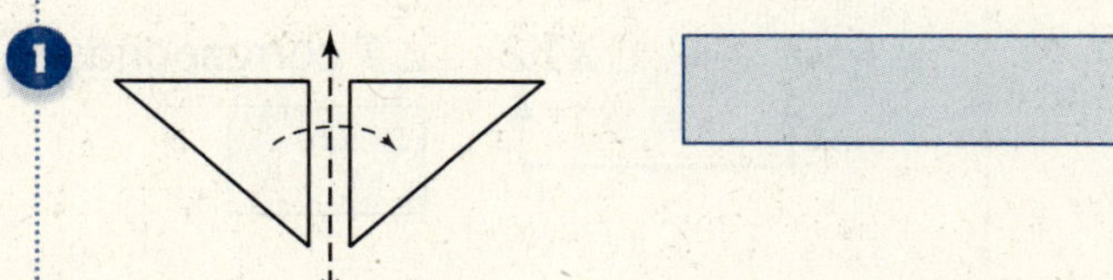

____

2 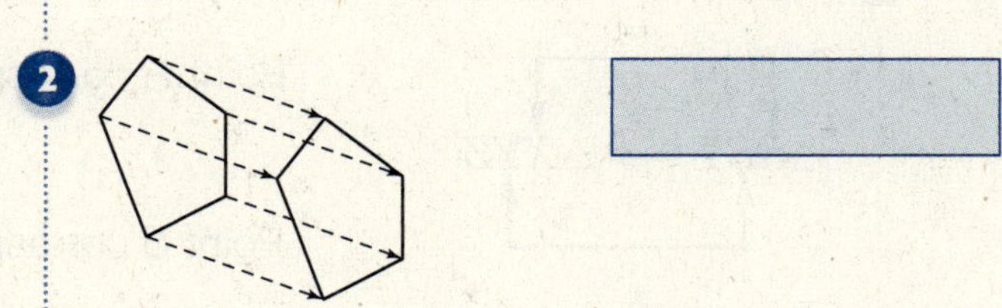

____

Your Turn **Identify each motion as a *translation*, *reflection*, or *rotation*.**

a.

____

b.

____

BUILD YOUR VOCABULARY (pages 88–89)

Pairing each point on the original figure, or [blank], with exactly one point on the [blank] is called **mapping**.

The moving of each [blank] of a preimage to a new figure called the image is a **transformation**.

The new figure in a [blank] is called the **image**.

In a transformation, the [blank] figure is called the **preimage**.

EXAMPLES

In the figure, $\triangle RST \rightarrow \triangle XYZ$ by a translation.

3 Name the image of $\angle T$.

$\triangle RST \rightarrow \triangle XYZ$ $\angle T$ corresponds to [blank].

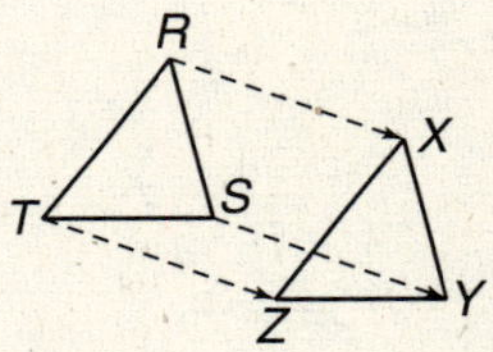

4 Name the side that corresponds to $\overline{XY}$.

$\triangle RST \rightarrow \triangle XYZ$

Point R corresponds to point [blank].

Point S corresponds to point [blank].

So, [blank] corresponds to [blank].

Your Turn **In the figure, $\triangle QRS \rightarrow \triangle DEF$ by a rotation.**

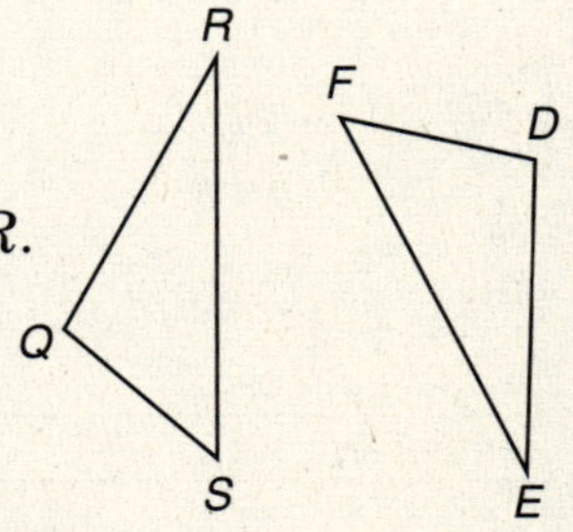

a. Name the angle that corresponds to $\angle R$.

[blank]

b. Name the side that corresponds to $\overline{QR}$.

[blank]

BUILD YOUR VOCABULARY (page 88)

Translations, reflections, and rotations are all **isometries** and do not change the ______ or ______ of the figure being moved.

EXAMPLE

5 **Identify the type(s) of transformations that were used to complete the work below.**

Some figures can be moved to ______ another without turning or flipping. Other figures have been turned around a ______ point with respect to the original.

Therefore, the transformations are ______ and ______.

Your Turn Identify the type(s) of transformations that were used to complete the work below.

HOMEWORK ASSIGNMENT

Page(s):

Exercises:

5–4 Congruent Triangles

WHAT YOU'LL LEARN

- Identify corresponding parts of congruent triangles.

BUILD YOUR VOCABULARY (page 88)

If a triangle can be translated, rotated, or reflected onto another triangle so that all of the ______ correspond, the triangles are said to be **congruent**.

The parts of congruent triangles that ______ are called **corresponding parts**.

KEY CONCEPT

Definition of Congruent Triangles If the corresponding parts of two triangles are congruent, then the two triangles are congruent. Likewise, if two triangles are congruent, then the corresponding parts of the two triangles are congruent.

EXAMPLES

1 If $\triangle ABC \cong \triangle FDE$, name the congruent angles and sides. Then draw the triangles, using arcs and slash marks to show congruent angles and sides.

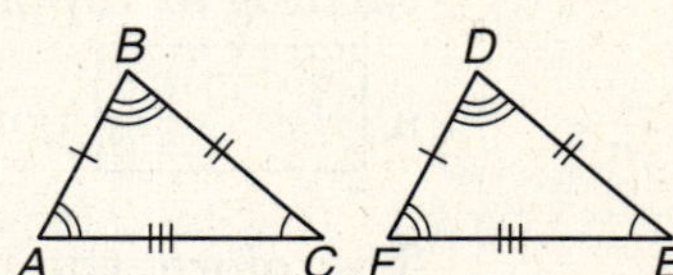

Name the three pairs of congruent angles by looking at the order of the vertices in the statement $\triangle ABC \cong \triangle FDE$.

$\angle A \cong$ ______, $\angle B \cong$ ______,

and $\angle C \cong$ ______.

Since A corresponds to ______, and B corresponds to ______,

______ $\cong$ ______.

Since B corresponds to D, and C corresponds to E,

______ $\cong$ ______.

Since ______ corresponds to F, and ______ corresponds to E,

______ $\cong \overline{FE}$.

REMEMBER IT

The order of the vertices in a congruence statement shows the corresponding parts of the congruent triangles.

2 **The corresponding parts of two congruent triangles are marked on the figure. Write a congruence statement for the two triangles.**

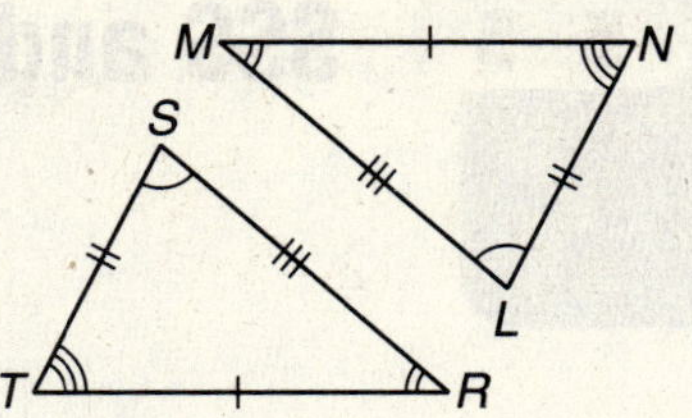

List the congruent angles and sides.

$\angle L \cong$ ______ $\angle M \cong \angle R$ $\angle N \cong$ ______

$\overline{LN} \cong \overline{ST}$ ______ $\cong \overline{TR}$ ______ $\cong \overline{SR}$

The congruence statement can be written by matching the ______ of the ______ angles. Therefore,

$\triangle LMN \cong$ ______.

Your Turn

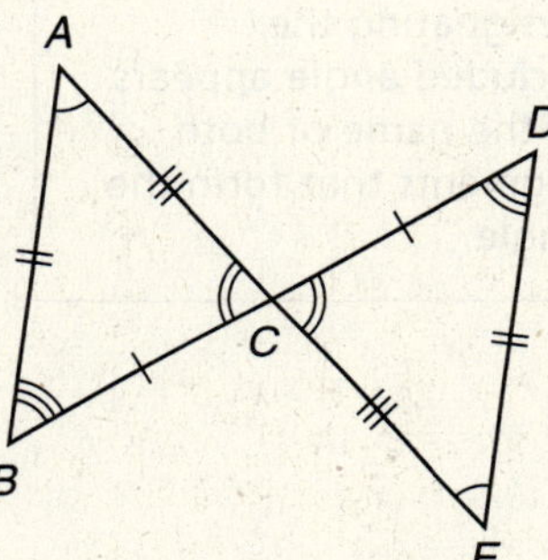

a. If $\triangle ACB \cong \triangle ECD$, name the congruent angles and sides. Then draw the triangles, using arcs and slash marks to show congruent angles and sides.

b. Write another congruence statement for the two triangles other than the one given above.

HOMEWORK ASSIGNMENT

Page(s):

Exercises:

5–5 SSS and SAS

WHAT YOU'LL LEARN

- Use the SSS and SAS tests for congruence.

Postulate 5-1 SSS Postulate
If three sides of one triangle are congruent to three corresponding sides of another triangle, then the triangles are congruent.

EXAMPLE

1 In two triangles, $\overline{DF} \cong \overline{UV}$, $\overline{FE} \cong \overline{VW}$, and $\overline{DE} \cong \overline{UW}$. Write a congruence statement for the two triangles.

REMEMBER IT

The letter designating the included angle appears in the name of both segments that form the angle.

Draw a pair of ______ triangles. Identify the congruent parts with ______. Label the vertices of one triangle.

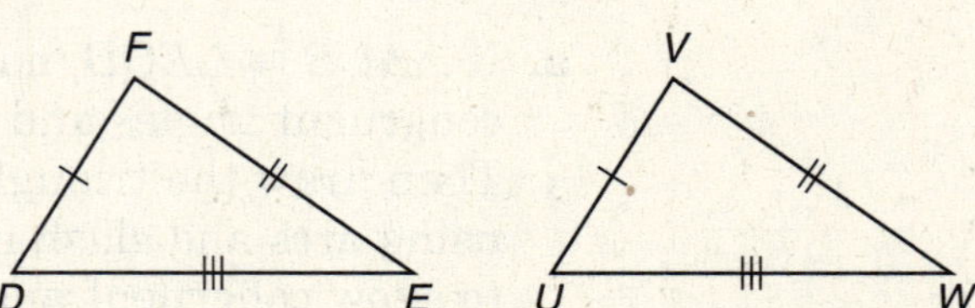

Use the given information to label the ______ in the second triangle.

By SSS, ______ ≅ ______.

Your Turn In two triangles, $\overline{CB} \cong \overline{EF}$, $\overline{CA} \cong \overline{ED}$, and $\overline{BA} \cong \overline{FD}$. Write a congruence statement for the two triangles.

______

BUILD YOUR VOCABULARY (page 88)

In a triangle, the ______ formed by two given ______ is the **included angle**.

Postulate 5-2 SAS Postulate
If two sides and the included angle of one triangle are congruent to the corresponding sides and included angle of another triangle, then the triangles are congruent.

WRITE IT

Explain the SSS and SAS tests for congruence in your own words. Give an example of each.

______
______
______
______
______
______
______
______
______
______

EXAMPLE

2 Determine whether the triangles shown at the right are congruent. If so, write a congruence statement and explain why the triangles are congruent. If not, explain why not.

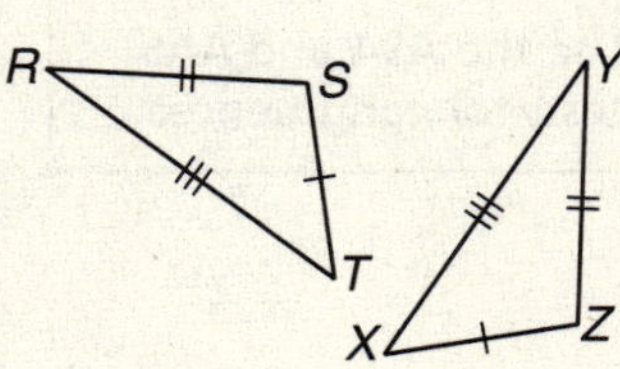

There are three pairs of ______ sides,

$\overline{RS} \cong$ ______, ______ $\cong \overline{ZX}$ and $\overline{RT} \cong$ ______.

Therefore, ______ $\cong$ ______ by ______.

Your Turn Determine whether the triangles to the right are congruent. If so, write a congruence statement and explain why the triangles are congruent. If not, explain why not.

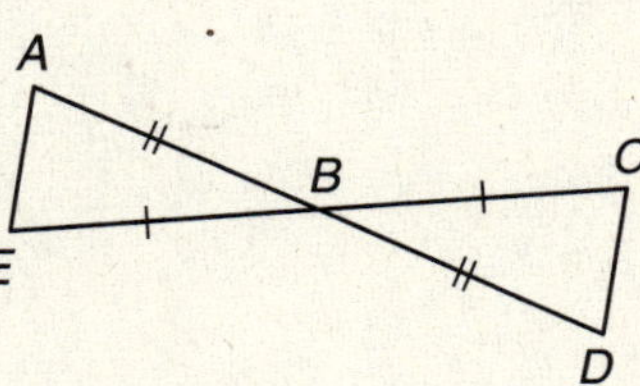

HOMEWORK ASSIGNMENT

Page(s):
Exercises:

5–6 ASA and AAS

WHAT YOU'LL LEARN

- Use the ASA and AAS tests for congruence.

BUILD YOUR VOCABULARY (page 88)

The ______ of the triangle that falls between two given ______ is called the **included side** and is the one common side to both angles.

Postulate 5-3 ASA Postulate
If two angles and the included side of one triangle are congruent to the corresponding angles and included side of another triangle, then the triangles are congruent.

EXAMPLE

1 In $\triangle DEF$ and $\triangle ABC$, $\angle D \cong \angle C$, $\angle E \cong \angle B$, and $\overline{DE} \cong \overline{CB}$. Write a congruence statement for the two triangles.

Draw a pair of ______ triangles. Mark the congruent parts with ______ and ______. Label the vertices of one triangle D, E, and F.

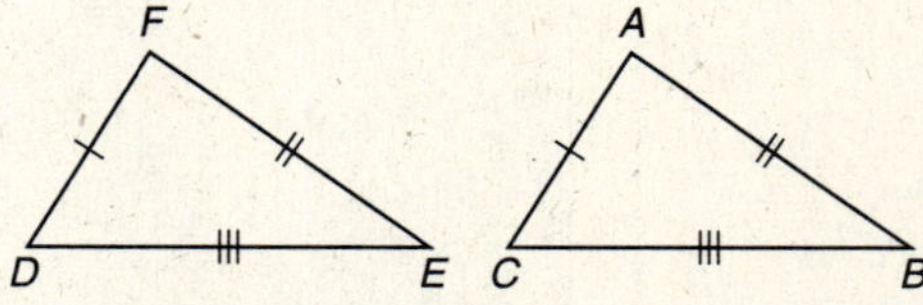

Locate C and B on the unlabeled triangle in the same positions as ______ and ______. The unassigned vertex is

______. Therefore, ______ $\cong$ ______.

Your Turn In $\triangle RST$ and $\triangle XYZ$, $\overline{ST} \cong \overline{XZ}$, $\angle S \cong \angle X$, and $\angle T \cong \angle Z$. Write a congruence statement for the two triangles.

Theorem 5-4 AAS Theorem
If two angles and a nonincluded side of one triangle are congruent to the corresponding two angles and nonincluded side of another triangle, then the triangles are congruent.

EXAMPLE

2 **$\triangle XYZ$ and $\triangle QRS$ each have one pair of sides and one pair of angles marked to show congruence. What other pair of angles needs to be marked so the two triangles are congruent by AAS?**

If $\angle Q$ and $\angle X$ are marked ______, and ______ $\cong$ ______, then ______ and ______ would have to be congruent for the triangles to be congruent by ______.

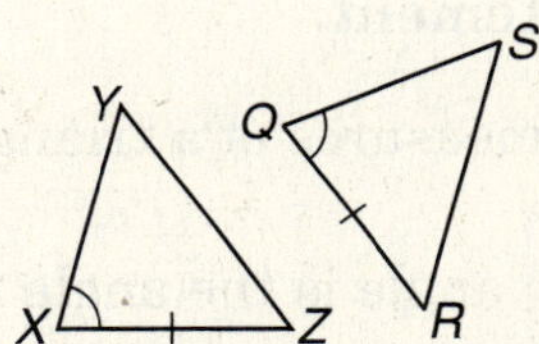

Your Turn $\triangle ACB$ and $\triangle FED$ each have one pair of sides and one pair of angles marked to show congruence. What other pair of angles needs to be marked so the two triangles are congruent by AAS?

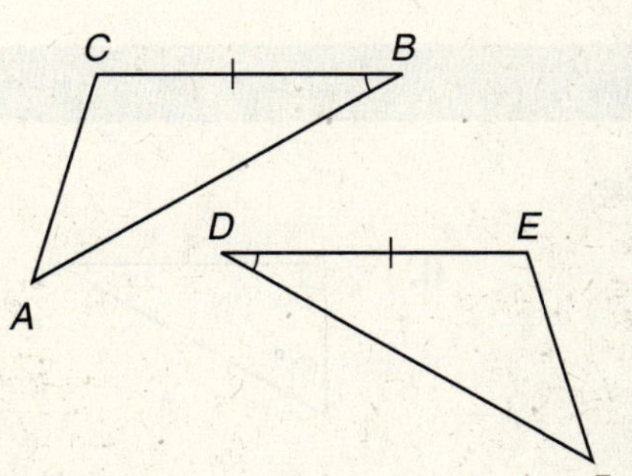

HOMEWORK ASSIGNMENT
Page(s):
Exercises:

CHAPTER 5

BRINGING IT ALL TOGETHER

STUDY GUIDE

FOLDABLES™	VOCABULARY PUZZLEMAKER	BUILD YOUR VOCABULARY
Use your **Chapter 5 Foldable** to help you study for your chapter test.	To make a crossword puzzle, word search, or jumble puzzle of the vocabulary words in Chapter 5, go to: www.glencoe.com/sec/math/t_resources/free/index.php	You can use your completed **Vocabulary Builder** (pages 88–89) to help you solve the puzzle.

5–1 Classifying Triangles

Complete each statement.

1. The sum of the measures of a triangle's interior angles is ______.
2. The ______ angle is the angle formed by two congruent sides of an isosceles triangle.
3. The ______ angles of a right triangle are complementary.
4. A triangle with no congruent sides is ______.

5–2 Angles of a Triangle

Find the value of each variable.

5. 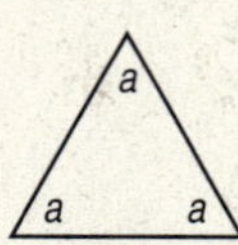

6. 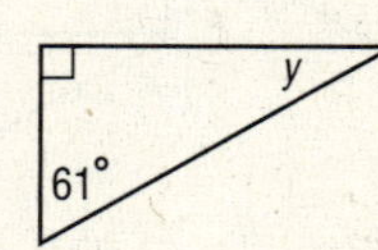

5–3 Geometry in Motion

Suppose $\triangle SRN \rightarrow \triangle CDA$.

7. Which angle corresponds to $\angle S$?

8. Name the preimage of $\overline{AD}$.

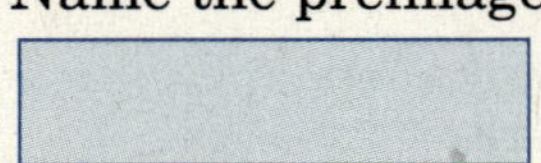

9. Identify the transformation that occurred in the mapping.

5–4 Congruent Triangles

If $\triangle ABC \cong \triangle QRS$, name the corresponding congruent parts.

10. $\angle B$

11. $\overline{AC}$

12. $\overline{RQ}$

13. $\angle C$

5–5 SSS and SAS

14. The pairs of triangles at the right are congruent. Write a congruence statement and the reason the triangles are congruent.

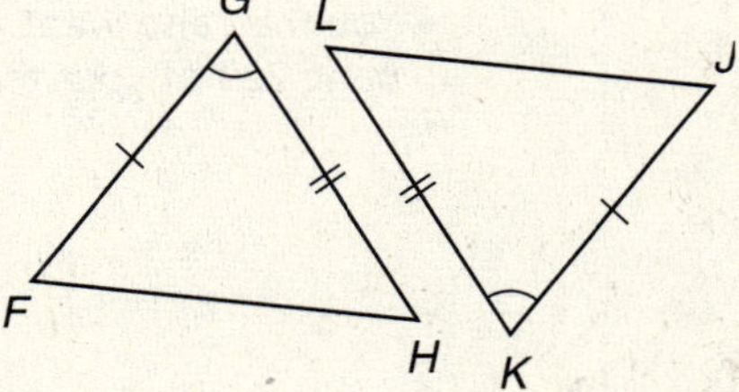

5-6 ASA and AAS

Underline the best term to make the statement true.

15. [Mapping/Congruence] of triangles is explained by SSS, SAS, ASA, and AAS.

16. AAS indicates two angles and their [included/nonincluded] side.

ARE YOU READY FOR THE CHAPTER TEST?

Visit **geomconcepts.com** to access your textbook, more examples, self-check quizzes, and practice tests to help you study the concepts in Chapter 5.

Check the one that applies. Suggestions to help you study are given with each item.

☐ **I completed the review of all or most lessons without using my notes or asking for help.**

- You are probably ready for the Chapter Test.
- You may want to take the Chapter 5 Practice Test on page 223 of your textbook as a final check.

☐ **I used my Foldable or Study Notebook to complete the review of all or most lessons.**

- You should complete the Chapter 5 Study Guide and Review on pages 220–222 of your textbook.
- If you are unsure of any concepts or skills, refer back to the specific lesson(s).
- You may also want to take the Chapter 5 Practice Test on page 223 of your textbook.

☐ **I asked for help from someone else to complete the review of all or most lessons.**

- You should review the examples and concepts in your Study Notebook and Chapter 5 Foldable.
- Then complete the Chapter 5 Study Guide and Review on pages 220–222 of your textbook.
- If you are unsure of any concepts or skills, refer back to the specific lesson(s).
- You may also want to take the Chapter 5 Practice Test on page 223 of your textbook.

Student Signature

Parent/Guardian Signature

Teacher Signature

More About Triangles

Use the instructions below to make a Foldable to help you organize your notes as you study the chapter. You will see Foldable reminders in the margin of this Interactive Study Notebook to help you in taking notes.

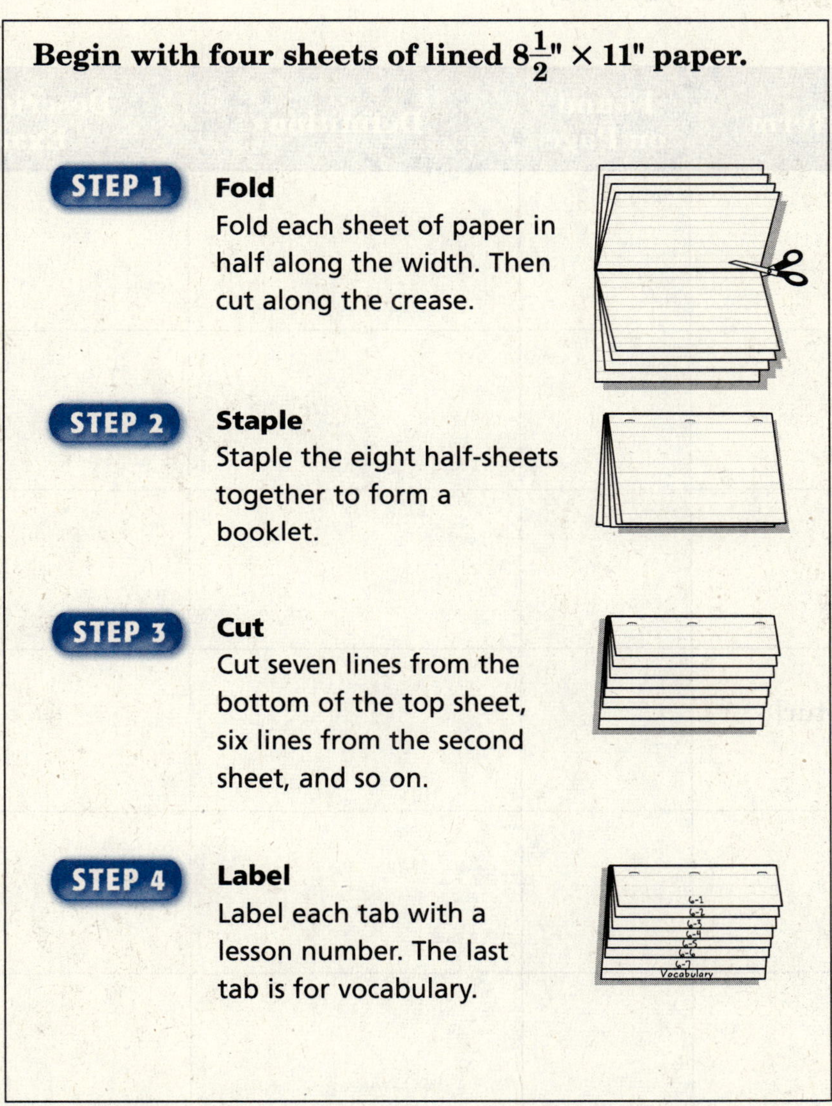

Begin with four sheets of lined $8\frac{1}{2}" \times 11"$ paper.

STEP 1 **Fold**
Fold each sheet of paper in half along the width. Then cut along the crease.

STEP 2 **Staple**
Staple the eight half-sheets together to form a booklet.

STEP 3 **Cut**
Cut seven lines from the bottom of the top sheet, six lines from the second sheet, and so on.

STEP 4 **Label**
Label each tab with a lesson number. The last tab is for vocabulary.

NOTE-TAKING TIP: As you read a lesson, take notes on the materials. Include definitions, concepts, and examples. After you finish each lesson, make an outline of what you learned.

BUILD YOUR VOCABULARY

This is an alphabetical list of new vocabulary terms you will learn in Chapter 6. As you complete the study notes for the chapter, you will see Build Your Vocabulary reminders to complete each term's definition or description on these pages. Remember to add the textbook page number in the second column for reference when you study.

Vocabulary Term	Found on Page	Definition	Description or Example
altitude			
angle bisector			
centroid			
circumcenter [SIR-kum-SEN-tur]			
concurrent			
Euler line			
hypotenuse [hi-PA-tin-oos]			

Vocabulary Term	Found on Page	Definition	Description or Example
incenter			
leg			
median			
nine-point circle			
orthocenter [OR-tho-SEN-tur]			
perpendicular bisector			
Pythagorean Theorem [puh-THA-guh-REE-uhn]			
Pythagorean triple			

6–1 Medians

WHAT YOU'LL LEARN

- Identify and construct medians in triangles.

BUILD YOUR VOCABULARY (page 109)

A **median** is a segment that joins a vertex of a triangle and the midpoint of the side opposite that vertex.

FOLDABLES™

ORGANIZE IT

Under the tab for Lesson 6-1, draw an example of a median. Label the congruent parts. Under the tab for Vocabulary, write the vocabulary words for this lesson.

EXAMPLE

1 **In $\triangle ABC$, $\overline{CE}$ and $\overline{AD}$ are medians.**

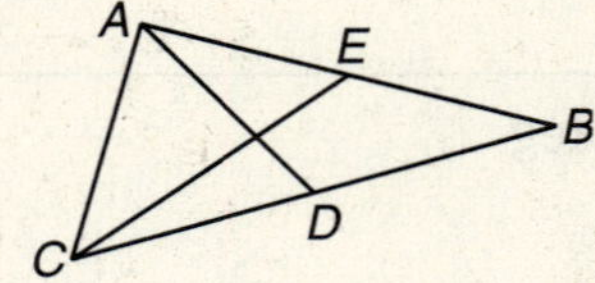

If $CD = 2x + 5$, $BD = 4x - 1$, and $AE = 5x - 2$, find BE.

Since $\overline{CE}$ and $\overline{AD}$ are medians, D and E are midpoints. Solve for x.

$CD = BD$ — Definition of median

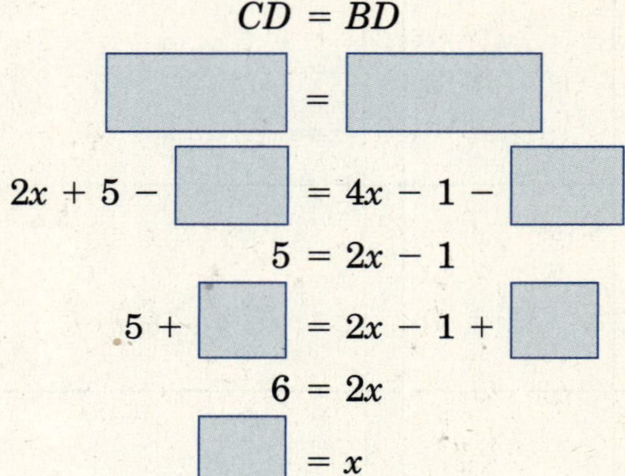

$\_\_\_ = \_\_\_$ — Substitution

$2x + 5 - \_\_\_ = 4x - 1 - \_\_\_$ — Subtract.

$5 = 2x - 1$

$5 + \_\_\_ = 2x - 1 + \_\_\_$ — Add.

$6 = 2x$ — Divide.

$\_\_\_ = x$

Use the values for x and AE to find BE.

$AE = BE$ — Definition of median

$5x - 2 = BE$ — Substitution

$5(\_\_\_) - 2 = BE$ — Substitution

$15 - 2 = BE$

$\_\_\_ = BE$

Your Turn In $\triangle OPS$, $\overline{ST}$ and $\overline{QP}$ are medians. If $PT = 3x - 1$, $OT = 2x + 1$, and $OQ = 4x - 2$, find SQ.

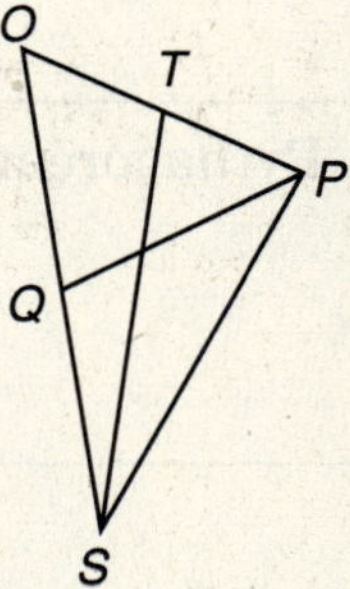

BUILD YOUR VOCABULARY (page 108)

The three [ ] of a triangle intersect at a common point known as the **centroid**.

When three or more lines or segments meet at the same point, they are said to be **concurrent**.

Theorem 6-1
The length of the segment from the vertex to the centroid is twice the length of the segment from the centroid to the midpoint.

EXAMPLES

In $\triangle XYZ$, $\overline{XP}$, $\overline{ZN}$, and $\overline{YM}$ are medians.

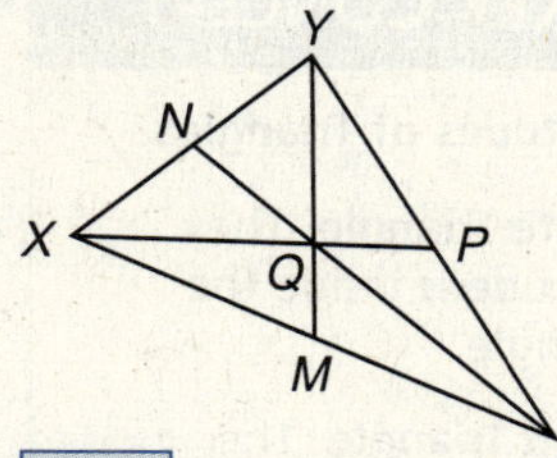

2 Find YQ if $QM = 4$.

Since $QM =$ [ ], $YQ = 2 \cdot$ [ ] or [ ].

3 If $QZ = 18$, what is ZN?

Since $QZ = 18$ and $QZ = \frac{2}{3} \cdot ZN$, solve the equation $18 = \frac{2}{3} \cdot ZN$ for ZN.

$$18 = \frac{2}{3} \cdot ZN$$

$$\frac{3}{2}(18) = \frac{3}{2}\left(\frac{2}{3}ZN\right)$$ Multiply each side by .

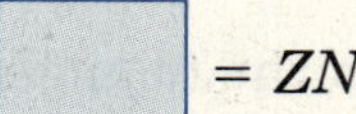 $= ZN$

Your Turn **In $\triangle EFG$, $\overline{FA}$, $\overline{GB}$, and $\overline{EC}$ are medians.**

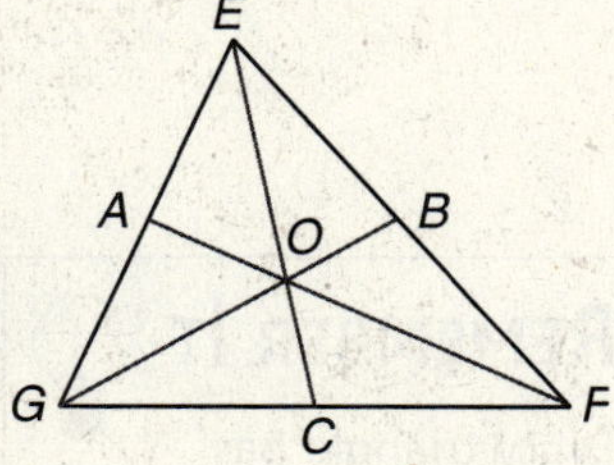

a. Find EO if $CO = 3$.

[ ]

b. If $FA = 18$, what are the measures of AO and OF?

[ ]

HOMEWORK ASSIGNMENT

Page(s):

Exercises:

6–2 Altitudes and Perpendicular Bisectors

What You'll Learn

- Identify and construct altitudes and perpendicular bisectors in triangles.

Build Your Vocabulary (page 108)

An **altitude** of a triangle is a perpendicular segment with one endpoint at a ______ and the other endpoint on the ______ opposite that vertex.

Key Concept

Altitudes of Triangles

Acute Triangle The altitude is inside the triangle.

Right Triangle The altitude is a side of the triangle.

Obtuse Triangle The altitude is outside the triangle.

Examples

1 Is $\overline{AD}$ an altitude of the triangle?

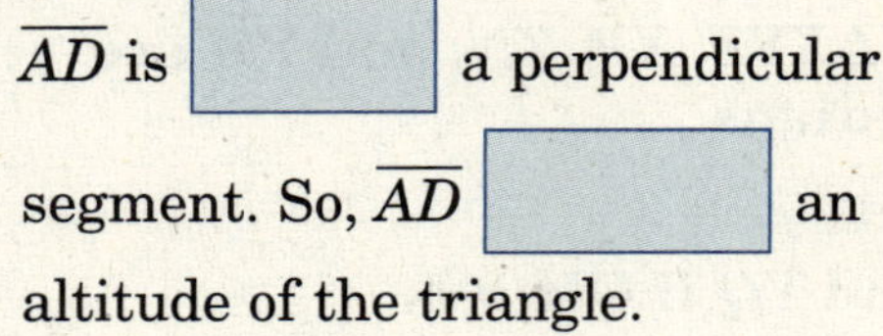

$\overline{AD}$ is ______ a perpendicular segment. So, $\overline{AD}$ ______ an altitude of the triangle.

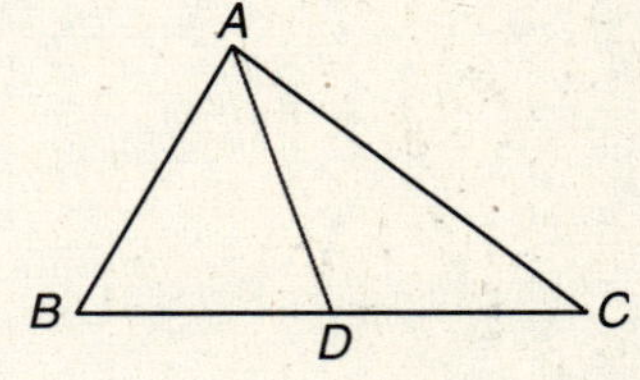

2 Is $\overline{GJ}$ an altitude of the triangle?

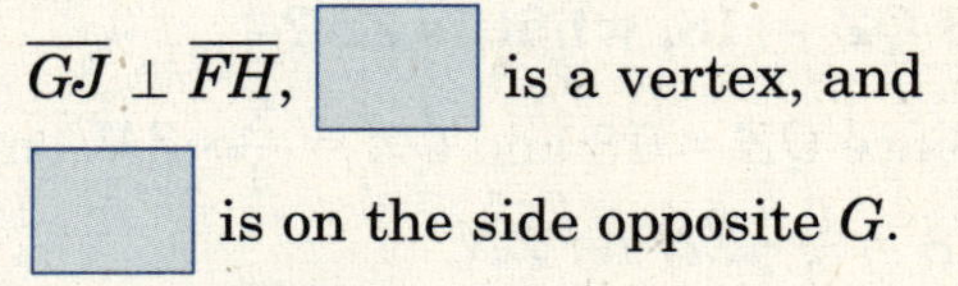

$\overline{GJ} \perp \overline{FH}$, ______ is a vertex, and ______ is on the side opposite G.

So, $\overline{GJ}$ ______ an altitude of the triangle.

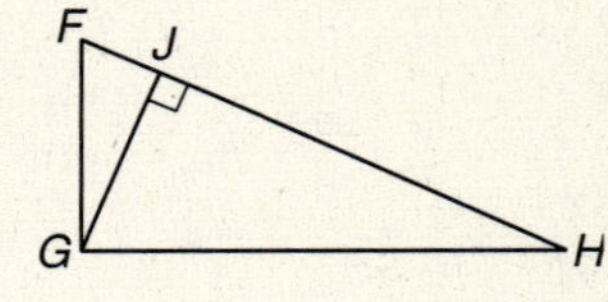

Your Turn

a. Is $\overline{BD}$ an altitude of the triangle?

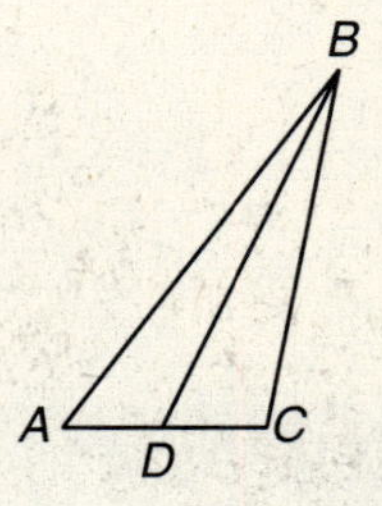

b. Is $\overline{XY}$ an altitude of the triangle?

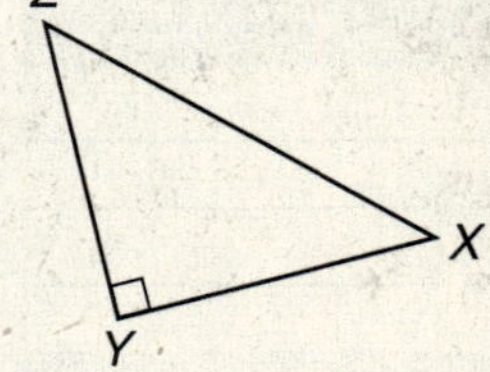

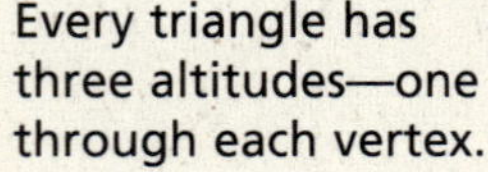

Remember It

Every triangle has three altitudes—one through each vertex.

BUILD YOUR VOCABULARY (page 109)

A ________ line or segment that ________ a side of a triangle is called the **perpendicular bisector** of that side.

EXAMPLES

FOLDABLES™

ORGANIZE IT

Under the tab for Lesson 6-2, draw one example of an altitude and one of a perpendicular bisector. Label congruent parts and right angles.

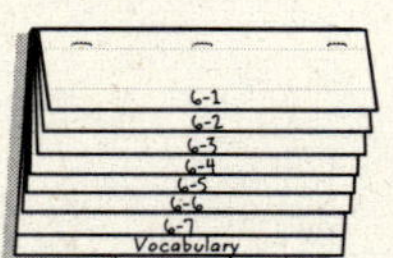

3 **Is $\overline{MN}$ a perpendicular bisector of a side of the triangle?**

Since N is the midpoint of $\overline{KL}$, $\overline{MN}$ is a bisector of side $\overline{KL}$. $\overline{MN}$ 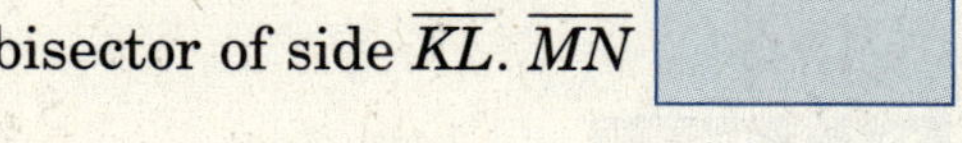 perpendicular to $\overline{KL}$, so $\overline{MN}$ is ________ a perpendicular bisector in $\triangle KLM$.

4 **Is $\overline{AD}$ a perpendicular bisector of a side of the triangle?**

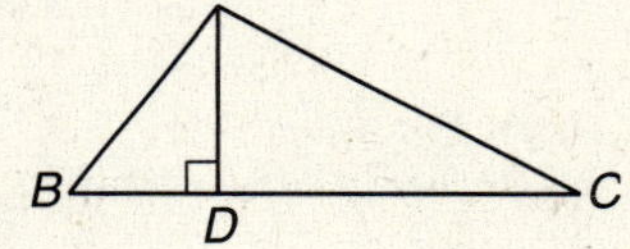

$\overline{AD} \perp \overline{BC}$ but D  the midpoint of $\overline{BC}$. So, $\overline{AD}$ ________ a perpendicular bisector of side $\overline{BC}$ in $\triangle ABC$.

Your Turn

a. Is $\overline{BD}$ a perpendicular bisector of the triangle?

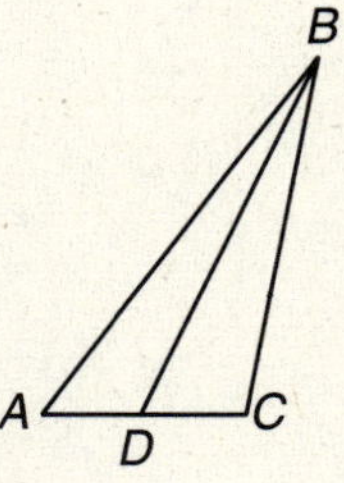

b. Is $\overline{LM}$ a perpendicular bisector of the triangle?

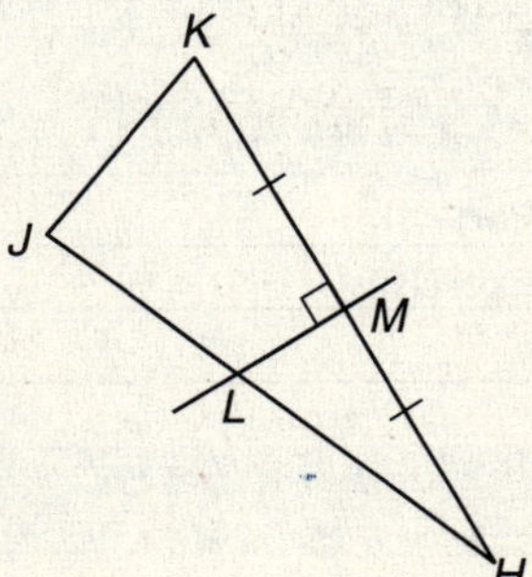

EXAMPLE

5 **Tell whether $\overline{MN}$ is an *altitude*, a *perpendicular bisector*, *both*, or *neither*.**

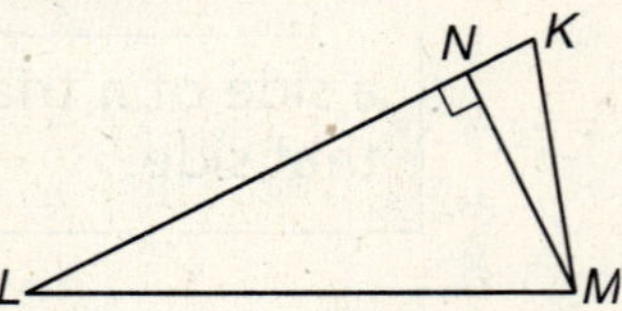

______; $\overline{MN} \perp \overline{KL}$ but N ______ the midpoint of $\overline{KL}$. So, $\overline{MN}$ ______ a ______ of side $\overline{KL}$ in $\triangle KLM$.

Your Turn Tell whether $\overline{XO}$ is an *altitude*, a *perpendicular bisector*, *both*, or *neither*.

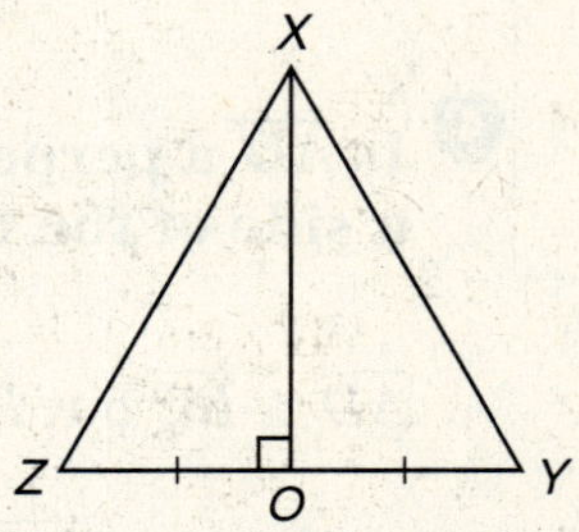

WRITE IT

How is a perpendicular bisector different from a median?

______

______

______

HOMEWORK ASSIGNMENT

Page(s):

Exercises:

6–3 Angle Bisectors of Triangles

What You'll Learn

- Identify and use angle bisectors in triangles.

Build Your Vocabulary (page 108)

An **angle bisector** of a triangle is a segment that separates an angle of the triangle into two ______ angles.

Foldables™

Organize It

Under the tab for Lesson 6-3, draw an example of an angle bisector. Label the congruent parts.

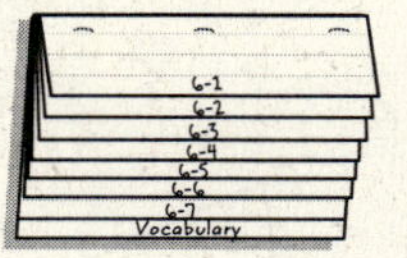

Examples

1 In $\triangle ABD$, $\overline{AC}$ bisects $\angle BAD$. If $m\angle 1 = 41$, find $m\angle 2$.

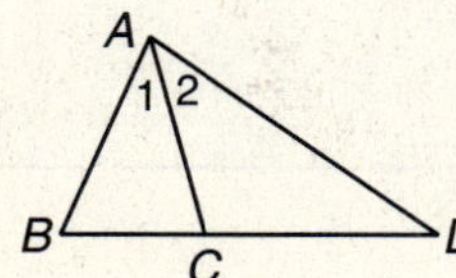

Since $\overline{AC}$ bisects $\angle BAD$, $m\angle 1 =$ ______.

Since $m\angle 1 =$ ______, $m\angle 2 =$ ______.

2 In $\triangle KMN$, $\overline{NL}$ bisects $\angle KNM$. If $\angle KNM$ is a right angle, find $m\angle 2$.

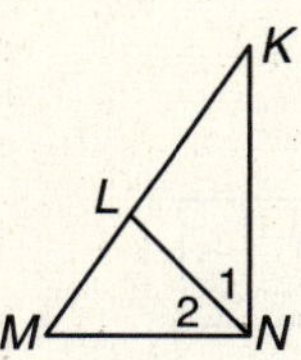

$m\angle 2 = \frac{1}{2}(m\angle KNM)$

$m\angle 2 = \frac{1}{2}(\quad)$

$m\angle 2 =$ ______

3 In $\triangle WYZ$, $\overline{ZX}$ bisects $\angle WZY$. If $m\angle 1 = 55$, find $m\angle WZY$.

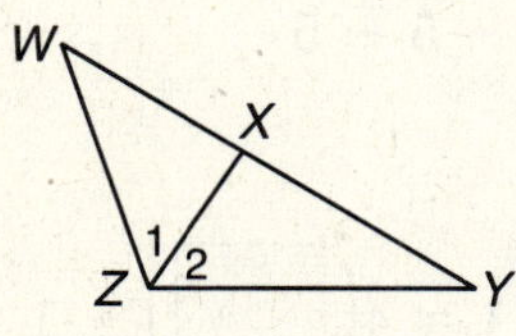

$m\angle WZY = 2(m\angle 1)$

$m\angle WZY = 2(\quad)$

$m\angle WZY =$ ______

Your Turn

a. In $\triangle XYZ$, $\overline{YW}$ bisects $\angle XYZ$. If $m\angle 2 = 33$, find $m\angle 1$.

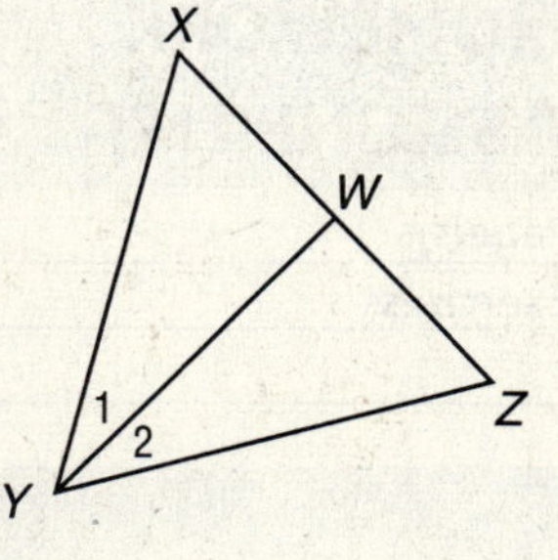

b. In $\triangle NOM$, $\overline{OP}$ bisects $\angle NOM$. If $\angle NOM = 85$, find $m\angle 4$.

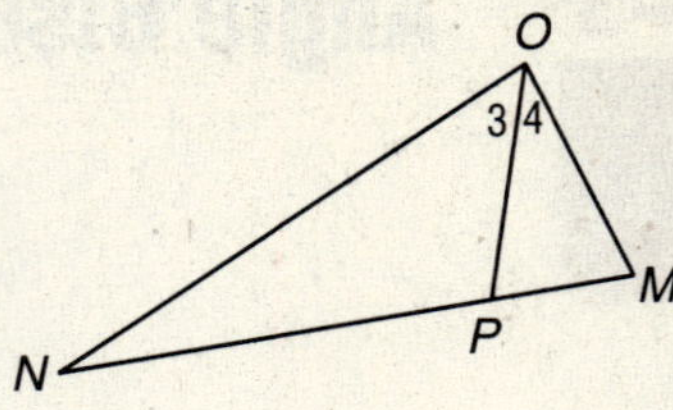

c. In $\triangle RST$, $\overline{SU}$ bisects $\angle RST$. If $m\angle 6 = 36.5$, find $m\angle RST$.

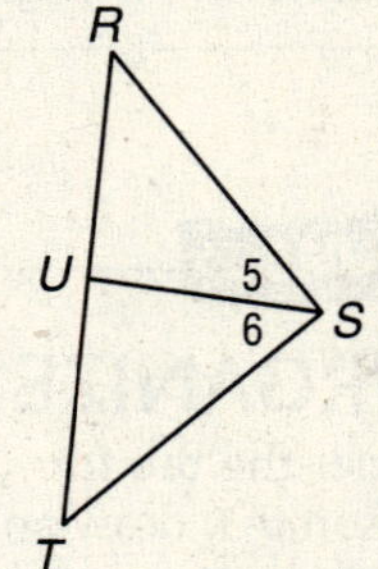

EXAMPLE

4 In $\triangle FHI$, $\overline{IG}$ is an angle bisector. Find $m\angle HIG$.

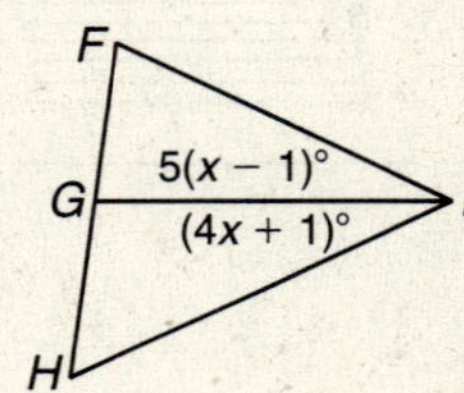

$m\angle HIG = m\angle FIG$

$\square = \square$

$4x + 1 = 5x - 5$ Distributive Property

$4x + 1 - 4x = 5x - 5 - 4x$ Subtract.

$1 = x - 5$

$1 + 5 = x - 5 + 5$ Add.

$\square = x$

$m\angle HIG = 4x + 1 = 4(\square) + 1 = \square + 1 = \square$

Your Turn In $\triangle JKL$, $\overline{KM}$ is an angle bisector. Find $m\angle JKM$.

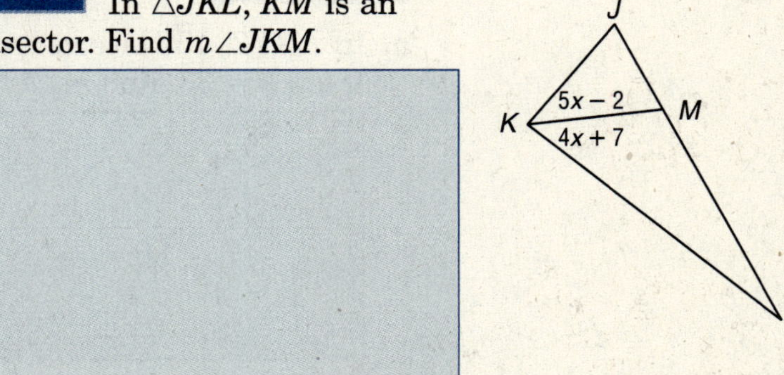

HOMEWORK ASSIGNMENT

Page(s):

Exercises:

6–4 Isosceles Triangles

WHAT YOU'LL LEARN

- Identify and use properties of isosceles triangles.

BUILD YOUR VOCABULARY (page 109)

A **leg** of an isosceles triangle is one of the two ______ sides.

Theorem 6-2 Isosceles Triangle Theorem
If two sides of a triangle are congruent, then the angles opposite those sides are congruent.

Theorem 6-3
The median from the vertex angle of an isosceles triangle lies on the perpendicular bisector of the base and the angle bisector of the vertex angle.

FOLDABLES

ORGANIZE IT

Under the tab for Lesson 6-4, draw an example of an isosceles triangle. Label the congruent parts, and the special names for sides and angles.

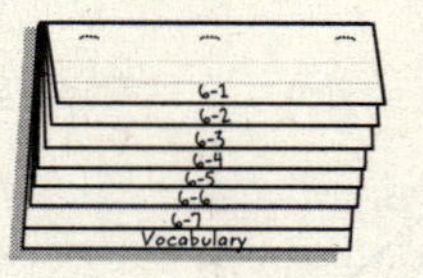

EXAMPLE

1 Find the values of the variables.

In the top triangle, find the value of base angle x. Since the triangle is isosceles, and one base angle = 35,

$x =$ ______.

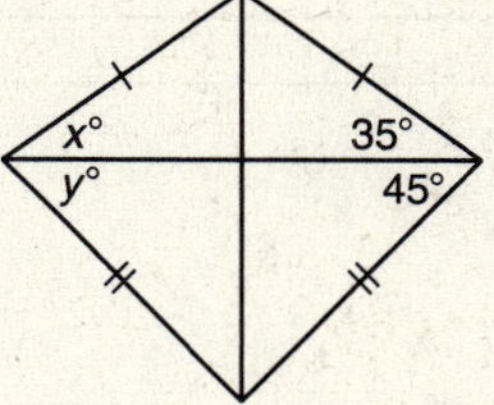

In the bottom triangle, find the value of base angle y. Since the other base angle = 45, $y =$ ______.

Your Turn Find the values of the variables.

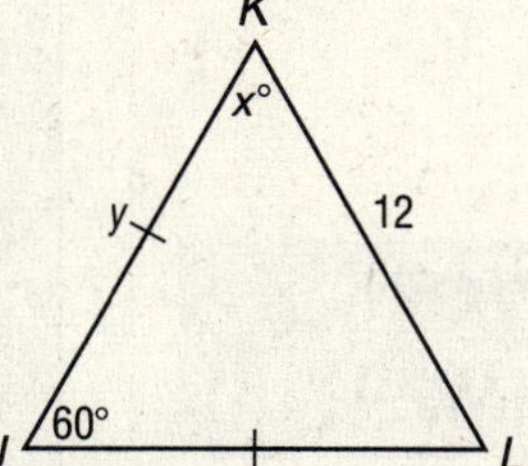

Theorem 6-4 Converse of Isosceles Triangle Theorem
If two angles of a triangle are congruent, then the sides opposite those angles are congruent.

EXAMPLE

2 **In $\triangle DEF$, $\angle 1 \cong \angle 2$ and $m\angle 1 = 28$. Find $m\angle F$, DF, and EF.**

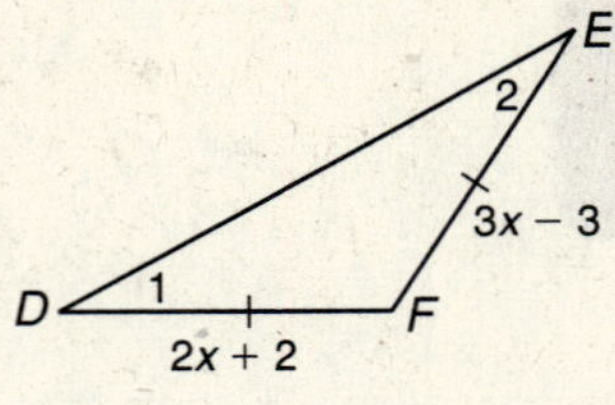

First, find $m\angle F$. You know that $m\angle 1 = 28$. Since $\angle 1 \cong \angle 2$, $m\angle 2 = 28$.

$m\angle 1 + m\angle 2 + m\angle F = 180$	Angle Sum Theorem
____ + ____ + $m\angle F = 180$	Replace $m\angle 1$ and $m\angle 2$.
____ + $m\angle F = 180$	
$56 + m\angle F - 56 = 180 - 56$	Subtract.
$m\angle F =$ ____	

Next, find DF. Since $\angle 1 \cong \angle 2$, Theorem 6-4 states that $\overline{DF} \cong \overline{EF}$.

$DF = EF$	Congruent segments
$2x + 2 = 3x - 3$	Replace *DF* and *EF*.
$2x + 2 -$ ____ $= 3x - 3 -$ ____	Subtract.
$2 = x - 3$	
$2 +$ ____ $= x - 3 +$ ____	Add.
____ $= x$	

By replacing x with 5, you find that $DF = 2x + 2 = 2(5) + 2 = 10 + 2$ or ____. $EF = 3x - 3 = 3(5) - 3 = 15 - 3$ or ____.

Your Turn Find the values of the variables.

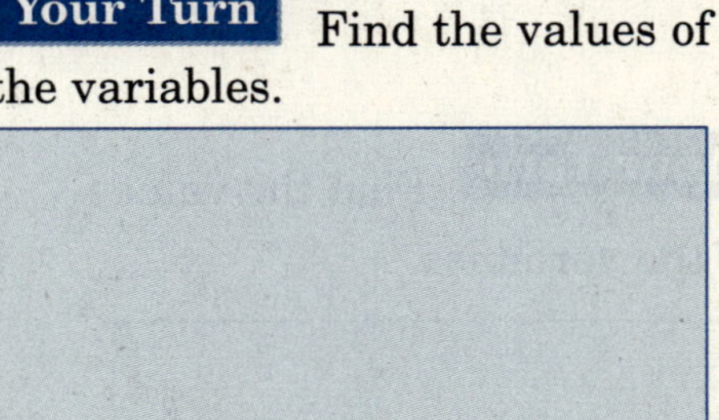

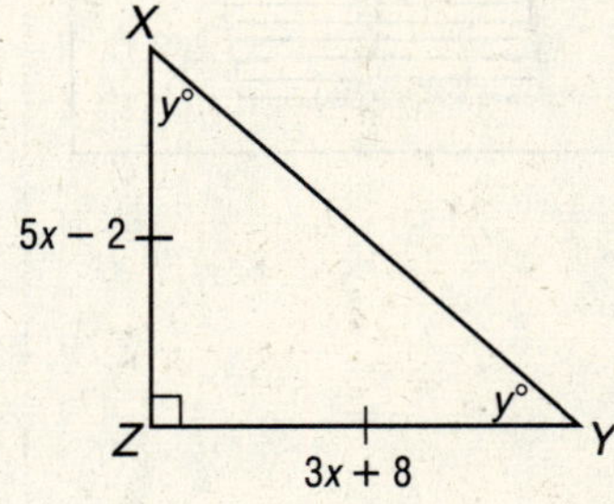

Theorem 6-5
A triangle is equilateral if and only if it is equiangular.

WRITE IT

Can an isosceles triangle be an equiangular triangle?

____

____

____

____

HOMEWORK ASSIGNMENT

Page(s):

Exercises:

6–5 Right Triangles

What You'll Learn

- Use tests for congruence of right triangles.

Build Your Vocabulary (pages 108–109)

In a ______ triangle the side opposite the ______ angle is known as the **hypotenuse**.

The two sides that form the ______ angle are called **legs**.

Theorem 6-6 LL Theorem
If two legs of one right triangle are congruent to the corresponding legs of another right triangle, then the triangles are congruent.

Theorem 6-7 HA Theorem
If the hypotenuse and an acute angle of one right triangle are congruent to the hypotenuse and corresponding angle of another right triangle, then the triangles are congruent.

Theorem 6-8 LA Theorem
If one leg and an acute angle of one right triangle are congruent to the corresponding leg and angle of another right triangle, then the triangles are congruent.

Postulate 6-1 HL Postulate
If the hypotenuse and a leg of one right triangle are congruent to the hypotenuse and corresponding leg of another right triangle, then the triangles are congruent.

Foldables™ Organize It

Under the tab for Lesson 6-5, draw an example of a right triangle. Label the special names for the sides of the triangle. Under the tab for Vocabulary, write the vocabulary words for this lesson.

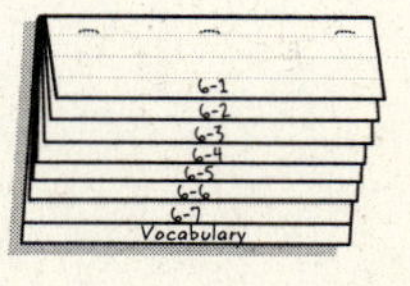

Examples

Determine whether each pair of right triangles is congruent by *LL*, *HA*, *LA*, or *HL*. If it is not possible to prove that they are congruent, write *not possible*.

1

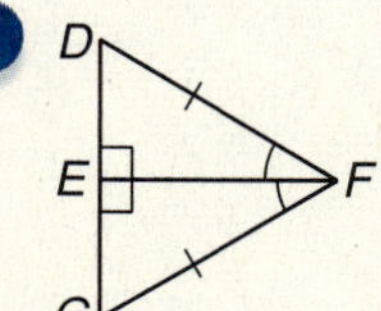

There is one pair of congruent ______ angles, $\angle DFE \cong \angle GFE$. The hypotenuses are congruent, $\overline{DF} \cong \overline{GF}$.

Therefore, $\triangle DEF \cong \triangle GEF$ by ______.

2

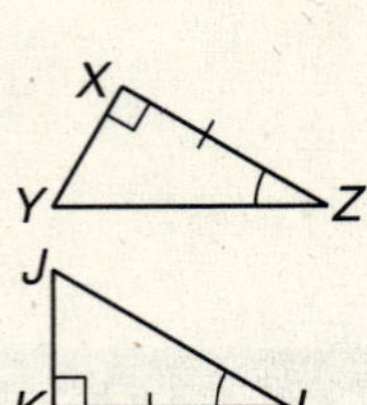

There is one pair of ______ acute angles, $\angle Z \cong \angle L$. There is one pair of ______, $\overline{XZ} \cong \overline{KL}$.

Therefore, $\triangle YXZ \cong \triangle JKL$ by ______.

WRITE IT

Which test for congruence is used to establish the LA Theorem? Explain.

Your Turn **Determine whether each pair of right triangles is congruent by *LL*, *HA*, *LA*, or *HL*. If it is not possible to prove that they are congruent, write *not possible*.**

a.

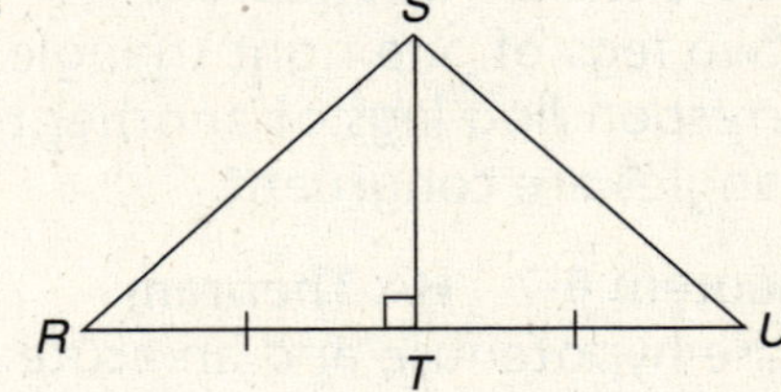

b.

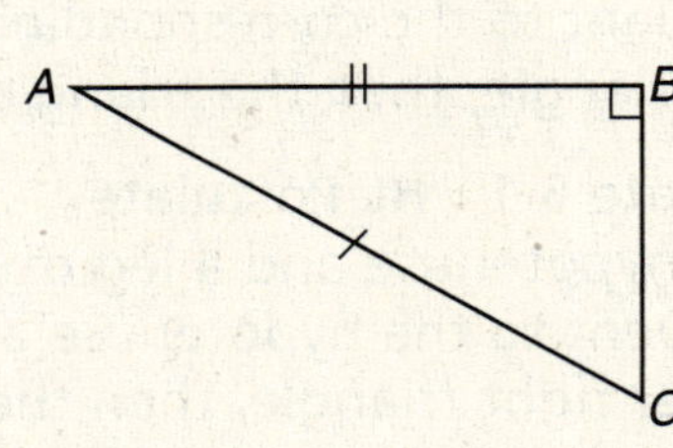

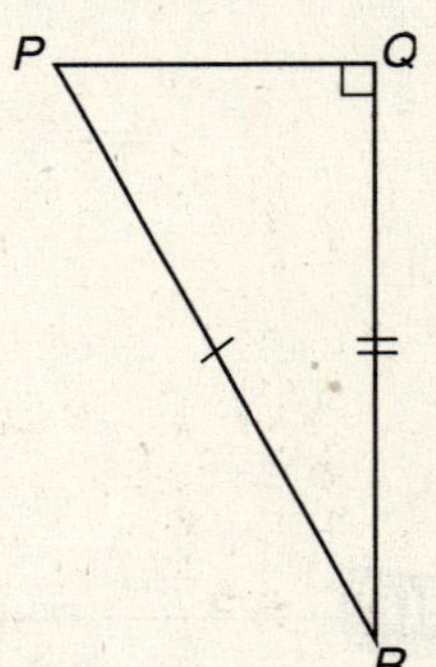

HOMEWORK ASSIGNMENT

Page(s):

Exercises:

6–6 The Pythagorean Theorem

What You'll Learn

- Use the Pythagorean Theorem and its converse.

Build Your Vocabulary (page 109)

The **Pythagorean Theorem** can be used to determine the lengths of the sides of a right triangle. It states that the ______ of the squares of the ______ of a right triangle equals the square of the hypotenuse.

Theorem 6-9 Pythagorean Theorem
In a right triangle, the square of the length of the hypotenuse *c* is equal to the sum of the squares of the lengths of the legs *a* and *b*.

Foldables™

Organize It

Under the tab for Lesson 6-5, write the Pythagorean Theorem. Draw a right triangle and label the legs *a* and *b*, and the hypotenuse *c*.

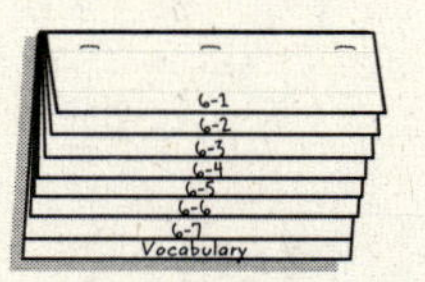

Example

1 Find the length of the hypotenuse of the right triangle.

Use the Pythagorean Theorem to find the length of the hypotenuse.

$c^2 = a^2 + b^2$

$c^2 = \square^2 + \square^2$ Replace *a* and *b*.

$c^2 = \square + \square$

$c^2 = 400$ Take the square root of each side.

$c = \square$ The length is ______.

Your Turn

a. Find the length of the hypotenuse of the right triangle.

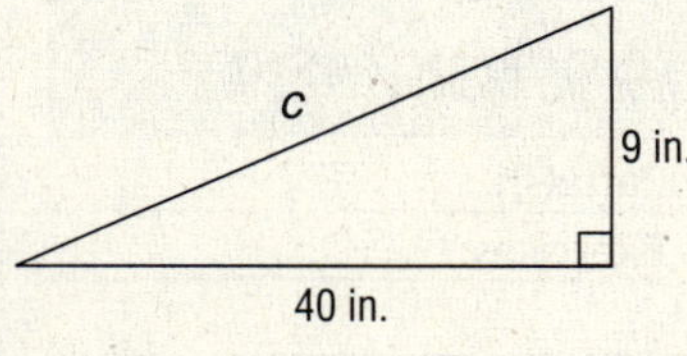

REMEMBER IT

Always check to see that c represents the length of the longest side.

b. Find the length of one leg of a right triangle if the length of the hypotenuse is 25 cm and the length of the other leg is 23 cm.

Theorem 6-10 Converse of the Pythagorean Theorem
If c is the measure of the longest side of a triangle, a and b are the lengths of the other two sides, and $c^2 = a^2 + b^2$, then the triangle is a right triangle.

EXAMPLE

2 The lengths of three sides of a triangle are 4, 5, and 6 meters. Determine whether this triangle is a right triangle.

Since the longest side is ______ meters, use ______ as c, the measure of the hypotenuse.

$c^2 = a^2 + b^2$ — Pythagorean Theorem

$6^2 \stackrel{?}{=} 4^2 + 5^2$ — Replace c with ______, a with ______, and b with ______.

$36 \stackrel{?}{=} 16 + 25$

$36 \neq 41$

Since c^2 ______ $a^2 + b^2$, the triangle ______ a right triangle.

Your Turn The lengths of three sides of a triangle are 5, 12, and 13 yards. Determine whether this triangle is a right triangle.

HOMEWORK ASSIGNMENT

Page(s):

Exercises:

6–7 Distance on the Coordinate Plane

WHAT YOU'LL LEARN

- Find the distance between two points on the coordinate plane.

Theorem 6-11 Distance Formula
If d is the measure of the distance between two points with coordinates (x_1, y_1) and (x_2, y_2), then
$d = \sqrt{(x_2 - x_1)^2 + (y_2 - y_1)^2}$.

EXAMPLE

1 **Use the Distance Formula to find the distance between $A(6, 2)$ and $B(4, -4)$. Round to the nearest tenth, if necessary.**

Use the Distance Formula. Replace (x_1, y_1) with (6, 2) and (x_2, y_2) with ______.

$d = \sqrt{(x_2 - x_1)^2 + (y_2 - y_1)^2}$ Distance Formula

$AB = \sqrt{(4 - \square)^2 + (\square - 2)^2}$ Substitution

$AB = \sqrt{(-2)^2 + (-6)^2}$

$AB = \sqrt{\square + \square}$

$AB = \sqrt{40}$

$AB \approx$ ______

FOLDABLES™

ORGANIZE IT

Under the tab for Lesson 6-7, write the Distance Formula. Then show an example to help you remember the main idea.

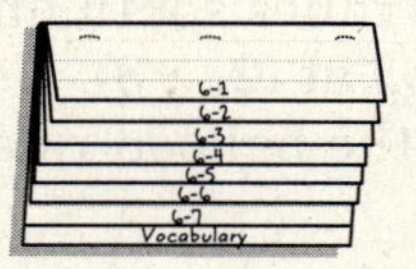

Your Turn

a. Use the Distance Formula to find the distance between $M(2, 2)$ and $N(-6, -4)$. Round to the nearest tenth, if necessary.

b. Determine whether $\triangle TRI$ with vertices $T(-4, 1)$, $R(2, 5)$, and $I(2, -2)$ is isosceles.

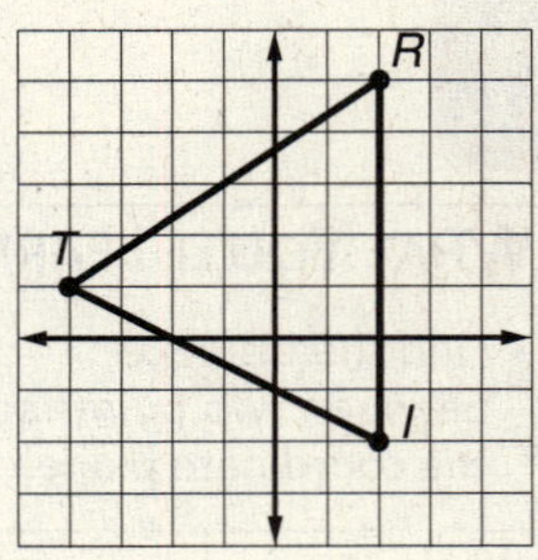

EXAMPLE

2 Akio took a ride in a hot-air balloon. The flight began 4 miles north of his house. The balloon landed 3 miles south and 2 miles east of his house. If the balloon traveled in a straight line between the starting and ending points of the flight, what was the length of Akio's balloon ride?

REMEMBER IT

Only use the positive square roots since distance is not negative.

Suppose Akio's house is located at the origin (0, 0). If the balloon ride began 4 miles north of his house, it began at (x_1, y_1), or (0, 4). Since the balloon landed 3 miles south and 2 miles east of his house, it landed at (x_2, y_2) at (2, −3). Use the Distance Formula to find the length of the balloon ride.

$$d = \sqrt{(x_2 - x_1)^2 + (y_2 - y_1)^2}$$
$$= \sqrt{(2 - 0)^2 + (-3 - 4)^2}$$
$$= \sqrt{2^2 + (-7)^2}$$
$$= \sqrt{4 + 49} = \_\_\_\_ \approx \_\_\_\_$$

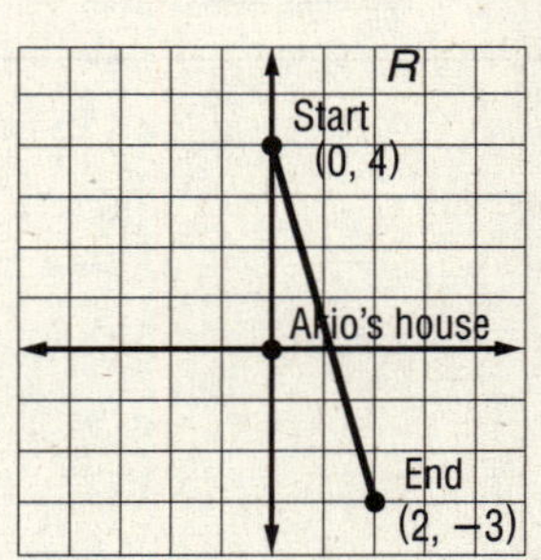

Akio's balloon ride was approximately ______ miles.

Your Turn Marcelle went to a friend's house to complete a homework project after school instead of going directly home. The school lies 2 blocks north of her home. Her friend's house is located 3 blocks west and 1 block north of her home. If Marcelle traveled in a straight path from school to her friend's home, what was the length of her walk?

HOMEWORK ASSIGNMENT

Page(s):

Exercises:

CHAPTER 6

BRINGING IT ALL TOGETHER

STUDY GUIDE

FOLDABLES™	VOCABULARY PUZZLEMAKER	BUILD YOUR VOCABULARY
Use your **Chapter 6 Foldable** to help you study for your chapter test.	To make a crossword puzzle, word search, or jumble puzzle of the vocabulary words in Chapter 6, go to: www.glencoe.com/sec/math/t_resources/free/index.php	You can use your completed **Vocabulary Builder** (pages 108–109) to help you solve the puzzle.

6-1

Medians

Complete the sentence.

1. The midpoint of a side of a triangle and the vertex of the opposite angle are endpoints of a ______.

2. A triangle's medians are ______ at the centroid.

3. In $\triangle ABC$, $\overline{BD}$ is a median and $BD = 6$. What is BE?

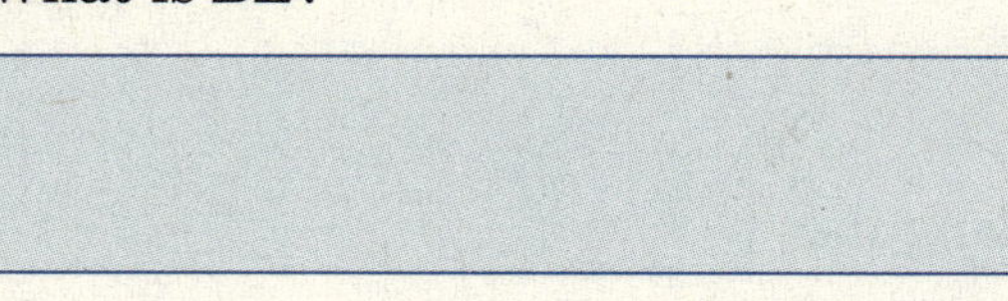

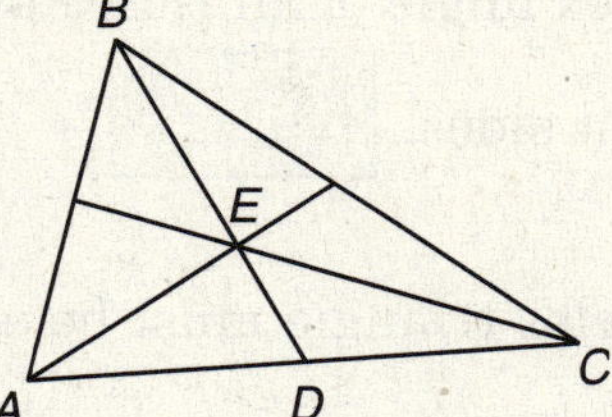

6-2

Altitudes and Perpendicular Bisectors

For the triangles shown, state whether *AB* is an *altitude*, a *perpendicular bisector*, *both*, or *neither*.

4.

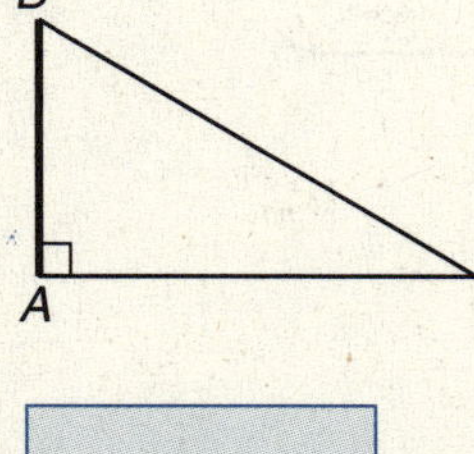

5.

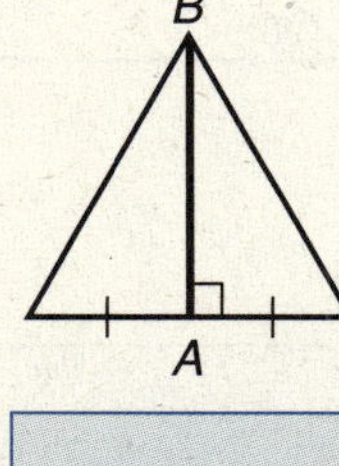

6.

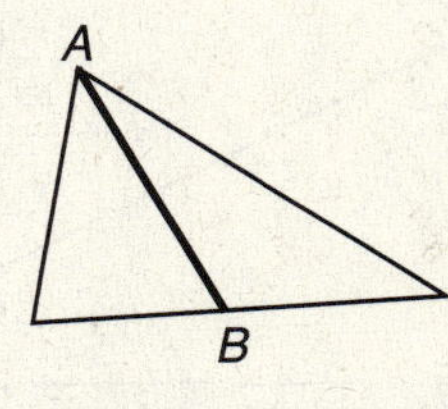

6-3

Angle Bisectors of Triangles

7. In $\triangle JKL$, $\overline{KH}$ bisects $\angle JKL$. If $m\angle 1 = 12$, find $m\angle JKL$.

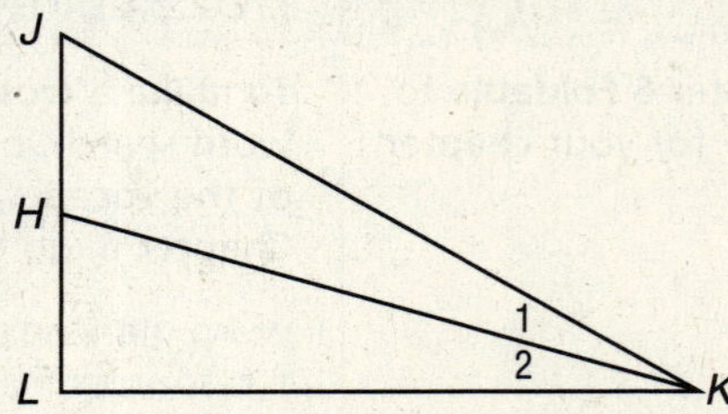

8. What is the value of x so that BD is an angle bisector?

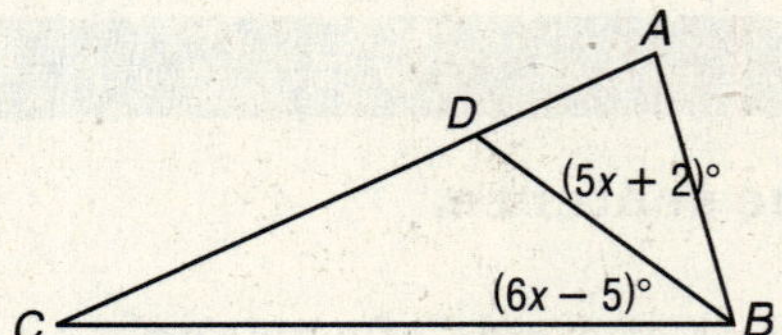

6-4

Isosceles Triangles

Indicate whether the statement is *true* or *false*.

9. The vertex angle of an isosceles triangle is opposite one of the congruent sides.

10. An isosceles triangle must be equiangular.

For each triangle, find the values of the variables.

11.

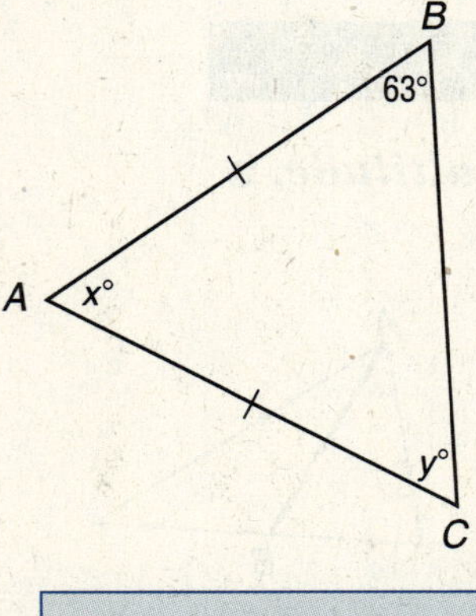

12.

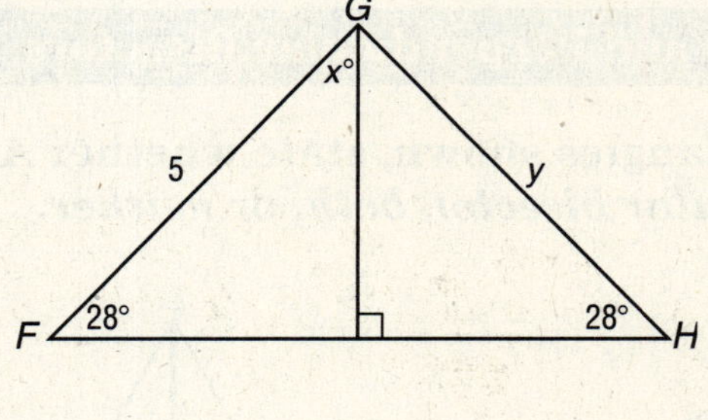

6-5

Right Triangles

Determine whether each pair of right triangles is congruent by *LL*, *HA*, *LA*, or *HL*. If it is not possible to prove that they are congruent, write *not possible*.

13.

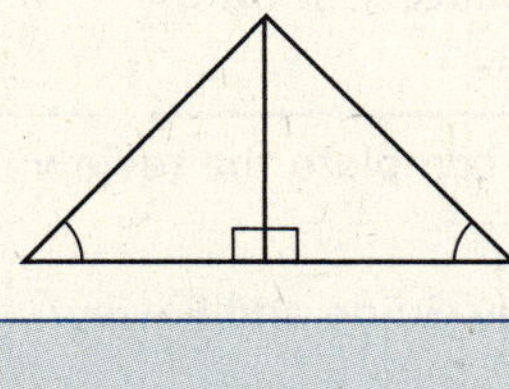

14.

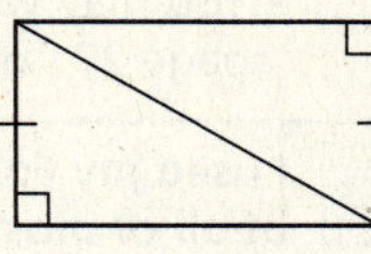

6-6

The Pythagorean Theorem

Find the missing measure in each right triangle. Round to the nearest tenth, if necessary.

15.

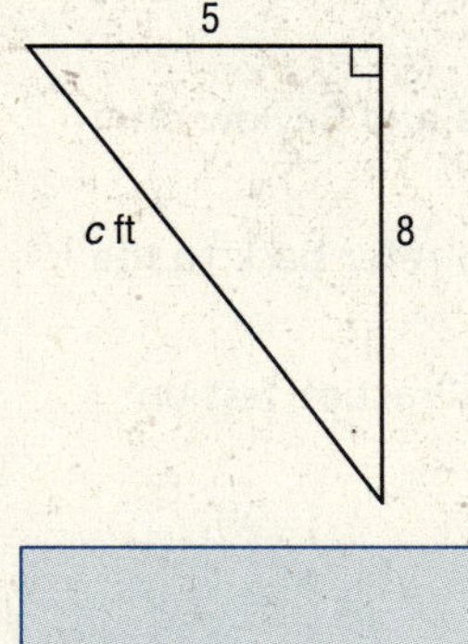

16.

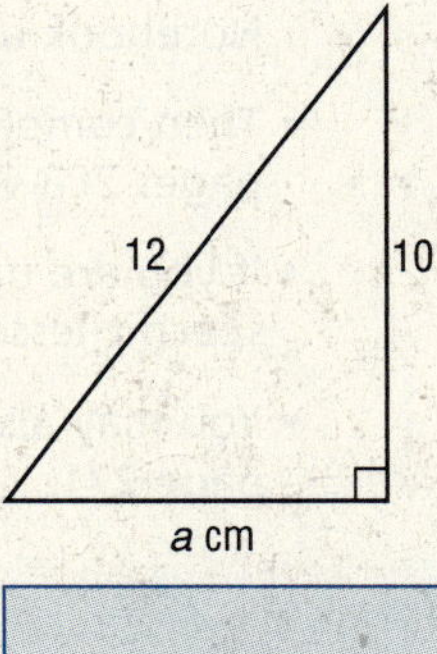

6-7

Distance on the Coordinate Plane

Use the Distance Formula to find the distance between each pair of points. Round to the nearest tenth, if necessary.

17. $G(-3, 1)$, $H(4, 5)$ **18.** $R(-1, 2)$, $S(5, -6)$ **19.** $A(12, 0)$, $B(0, 5)$

20. Andre walked 2 blocks west of his home to school. After school, he walked to the store which is 1 block east and 1 block north of his home. About how far apart are the school and the store?

ARE YOU READY FOR THE CHAPTER TEST?

Visit **geomconcepts.com** to access your textbook, more examples, self-check quizzes, and practice tests to help you study the concepts in Chapter 6.

Check the one that applies. Suggestions to help you study are given with each item.

☐ **I completed the review of all or most lessons without using my notes or asking for help.**

- You are probably ready for the Chapter Test.
- You may want to take the Chapter 6 Practice Test on page 271 of your textbook as a final check.

☐ **I used my Foldable or Study Notebook to complete the review of all or most lessons.**

- You should complete the Chapter 6 Study Guide and Review on pages 268–270 of your textbook.
- If you are unsure of any concepts or skills, refer back to the specific lesson(s).
- You may also want to take the Chapter 6 Practice Test on page 271.

☐ **I asked for help from someone else to complete the review of all or most lessons.**

- You should review the examples and concepts in your Study Notebook and Chapter 6 Foldable.
- Then complete the Chapter 6 Study Guide and Review on pages 268–270 of your textbook.
- If you are unsure of any concepts or skills, refer back to the specific lesson(s).
- You may also want to take the Chapter 6 Practice Test on page 271.

Student Signature

Parent/Guardian Signature

Teacher Signature

Triangle Inequalities

Use the instructions below to make a Foldable to help you organize your notes as you study the chapter. You will see Foldable reminders in the margin of this Interactive Study Notebook to help you in taking notes.

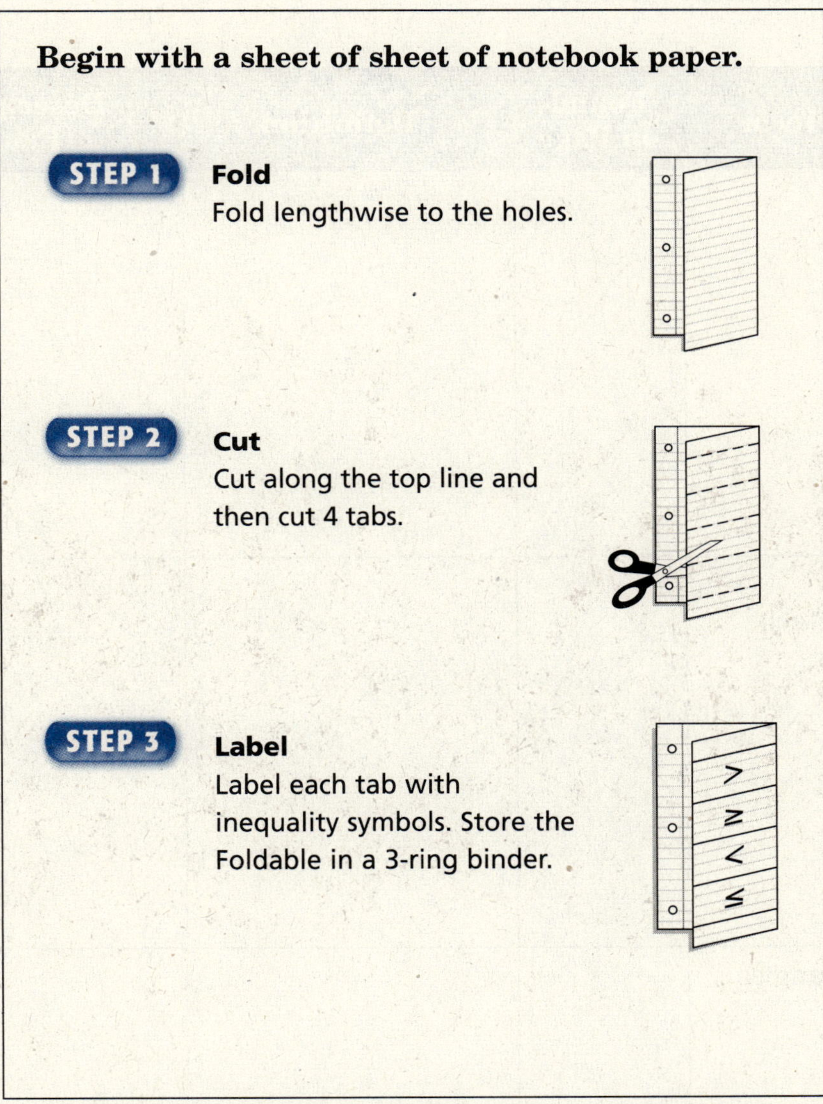

Begin with a sheet of sheet of notebook paper.

STEP 1 **Fold**
Fold lengthwise to the holes.

STEP 2 **Cut**
Cut along the top line and then cut 4 tabs.

STEP 3 **Label**
Label each tab with inequality symbols. Store the Foldable in a 3-ring binder.

NOTE-TAKING TIP: When you take notes, define new vocabulary words, describe new ideas, and write examples that help you remember the meanings of the words and ideas.

BUILD YOUR VOCABULARY

This is an alphabetical list of new vocabulary terms you will learn in Chapter 7. As you complete the study notes for the chapter, you will see Build Your Vocabulary reminders to complete each term's definition or description on these pages. Remember to add the textbook page number in the second column for reference when you study.

Vocabulary Term	Found on Page	Definition	Description or Example
exterior angle			
inequality [IN-ee-KWAL-a-tee]			
remote interior angles			

7-1 Segments, Angles, and Inequalities

What You'll Learn

- Apply inequalities to segment and angle measurements.

Build Your Vocabulary (page 130)

Statements that contain the symbols ______ or ______ compare quantities or measures that do not have the same value and are called **inequalities**.

Foldables

Organize It

Write words under each tab to describe each symbol on your foldable.

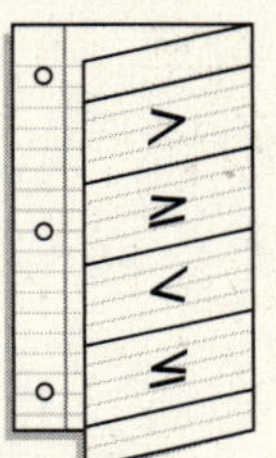

Postulate 7-1 Comparison Property
For any two real numbers a and b, exactly one of the following statements is true: $a < b$, $a = b$, or $a > b$.

EXAMPLE

1 **Refer to the number line and replace ● in *DR* ● *LN* with <, >, or = to make a true sentence.**

$DR \bullet LN$

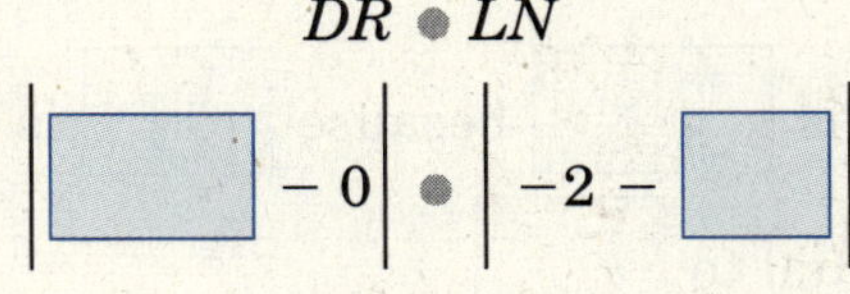

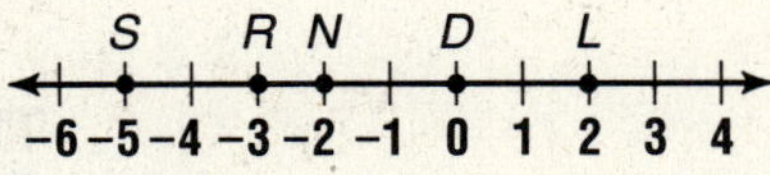

$|-3| \bullet |-4|$

$3 \bullet 4$

3 ______ 4

Your Turn Refer to the number line and replace ● in $PR \bullet QS$ with <, >, or = to make a true sentence.

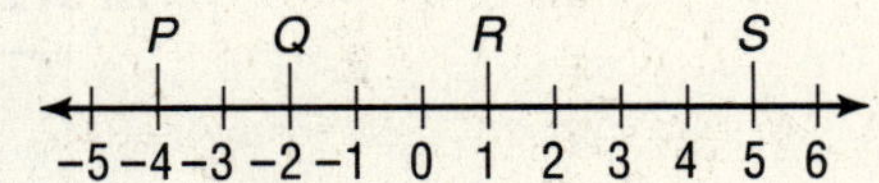

Theorem 7-1
If point C is between points A and B, and A, C, and B are collinear, then $AB > AC$ and $AB > CB$.

Theorem 7-2
If $\overrightarrow{EP}$ is between $\overrightarrow{ED}$ and $\overrightarrow{EF}$, then $m\angle DEF > m\angle DEP$ and $m\angle DEF > m\angle PEF$.

EXAMPLES

Refer to the figure. Determine whether each statement is *true* or *false*.

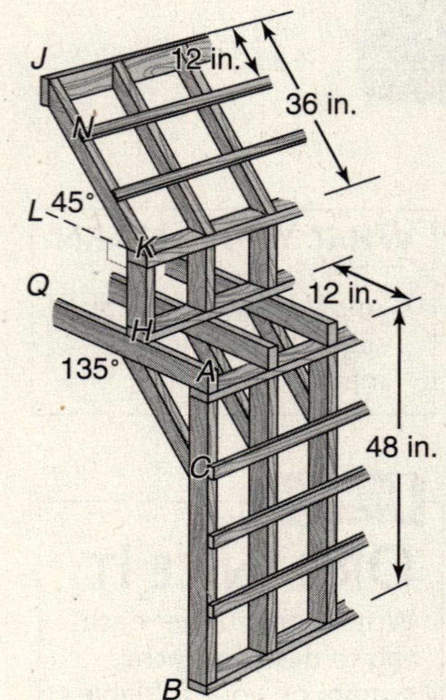

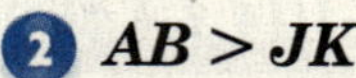

2 $AB > JK$

$AB =$ ____ and $JK =$ ____

$48 >$ ____ Substitution

This is ____ because ____ is greater than ____.

3 $m\angle AHC \not\geq m\angle HKL$

$m\angle AHC =$ ____ and $m\angle HKL =$ ____

$45 \not\geq$ ____ Substitution

This is ____ because ____ is not greater than or equal to ____.

Your Turn **Refer to the figure. Determine whether each statement is *true* or *false*.**

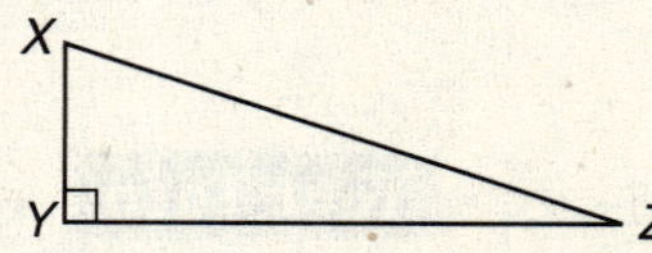

a. $XY < XZ$

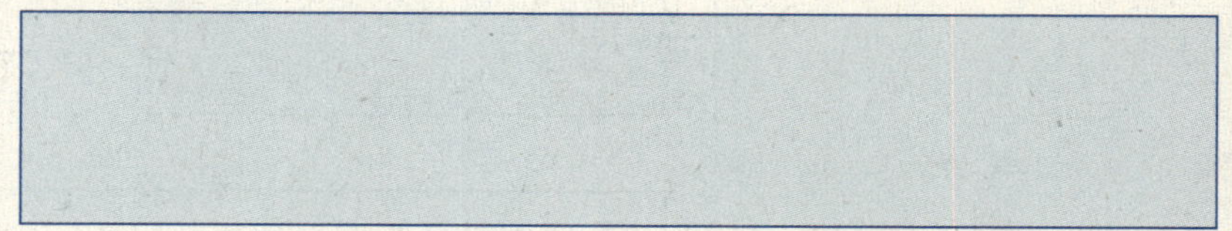

b. $m\angle XYZ < m\angle ZXY$

KEY CONCEPTS

Transitive Property
For any numbers a, b, and c,

1. If $a < b$ and $b < c$, then $a < c$.
2. If $a > b$ and $b > c$, then $a > c$.

Addition and Subtraction Properties For any numbers a, b, and c,

1. If $a < b$, then $a + c < b + c$ and $a - c < b - c$.
2. If $a > b$, then $a + c > b + c$ and $a - c > b - c$.

Multiplication and Division Properties
For any numbers a, b, and c,

1. If $c > 0$, and $a < b$ then $ac < bc$ and $\frac{a}{c} < \frac{b}{c}$.
2. If $c > 0$ and $a > b$ then $ac > bc$ and $\frac{a}{c} > \frac{b}{c}$.

EXAMPLE

4 **In the figure, $m\angle C > m\angle A$. If each of these measures were divided by 5, would the inequality still be true?**

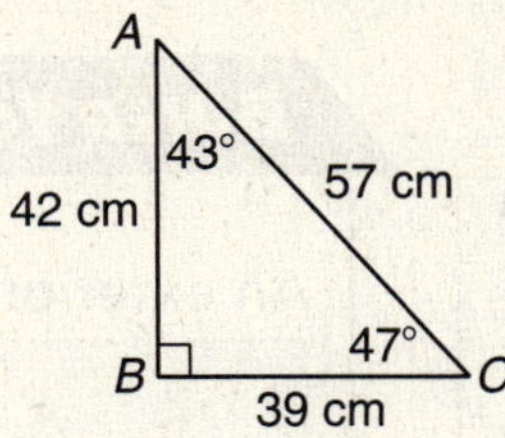

$m\angle C > m\angle A$

47 > ______ Replace $m\angle C$ with ______

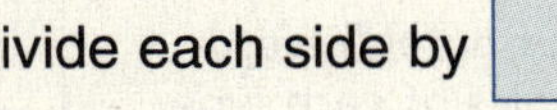

and $m\angle A$ with ______.

47 ÷ ______ > 43 ÷ ______ Divide each side by ______.

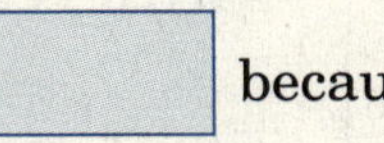

______ > ______

The inequality still holds ______ because ______ is greater than ______.

Your Turn In $\triangle XYZ$, $m\angle X > m\angle Z$. If each of these measures doubled, would this inequality still hold true?

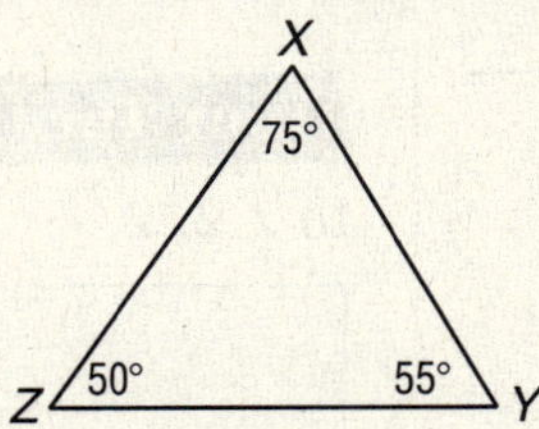

HOMEWORK ASSIGNMENT

Page(s):

Exercises:

7–2 Exterior Angle Theorem

WHAT YOU'LL LEARN

- Identify exterior angles and remote interior angles of a triangle and use the Exterior Angle Theorem.

FOLDABLES™

ORGANIZE IT

In your notes, record examples of each type of inequality under the appropriate tab. Be sure to write about the relationships between sides and angles of a triangle.

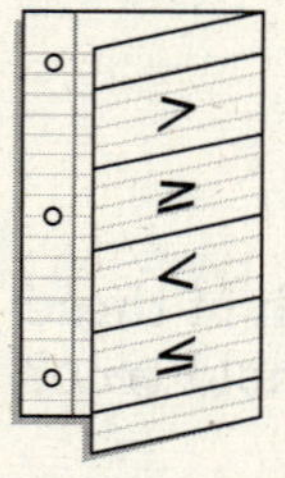

BUILD YOUR VOCABULARY (page 130)

An **exterior angle** of a triangle is an angle that forms a ______ pair with one of the angles of the triangle.

Remote interior angles of a triangle are the ______ angles that *do not* form a linear pair with the ______ angle.

EXAMPLE

1 Name the remote interior angles with respect to ∠4.

Angle ______ forms a ______ with ∠2. Therefore, ______ and ∠3 are remote ______ angles with respect to ∠4.

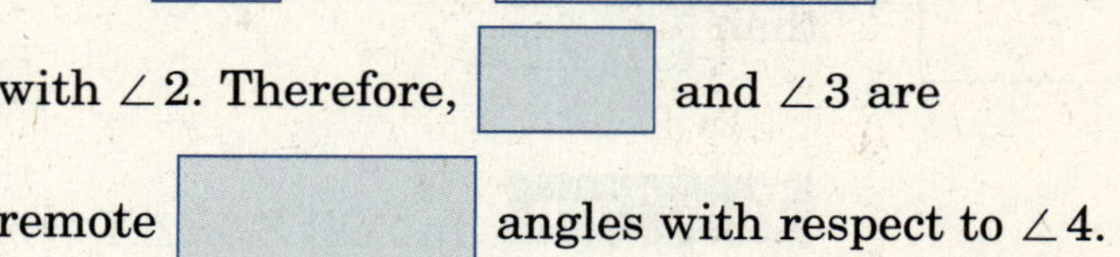

Your Turn Name the remote interior angles with respect to ∠2.

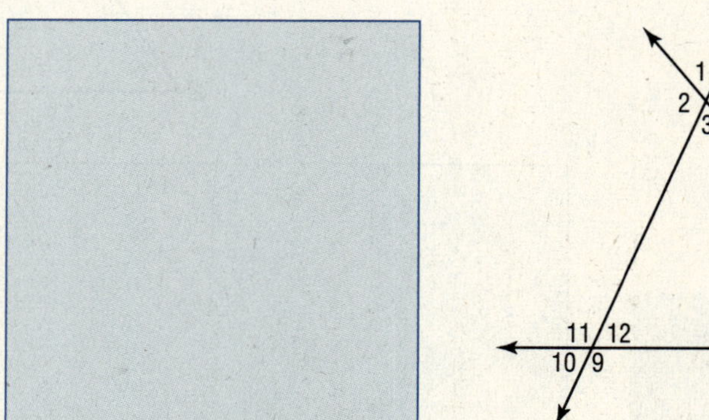

Theorem 7-3 Exterior Angle Theorem
The measure of an exterior angle of a triangle is equal to the sum of the measures of its two remote interior angles.

FOLDABLES™

ORGANIZE IT

Under the tab labeled with a greater than sign, summarize Theorem 7-4.

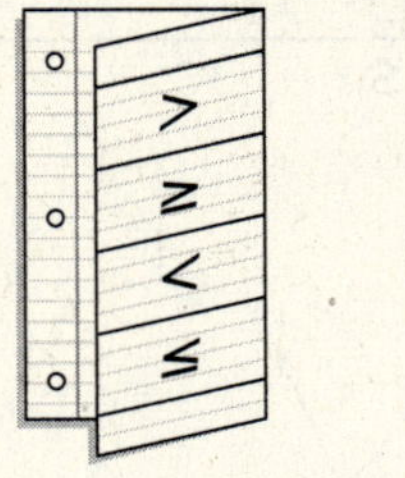

Theorem 7-4 Exterior Angle Inequality Theorem
The measure of an exterior angle of a triangle is greater than the measure of either of its two remote interior angles.

EXAMPLES

2 **In the figure, if $m\angle 1 = 145$ and $m\angle 5 = 82$, what is $m\angle 3$?**

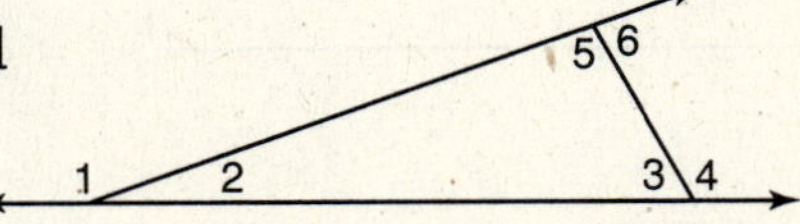

$m\angle 1 = m\angle 5 +$ ______ Exterior Angle Theorem

$145 =$ ______ $+ m\angle 3$ Replace $m\angle 1$ with 145 and $m\angle 5$ with 82.

$145 -$ ______ $= 82 + m\angle 3 -$ ______ Subtract ______ from each side.

______ $= m\angle 3$

3 **In the figure, if $m\angle 6 = 8x$, $m\angle 3 = 12$, and $m\angle 2 = 4(x + 5)$, find the value of x.**

$m\angle 6 = m\angle 3 +$ ______ Exterior Angle Theorem

______ $=$ ______ $+ 4(x + 5)$ Replace $m\angle 6$ with $8x$, $m\angle 3$ with 12 and $m\angle 2$ with $4(x + 5)$.

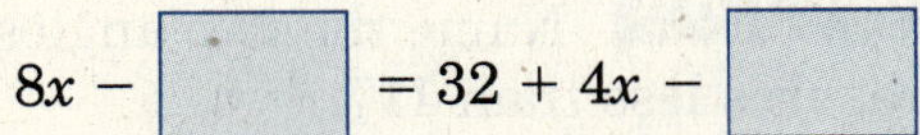

$8x = 12 +$ ______ $+$ ______

$8x =$ ______ $+ 4x$ Combine ______ terms.

$8x -$ ______ $= 32 + 4x -$ ______ Subtract ______ from each side.

______ $= 32$

$\frac{4x}{\_\_\_} = \frac{32}{\_\_\_}$ Divide each side by ______.

$x =$ ______

REMEMBER IT

The measures of the angles in any triangle have a sum of 180 degrees.

Your Turn

a. Find the measure of $\angle 1$ in the figure.

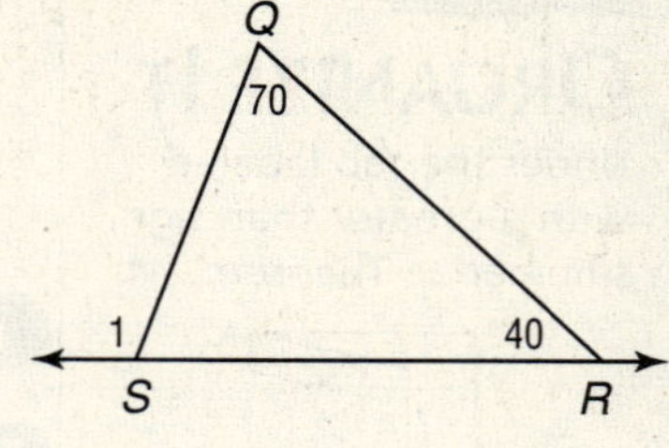

b. In the figure, if $m\angle 6 = 10x + 3$, $m\angle 3 = 6x - 6$, and $m\angle 12 = 49$, find the value of x.

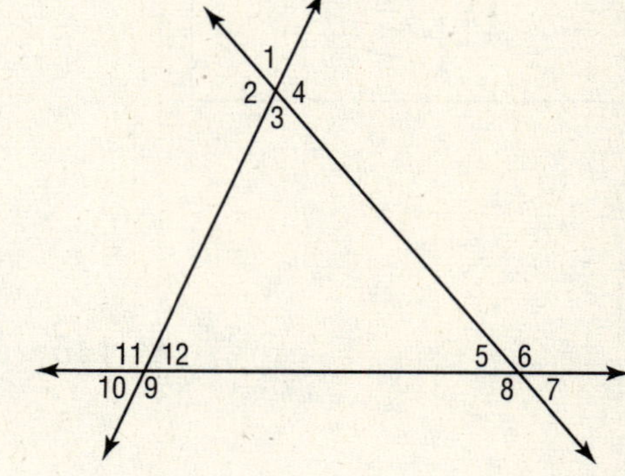

EXAMPLE

4 Name the two angles in $\triangle CDE$ that have measures less than 82.

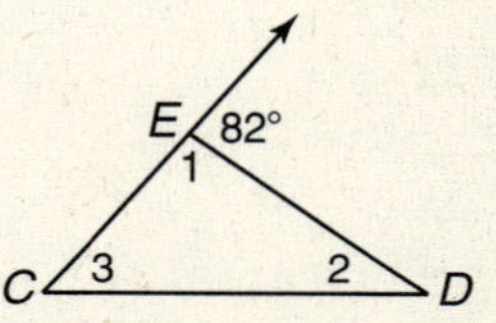

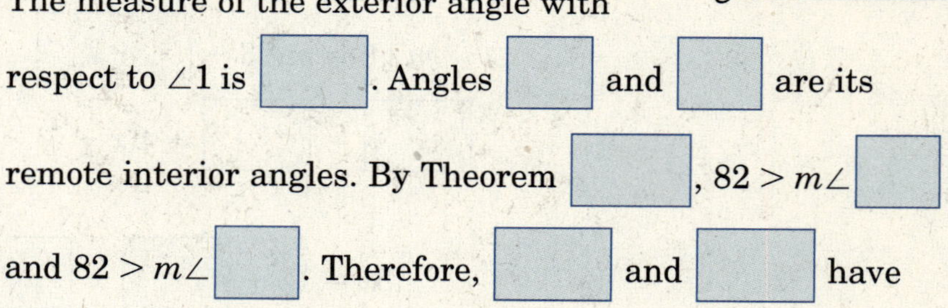

The measure of the exterior angle with respect to $\angle 1$ is ______. Angles ______ and ______ are its remote interior angles. By Theorem ______, $82 > m\angle$ ______ and $82 > m\angle$ ______. Therefore, ______ and ______ have measures less than 82.

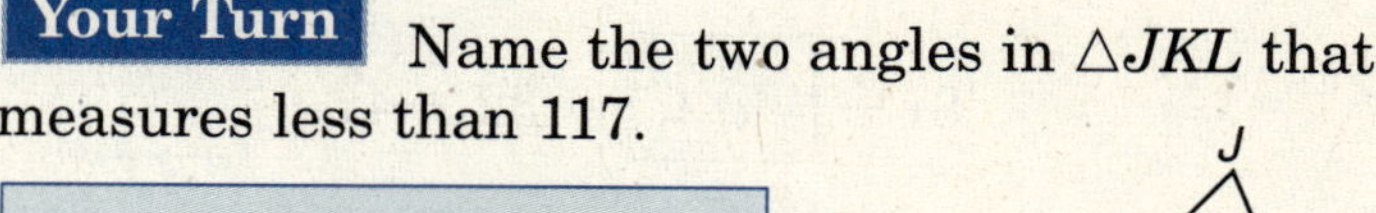

Your Turn Name the two angles in $\triangle JKL$ that have measures less than 117.

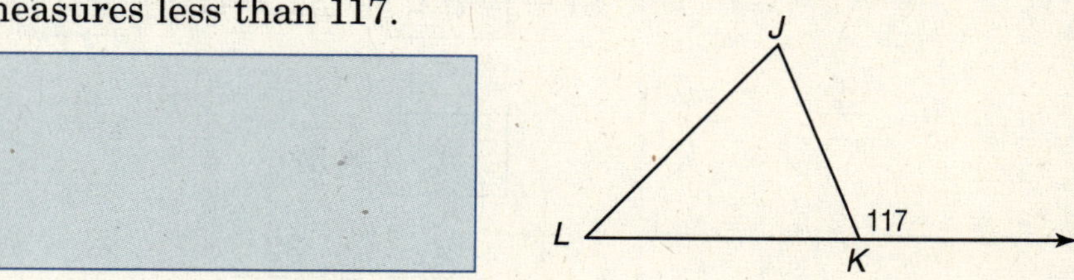

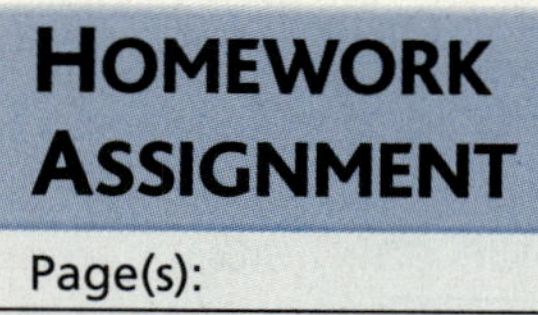

HOMEWORK ASSIGNMENT

Page(s):

Exercises:

Theorem 7-5

If a triangle has one right angle, then the other two angles must be acute.

7–3 Inequalities Within a Triangle

WHAT YOU'LL LEARN

- Identify the relationships between the sides and angles of a triangle.

Theorem 7-6
If the measures of three sides of a triangle are unequal, then the measures of the angles opposite those sides are unequal in the same order.

Theorem 7-7
If the measures of three angles of a triangle are unequal, then the measures of the sides opposite those angles are unequal in the same order.

EXAMPLE

1 In $\triangle KLM$, list the angles in order from least to greatest measure.

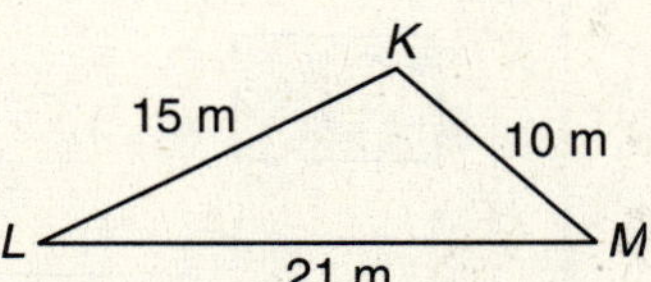

Write the segment measures in order from  to greatest. Then, use Theorem ______ to write the measures of the angles opposite those sides in the same order.

KM < KL < LM

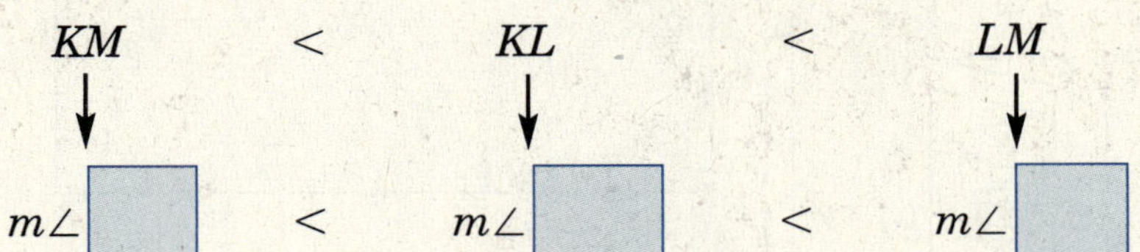

Therefore, the angles in order from least to greatest are ∠ ______, ∠ ______, and ∠ ______.

Your Turn In $\triangle QPS$, list the angles in order from least to greatest measure.

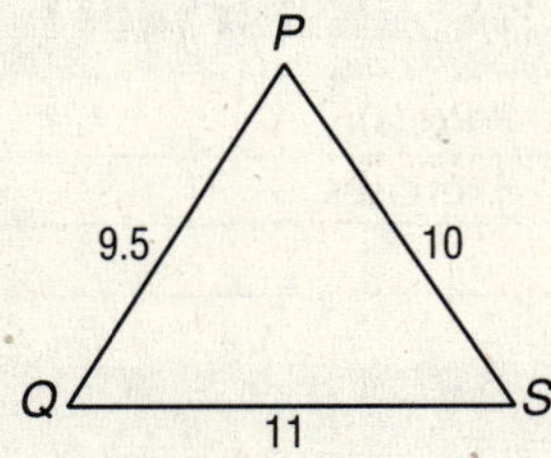

EXAMPLE

2 Identify the side of △*KLM* with the greatest measure.

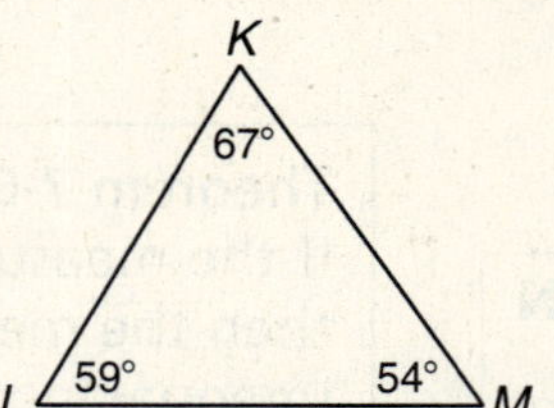

Write the angle measures in order from least to ______.

Then, use Theorem ______ to write the measures of the sides opposite those angles in the same order.

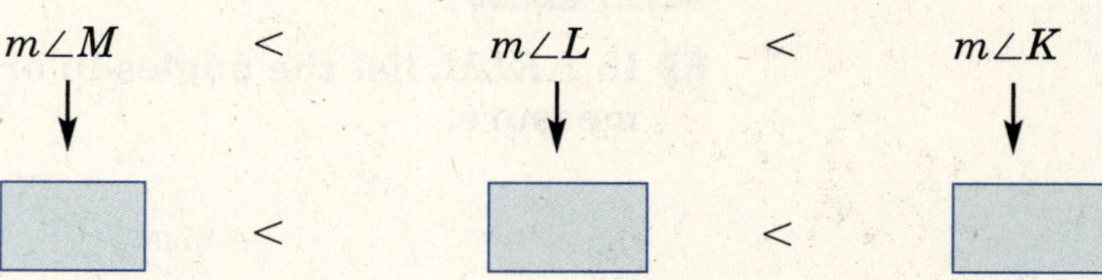

Therefore, ______ has the greatest measure.

Your Turn In △*XYZ*, list the sides in order from least to greatest measure.

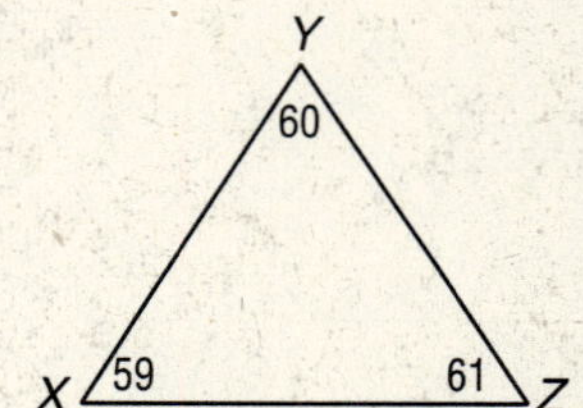

Theorem 7-8
In a right triangle, the hypotenuse is the side with the greatest measure.

FOLDABLES™

ORGANIZE IT

Under the tab labeled with a greater than sign, summarize Theorem 7-8 using the words "greater than".

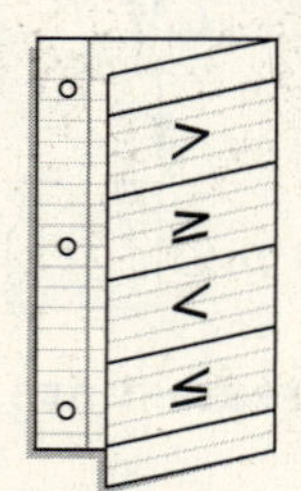

HOMEWORK ASSIGNMENT

Page(s):

Exercises:

7–4 Triangle Inequality Theorem

What You'll Learn

- Identify and use the Triangle Inequality Theorem.

Theorem 7-9 Triangle Inequality Theorem
The sum of the measures of any two sides of a triangle is greater than the measure of the third side.

Foldables

Organize It

Under the tab labeled with a greater than sign, summarize Theorem 7-9.

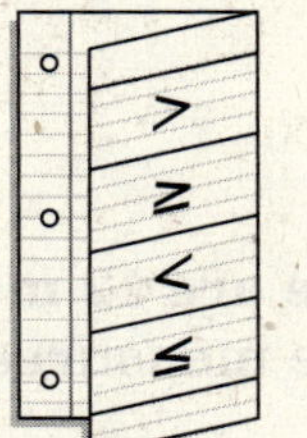

EXAMPLES

1 Determine if the three numbers can be the measures of the sides of a triangle.

6, 7, 9

$6 + 7 > 9$ _____

$6 + 9 > 7$ _____

$7 + 9 > 6$ _____

All possible cases _____ true. Sides with these measures _____ form a triangle.

2 1, 7, 8

$7 + 8 > 1$ _____

$8 + 1 > 7$ _____

$1 + 7 > 8$ _____

All possible cases _____ true. Sides with these measures _____ form a triangle.

Your Turn **Determine if the three numbers can be the measures of the sides of a triangle.**

a. 15, 40, 19 _____

b. 4, 18, 21 _____

EXAMPLES

3 **What are the greatest and least possible whole-number measures for a side of a triangle whose other two sides measure 4 feet and 6 feet?**

Let x be the measure of the third side of the triangle. x is greater than the difference of the measures of the two other sides.

$x > 6 -$ ______

$x >$ ______

x is less than the sum of the measures of the two other sides.

$x < 6 +$ ______

$x <$ ______

Therefore, ______ $< x <$ ______.

WRITE IT

In your own words, explain why two sides of a triangle, when added together, cannot equal the length of the third side.

______

______

______

______

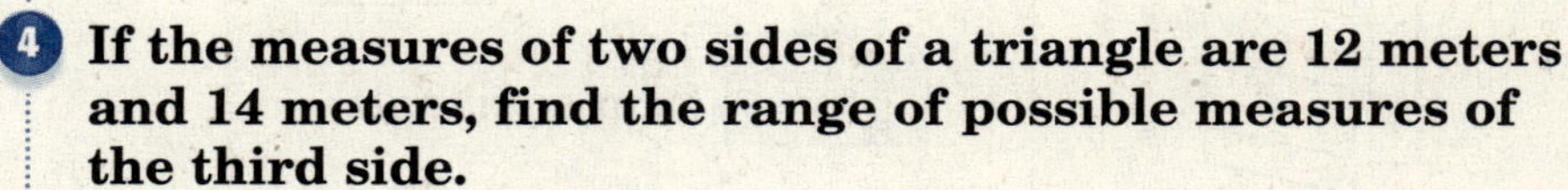

4 **If the measures of two sides of a triangle are 12 meters and 14 meters, find the range of possible measures of the third side.**

Let x be the measure of the third side of the triangle. x is greater than the difference of the measures of the two other sides.

$x > 14 -$ ______

$x >$ ______

x is less than the sum of the measures of the two other sides.

$x < 14 +$ ______

$x <$ ______

Therefore, ______ $< x <$ ______.

Your Turn

a. What are the greatest and least possible whole-number measures for a side of a triangle whose other two sides measure 23 cm and 29 cm?

b. If the measures of two sides of a triangle are 11 inches and 3 inches, find the range of possible measures of the third side.

HOMEWORK ASSIGNMENT

Page(s):

Exercises:

CHAPTER 7 BRINGING IT ALL TOGETHER

STUDY GUIDE

FOLDABLES™	VOCABULARY PUZZLEMAKER	BUILD YOUR VOCABULARY
Use your **Chapter 7 Foldable** to help you study for your chapter test.	To make a crossword puzzle, word search, or jumble puzzle of the vocabulary words in Chapter 7, go to: www.glencoe.com/sec/math/t_resources/free/index.php	You can use your completed **Vocabulary Builder** (page 130) to help you solve the puzzle.

7-1 Segments, Angles, and Inequalities

Replace ● with <, >, or = to make a true sentence.

1. JK ● KX

2. LM ● JL

3. KM ● JL

4. KL ● XM

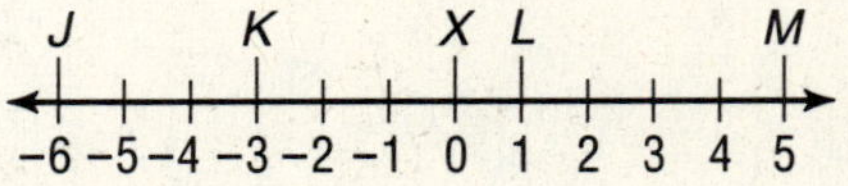

5. $m\angle BCD$ ● $m\angle BDE$

6. $m\angle CBE$ ● $m\angle EDC$

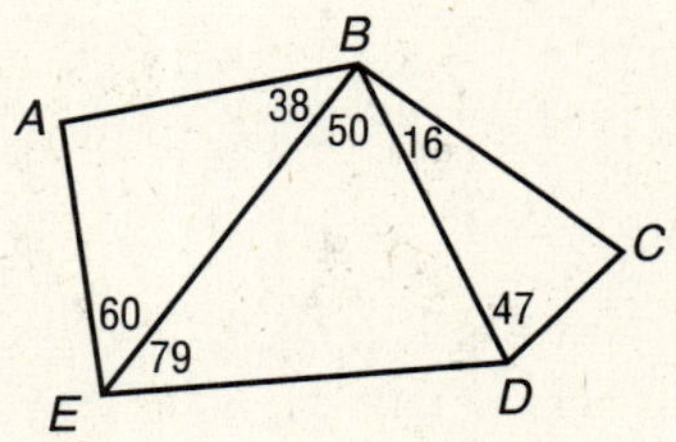

7-2 Exterior Angle Theorem

7. Name the remote interior angles of $\triangle ABC$ with respect to $\angle 5$.

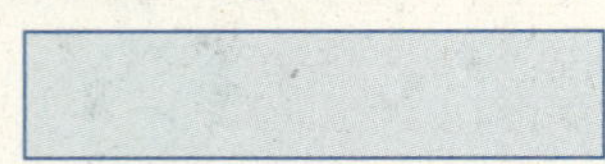

8. $\overline{BD} \perp \overline{AC}$ and $m\angle 15 = 139$.

What is $m\angle 10$?

9. If $m\angle 1 = 19x$, $m\angle 16 = 6x$, and $m\angle DAB = 91$, find the value of x.

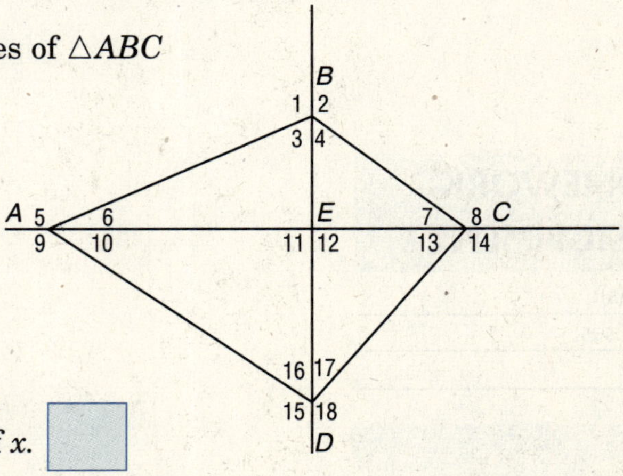

7-3 Inequalities Within a Triangle

In each triangle, list the angles from least to greatest.

10.

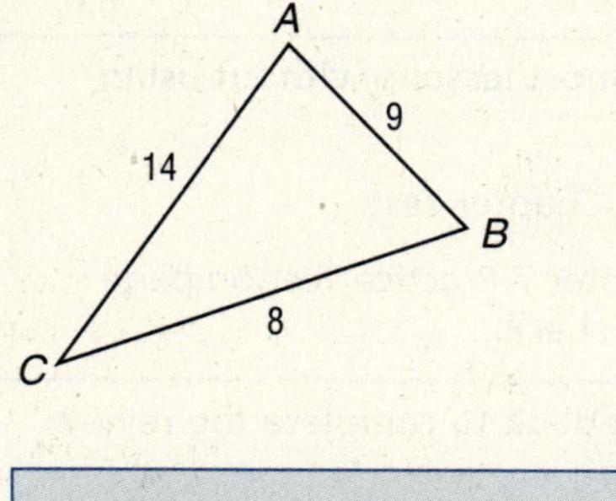

11.

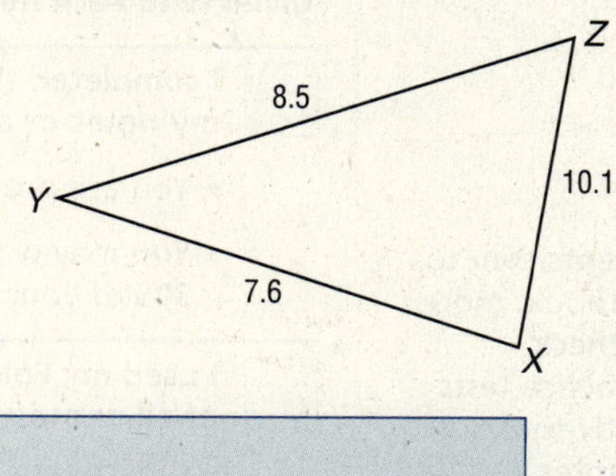

In each triangle, list the sides measuring least to greatest.

12.

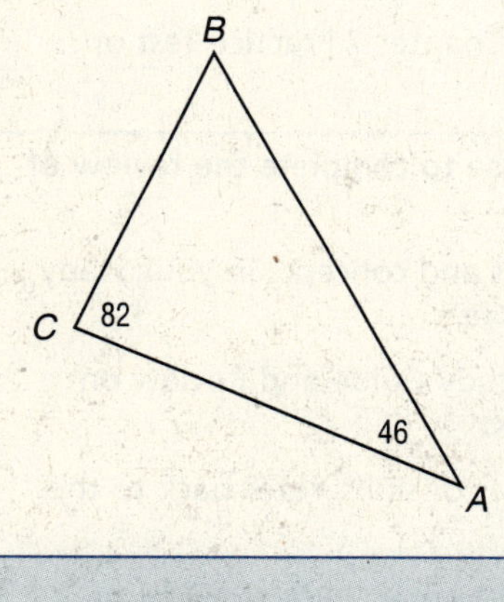

13.

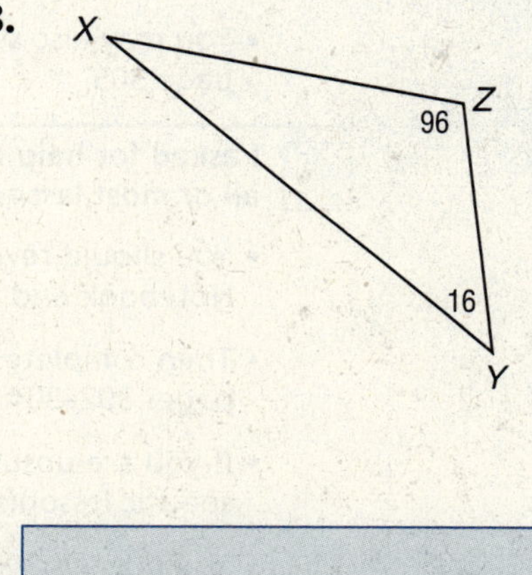

7-4 Triangle Inequality Theorem

Determine if the numbers given can be measures of the sides of a triangle.

14. 7.7, 16.8, 11.3

15. 36, 12, 28

16. 7, 9, 16

Find the range of possible values for the third side of the triangle.

17. 16, 7

18. 12, 10

19. 5, 9

ARE YOU READY FOR THE CHAPTER TEST?

Math Online

Visit **geomconcepts.com** to access your textbook, more examples, self-check quizzes, and practice tests to help you study the concepts in Chapter 7.

Check the one that applies. Suggestions to help you study are given with each item.

☐ **I completed the review of all or most lessons without using my notes or asking for help.**

- You are probably ready for the Chapter Test.
- You may want to take the Chapter 7 Practice Test on page 305 of your textbook as a final check.

☐ **I used my Foldable or Study Notebook to complete the review of all or most lessons.**

- You should complete the Chapter 7 Study Guide and Review on pages 302–304 of your textbook.
- If you are unsure of any concepts or skills, refer back to the specific lesson(s).
- You may also want to take the Chapter 7 Practice Test on page 305.

☐ **I asked for help from someone else to complete the review of all or most lessons.**

- You should review the examples and concepts in your Study Notebook and Chapter 7 Foldable.
- Then complete the Chapter 7 Study Guide and Review on pages 302–304 of your textbook.
- If you are unsure of any concepts or skills, refer back to the specific lesson(s).
- You may also want to take the Chapter 7 Practice Test on page 305.

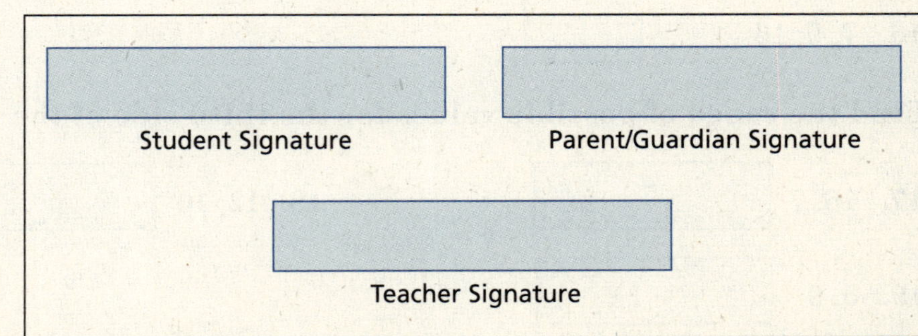

Quadrilaterals

Use the instructions below to make a Foldable to help you organize your notes as you study the chapter. You will see Foldable reminders in the margin of this Interactive Study Notebook to help you in taking notes.

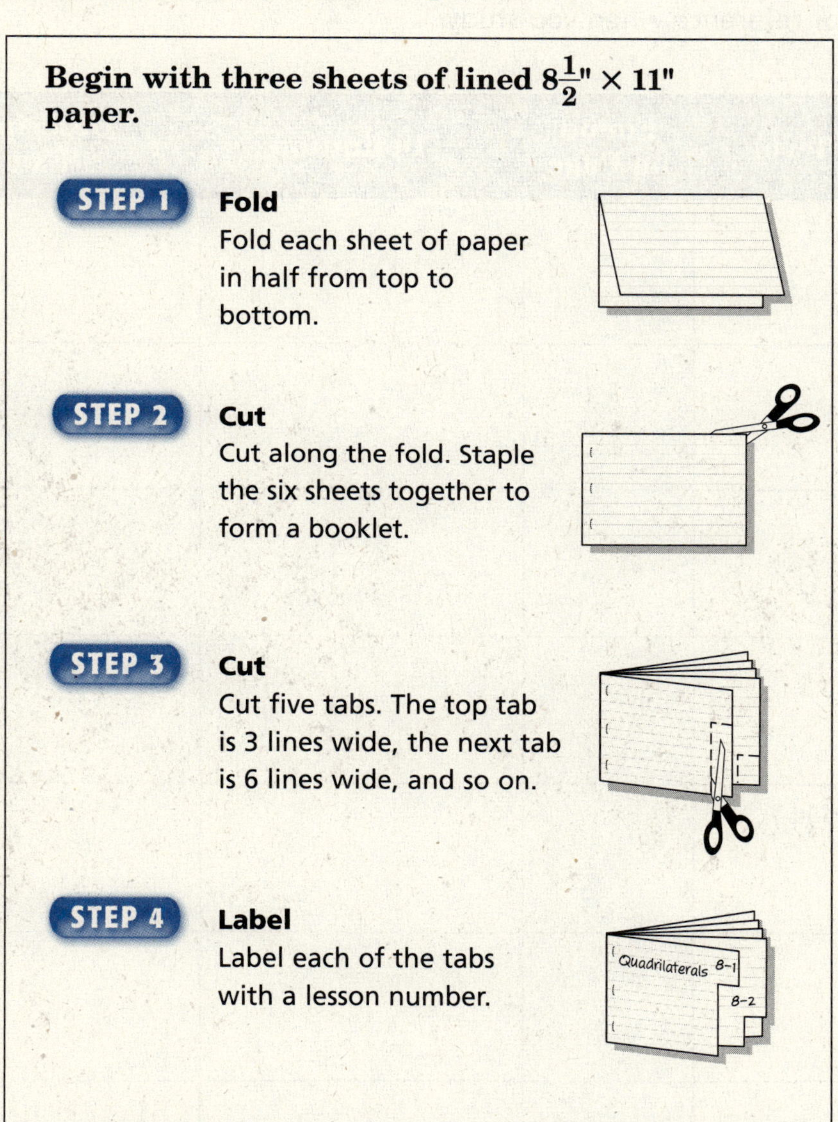

NOTE-TAKING TIP: When you read and learn new concepts, help yourself remember these concepts by taking notes, writing definitions and explanations, and drawing models as needed.

BUILD YOUR VOCABULARY

This is an alphabetical list of new vocabulary terms you will learn in Chapter 8. As you complete the study notes for the chapter, you will see Build Your Vocabulary reminders to complete each term's definition or description on these pages. Remember to add the textbook page number in the second column for reference when you study.

Vocabulary Term	Found on Page	Definition	Description or Example
base angles			
bases			
consecutive [con-SEK-yoo-tiv]			
diagonals			
isosceles trapeziod			
kite			
legs			
median			

Vocabulary Term	Found on Page	Definition	Description or Example
midsegment			
nonconsecutive			
parallelogram			
quadrilateral			
rectangle			
rhombus [ROM-bus]			
square			
trapezoid [TRAP-a-ZOYD]			

8–1 Quadrilaterals

What You'll Learn

- Identify parts of quadrilaterals and find the sum of the measures of the interior angles of a quadrilateral.

FOLDABLES™

Organize It

Under the tab for Lesson 8-1, write the rules for classifying quadrilaterals. Draw a quadrilateral and label the consecutive and nonconsecutive sides, as well as the diagonals.

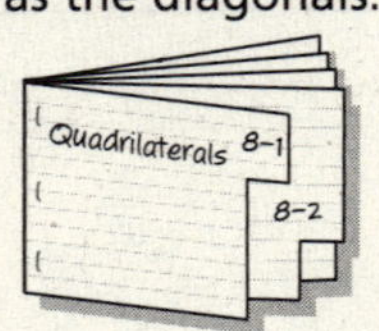

Build Your Vocabulary (pages 146–147)

A **quadrilateral** is a ______ geometric figure with ______ sides and ______ vertices.

Any two sides, vertices, or angles of a quadrilateral are either **consecutive** or **nonconsecutive**.

Segments that join ______ vertices are called **diagonals**.

Examples

Refer to quadrilateral *DEFG*.

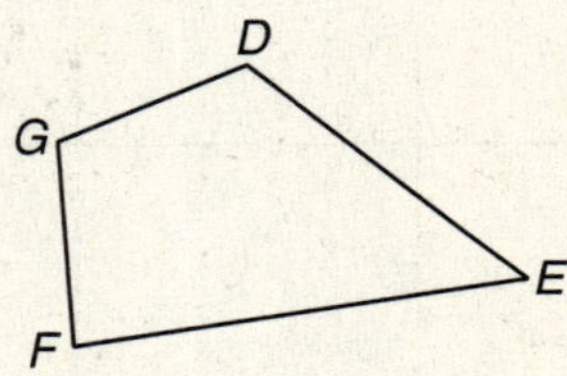

1 Name all pairs of consecutive angles.

______ and $\angle G$, ______ and $\angle F$, ______ and $\angle E$, and ______ and $\angle D$ are consecutive angles.

2 Name all pairs of nonconsecutive vertices.

D and ______ are nonconsecutive vertices. ______ and *G* are nonconsecutive vertices.

3 Name all pairs of consecutive sides.

$\overline{DG}$ and ______, $\overline{FG}$ and ______, $\overline{EF}$ and ______, and $\overline{DE}$ and ______ are pairs of consecutive sides.

Remember It

Consecutive sides share a vertex; nonconsecutive sides do not.

Consecutive vertices are the endpoints of a side while nonconsecutive vertices are not.

Consecutive angles share a side of the quadrilateral while nonconsecutive angles do not.

Your Turn **Refer to quadrilateral *WXYZ*.**

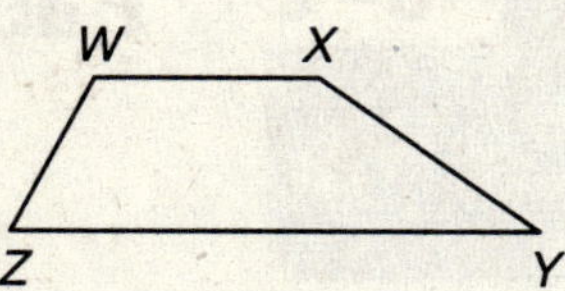

a. Name all pairs of consecutive angles.

b. Name all pairs of nonconsecutive vertices.

c. Name all pairs of consecutive sides.

Theorem 8-1
The sum of the measures of the angles of a quadrilateral is 360.

EXAMPLE

Remember It

In a quadrilateral, nonconsecutive sides, vertices, or angles are also called *opposite* sides, vertices, or angles.

4 **Find the missing measure if three of the four angle measures in quadrilateral *ABCD* are 90, 120, and 40.**

$m\angle A + m\angle B + m\angle C + m\angle D = 360$ Theorem 8-1

$\_\_\_ + \_\_\_ + \_\_\_ + m\angle D = 360$ Substitution

$\_\_\_ + m\angle D = 360$

$250 + m\angle D - 250 = 360 - 250$ Subtract.

$m\angle D = \_\_\_$

Homework Assignment

Page(s):

Exercises:

Your Turn Find the missing measure if three of the four angle measures in quadrilateral *RMSQ* are 115, 75, and 50.

8–2 Parallelograms

What You'll Learn

- Identify and use the properties of parallelograms.

Build Your Vocabulary (page 147)

A **parallelogram** is a ______ with two pairs of ______ sides.

Theorem 8-2
Opposite angles of a parallelogram are congruent.

Theorem 8-3
Opposite sides of a parallelogram are congruent.

Theorem 8-4
The consecutive angles of a parallelogram are supplementary.

Foldables

Organize It

Under the tab for Lesson 8-2, write the definitions and theorems to help you classify parallelograms. Draw a parallelogram and label the congruent sides and angles, as well as properties of the diagonals.

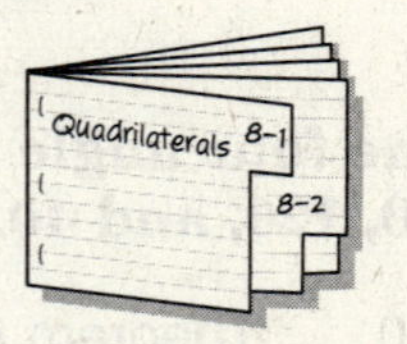

Examples

In parallelogram *KLMN*, $KL = 23$, $KN = 15$, and $m\angle K = 105$.

1 Find *LM* and *MN*.

$\overline{KL} \cong \overline{MN}$ and $\overline{KN} \cong \overline{LM}$ — Theorem 8-3

$KL =$ ______ and $KN =$ ______ — Definition of congruent segments

______ $= MN$ and ______ $= LM$ — Replace *KL* with ______ and *KN* with ______.

2 Find $m\angle M$.

$\angle M \cong \angle K$ — Theorem 8-2

$m\angle M =$ ______ — Definition of congruent angles

$m\angle M =$ ______ — Replace $m\angle K$ with ______.

3 **Find $m\angle L$.**

$m\angle L + m\angle K = 180$ Theorem 8-4

$m\angle L +$ ______ $= 180$ Replace $m\angle K$ with .

$m\angle L + 105 - 105 = 180 - 105$ Subtract.

$m\angle L =$ ______

Your Turn **In parallelogram $ABCD$, $AB = 8$, $BC = 3$, and $m\angle C = 115$.**

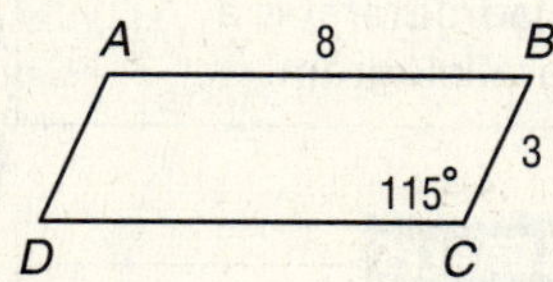

a. Find AD and CD.

b. Find $m\angle A$.

c. Find $m\angle B$.

Theorem 8-5
The diagonals of a parallelogram bisect each other.

Theorem 8-6
A diagonal of a parallelogram separates it into two congruent triangles.

EXAMPLE

4 **In parallelogram $PQRS$, if $PR = 32$, find PL.**

Theorem 8-5 states that the diagonals of a parallelogram bisect each other. Therefore, $\overline{PL} \cong \overline{LR}$ or $PL = \frac{1}{2}(PR)$.

$PL = \frac{1}{2}(PR)$

$PL = \frac{1}{2}(32)$ or ______ Replace PR with 32.

Homework Assignment

Page(s):

Exercises:

Your Turn In parallelogram $PARL$, if $LA = 48$, find LO.

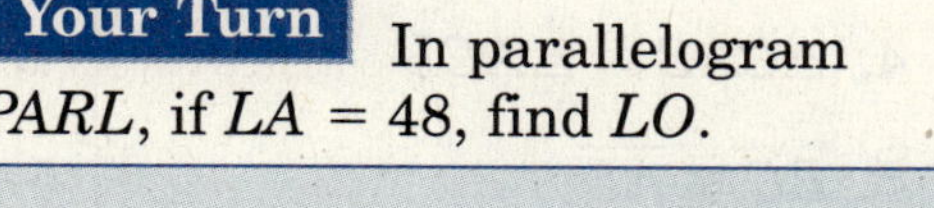

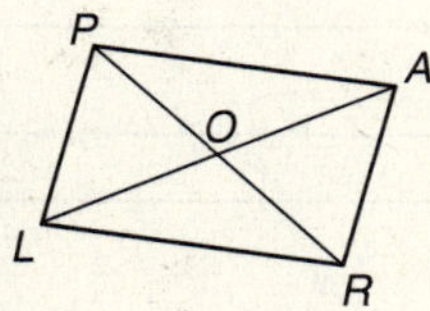

8–3 Tests for Parallelograms

What You'll Learn

- Identify and use tests to show that a quadrilateral is a parallelogram.

Foldables™

Organize It

Under the tab for Lesson 8-3, write the tests for parallelograms. Remember to include the definition of a parallelogram. Draw pictures to accompany each theorem.

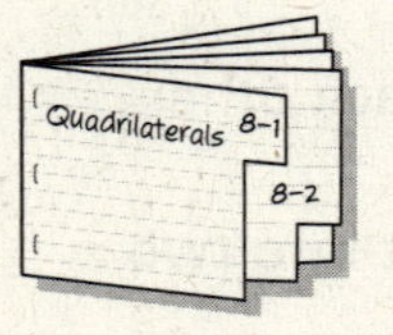

Review It

What does CPCTC represent? (*Lesson 5-4*)

______

______

______

______

Theorem 8-7
If both pairs of opposite sides of a quadrilateral are congruent, then the quadrilateral is a parallelogram.

EXAMPLE

1 In quadrilateral $WXYZ$, if $\triangle WYZ \cong \triangle YWX$, how could you prove that $WXYZ$ is a parallelogram?

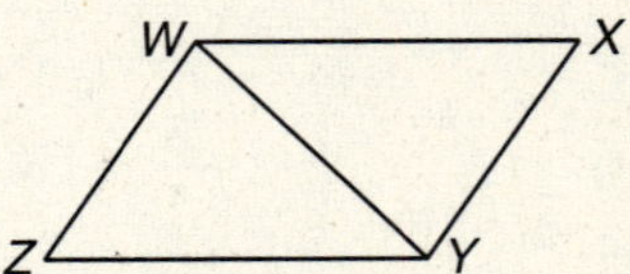

Show that both pairs of opposite sides are congruent.

Statement	Reason
1. $\triangle WYZ \cong \triangle YWX$	1. Given
2. $\overline{YZ} \cong \overline{WX}$	2.
3. $\overline{WZ} \cong \overline{YX}$	3. CPCTC
4. $WXYZ$ is a parallelogram.	4.

Your Turn In quadrilateral $ABCD$, $\angle CAB \cong \angle ACD$ and $\overline{AB} \cong \overline{CD}$. Show that $ABCD$ is a parallelogram by providing a reason for each step.

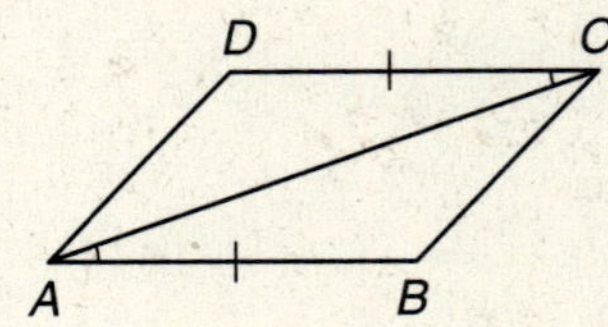

Statement	Reason
1. $\angle CAB \cong \angle ACD$	1. Given
2. $\overline{AB} \cong \overline{CD}$	2. Given
3. $\overline{AC} \cong \overline{AC}$	3.
4. $\triangle CAB \cong \triangle ACD$	4. SAS
5. $\overline{BC} \cong \overline{AD}$	5.
6. $ABCD$ is a parallelogram.	6.

Theorem 8-8
If one pair of opposite sides of a quadrilateral is parallel and congruent, then the quadrilateral is a parallelogram.

Theorem 8-9
If the diagonals of a quadrilateral bisect each other, then the quadrilateral is a parallelogram.

EXAMPLES

Determine whether each quadrilateral is a parallelogram. If the figure is a parallelogram, give a reason for your answer.

The figure has one pair of opposite sides that are ________ and congruent. Therefore, the quadrilateral is a ________ by Theorem 8-8.

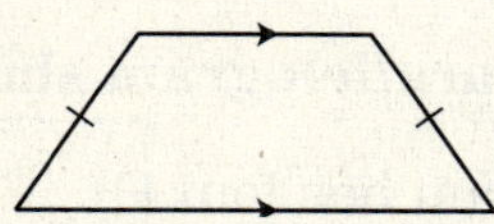

One pair of opposite sides is congruent but ________. The other pair of opposite sides is ________ but not ________. Therefore, the quadrilateral ________ a parallelogram.

WRITE IT

Explain how alternate interior angles could be used to show that opposite angles in a parallelogram are congruent.

____

____

____

____

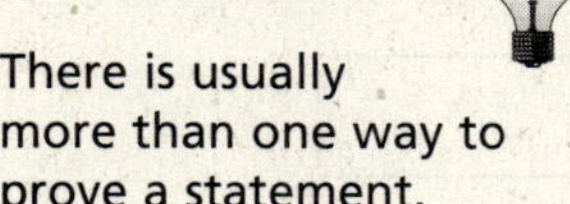

REMEMBER IT

There is usually more than one way to prove a statement.

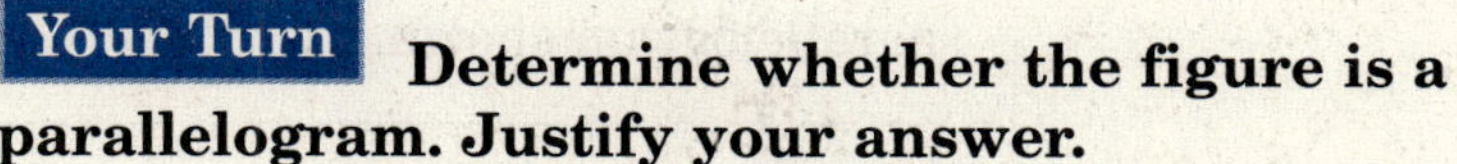

Your Turn **Determine whether the figure is a parallelogram. Justify your answer.**

a.

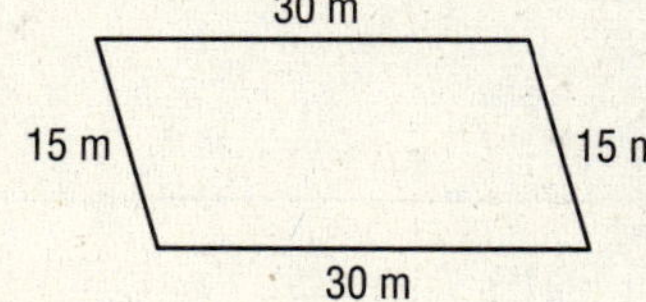

b.

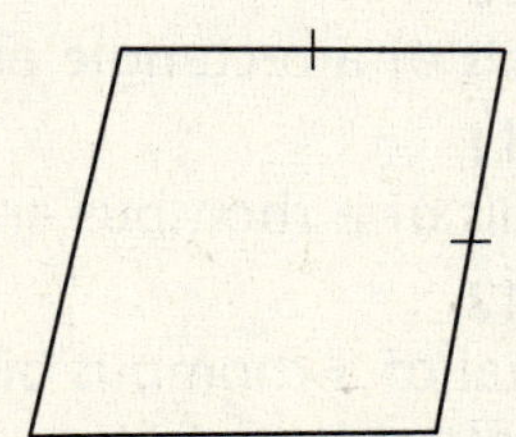

HOMEWORK ASSIGNMENT

Page(s):

Exercises:

8–4 Rectangles, Rhombi, and Squares

What You'll Learn

- Identify and use properties of rectangles, rhombi, and squares.

Build Your Vocabulary (page 147)

A **rectangle** is a ______ with four ______ angles.

A parallelogram with ______ congruent sides is a **rhombus**.

A parallelogram with ______ sides and four ______ angles is a **square**.

Foldables™

Organize It

Under the tab for Lesson 8-4, draw the diagram for classifying rectangles, rhombi, and squares. Write notes and theorems to help you remember the main idea.

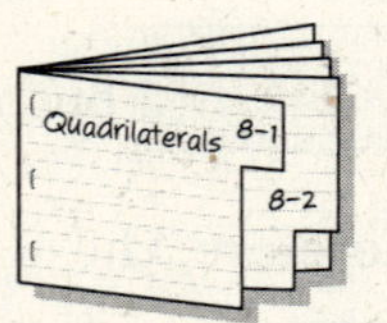

EXAMPLE

1 Identify the parallelogram shown.

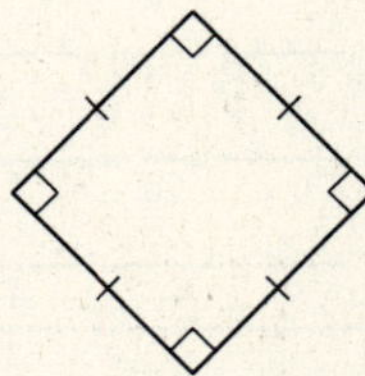

The parallelogram has four ______ sides and ______ right angles. It is a ______.

Your Turn Identify the parallelogram shown.

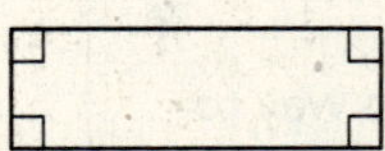

______

Remember It

Rhombi is the plural of *rhombus*.

Theorem 8-10
The diagonals of a rectangle are congruent.

Theorem 8-11
The diagonals of a rhombus are perpendicular.

Theorem 8-12
Each diagonal of a rhombus bisects a pair of opposite angles.

EXAMPLES

Refer to rhombus ABCD.

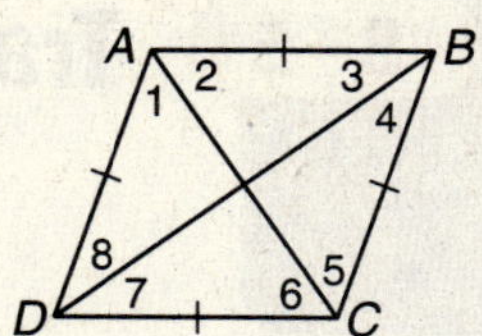

2 **Which angles are congruent to ∠1?**

Theorem 8-12 states the diagonals of a rhombus ______ opposite ______. Therefore, ______ is congruent to ∠2, ______, and ∠6.

3 **If $m\angle 7 = 35$, find $m\angle ADC$.**

Theorem 8-12 states the diagonals of a ______ bisect ______ angles.

Therefore, $m\angle 7 = \frac{1}{2}(m\angle ADC)$.

$\_\_\_ = \frac{1}{2}(m\angle ADC)$ $m\angle 7 =$ ______.

$\_\_\_ \cdot \_\_\_ = \_\_\_ \cdot \frac{1}{2}(m\angle ADC)$ Multiply each side.

$\_\_\_ = m\angle ADC$

WRITE IT

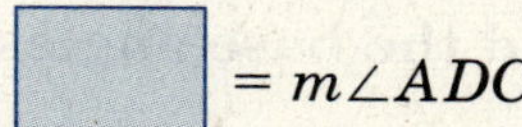

Explain how squares can be rhombi, rectangles, and parallelograms.

______

______

______

______

______

______

Your Turn **Refer to the figure.**

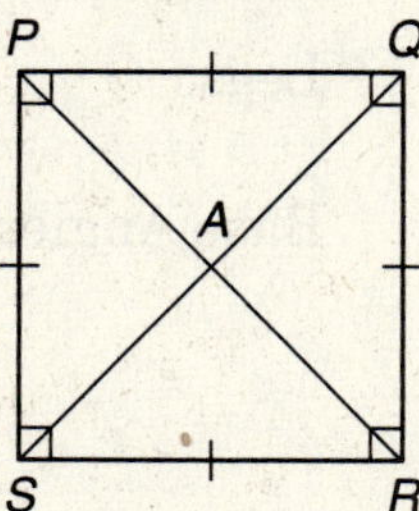

a. Which angles are congruent to $\angle PSQ$ in square $PQRS$?

______

b. If $AP = 7$, find QS.

______

HOMEWORK ASSIGNMENT

Page(s):

Exercises:

8–5 Trapezoids

What You'll Learn

- Identify and use properties of trapezoids and isosceles trapezoids.

Build Your Vocabulary (pages 146–147)

A **trapezoid** is a quadrilateral with exactly ____ pair of ____ sides.

The ____ sides are the **bases** of the trapezoid.

The ____ sides of the trapezoid are known as **legs**.

Each trapezoid has ____ pairs of **base angles**.

Foldables™ Organize It

Under the tab for Lesson 8-5, write the definition of a trapezoid and of an isosceles trapezoid. Draw a trapezoid and label the bases, legs, and base angles. Label the congruent angles, and draw the median.

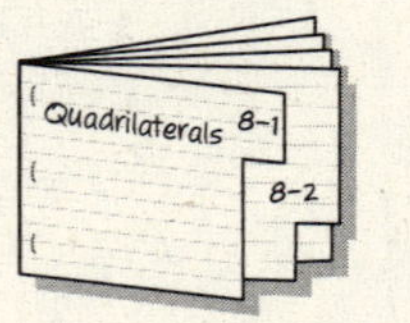

Example

1 In trapezoid *ABCD*, name the bases, legs, and the base angles.

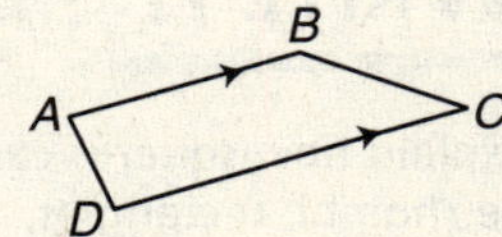

Bases: $\overline{AB}$ and ____ are parallel segments.

Legs: $\overline{AD}$ and ____ are nonparallel segments.

Base Angles: $\angle A$ and ____ form one pair of base angles, while $\angle C$ and ____ are the other pair of base angles.

Your Turn In trapezoid *WXYZ*, name the bases, legs, and the base angles.

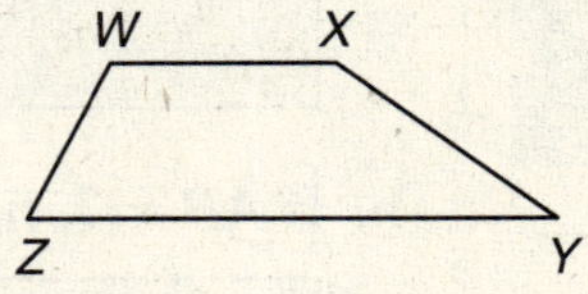

BUILD YOUR VOCABULARY (pages 146–147)

The **median** of a trapezoid is the segment that joins the ______ of the legs.

Another name for the median is the **midsegment**.

If the legs of the trapezoid are ______, then the trapezoid is an **isosceles trapezoid**.

Theorem 8-13
The median of a trapezoid is parallel to the bases, and the length of the median equals one-half the sum of the lengths of the bases.

EXAMPLE

2 **Find the length of the median *KL* in trapezoid *EFGH* if *EF* = 35 and *GH* = 40.**

$KL = \frac{1}{2}(EF + GH)$ Theorem 8-13

$KL = \frac{1}{2}(\_\_\_\_ + \_\_\_\_)$ Replace *EF* and *GH*.

$KL = \frac{1}{2}(\_\_\_\_)$ or ______

REVIEW IT

The word "isosceles" is used for classifying triangles and trapezoids. What similarities do isosceles triangles and isosceles trapezoids have? (*Lesson 6-4*)

______

______

______

______

______

Your Turn Find the length of the median *NO* in trapezoid *JKLM* if *JK* = 22 and *LM* = 26.

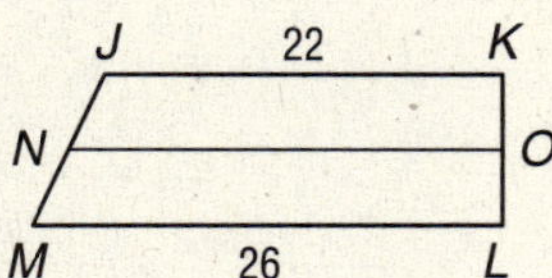

Theorem 8-14
Each pair of base angles in an isosceles trapezoid is congruent.

REMEMBER IT

Trapezoids and parallelograms are both quadrilaterals, but no quadrilateral can be both a trapezoid and a parallelogram.

EXAMPLE

3 The measure of one angle in an isosceles trapezoid is 55. Find the measures of the other three angles.

Let $\angle 1$ be the given angle, and let $\angle 2$ be the base angle congruent to $\angle 1$.

$\angle 2 \cong \angle 1$ Theorem 8-14

$m\angle 2 =$

$m\angle 2 =$ Replace $m\angle 1$ by 55.

Let $\angle 3$ and $\angle 4$ be the other pair of base angles, with $\angle 3$ adjacent to $\angle 1$.

$m\angle 3 + m\angle 1 =$ Consecutive interior angles are supplementary.

$m\angle 3 +$ $= 180$ Replace $m\angle 1$ with .

$m\angle 3 + 55 -$ $= 180 -$ Subtract 55 from each side.

$m\angle 3 =$

Since $\angle 3$ and $\angle 4$ are congruent base angles, $m\angle 4 =$.

The measures of the three missing angles are 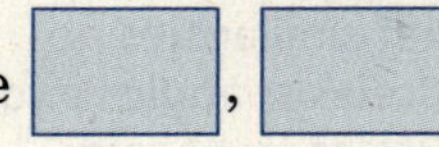,

and .

Your Turn The measure of one angle in an isosceles trapezoid is 76. Find the measures of the other three angles.

HOMEWORK ASSIGNMENT

Page(s):

Exercises:

CHAPTER 8

BRINGING IT ALL TOGETHER

STUDY GUIDE

FOLDABLES™	VOCABULARY PUZZLEMAKER	BUILD YOUR VOCABULARY
Use your **Chapter 8 Foldable** to help you study for your chapter test.	To make a crossword puzzle, word search, or jumble puzzle of the vocabulary words in Chapter 8, go to: www.glencoe.com/sec/math/t_resources/free/index.php	You can use your completed **Vocabulary Builder** (pages 146–147) to help you solve the puzzle.

8-1

Quadrilaterals

1. Name the side opposite $\overline{WZ}$.

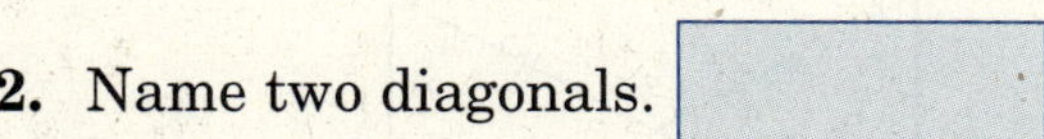

2. Name two diagonals.

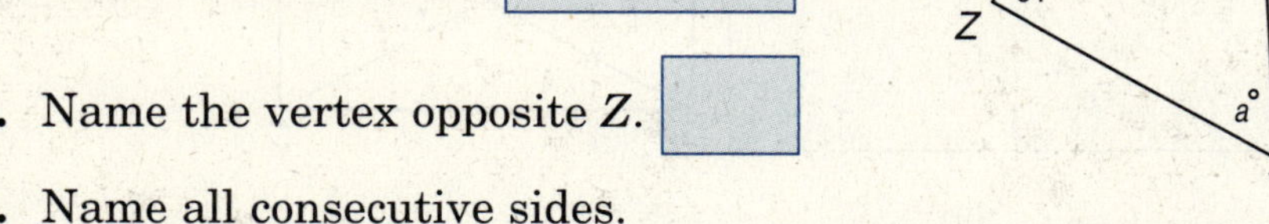

3. Name the vertex opposite Z.

4. Name all consecutive sides.

5. Find $m\angle X$ and $m\angle Y$.

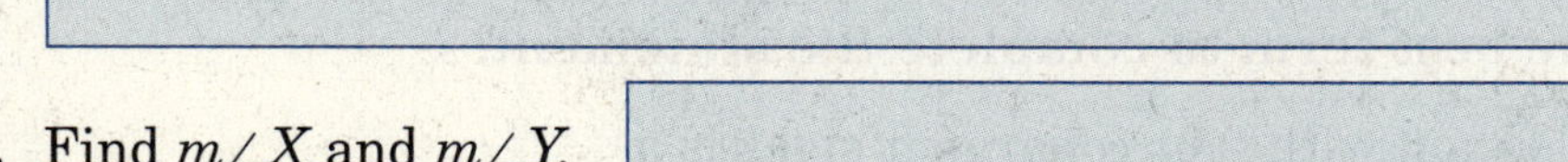

8-2

Parallelograms

Given that *JKLM* is a parallelogram, find the missing measures.

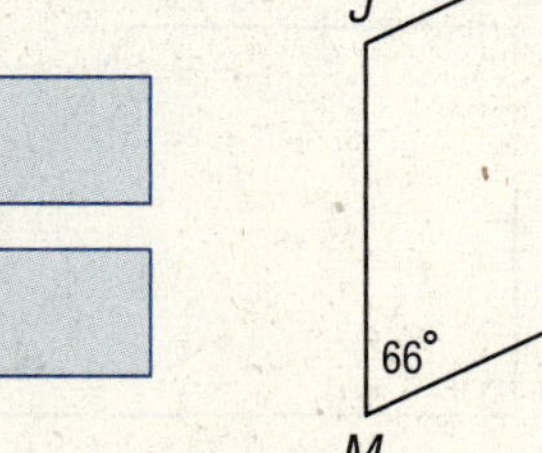

6. $m\angle L$ 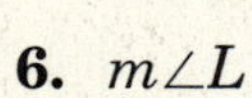

7. $m\angle J$ 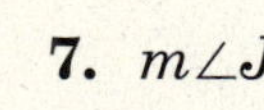

8. LM 

9. $m\angle K$ 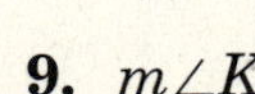

10. If the measure of one angle of parallelogram $PQRS$ is 79, what are the measures of the other three interior angles?

8-3 Tests for Parallelograms

State whether each figure is a parallelogram. Justify your reason.

11.

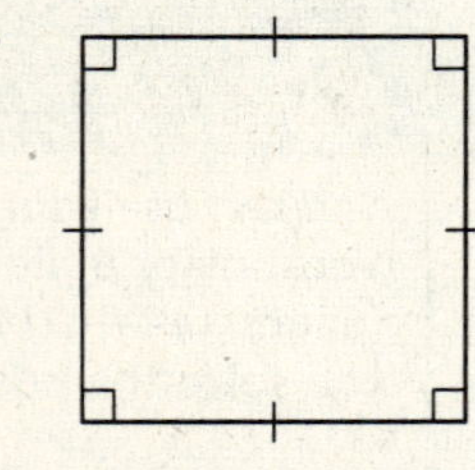

12.

13. Explain why quadrilateral *ABCD* is a parallelogram.

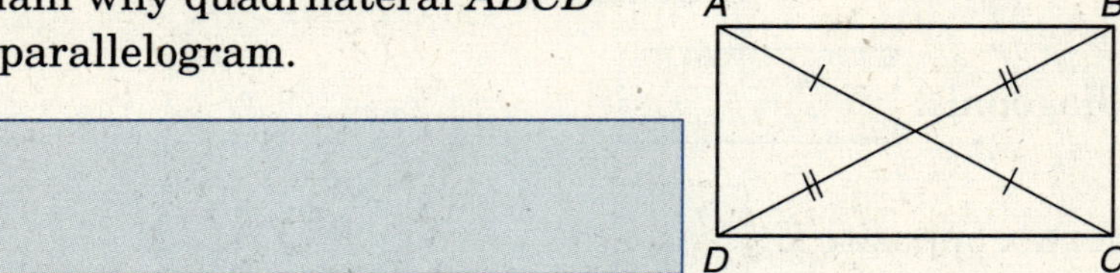

8-4 Rectangles, Rhombi, and Squares

Underline the best term to complete the statement.

14. A parallelogram with four congruent sides is a [rhombus/rectangle].

Identify each figure with as many terms as possible. Indicate if no term applies.

Quadrilateral *Parallelogram* *Square* *Rhombus* *Rectangle*

15.

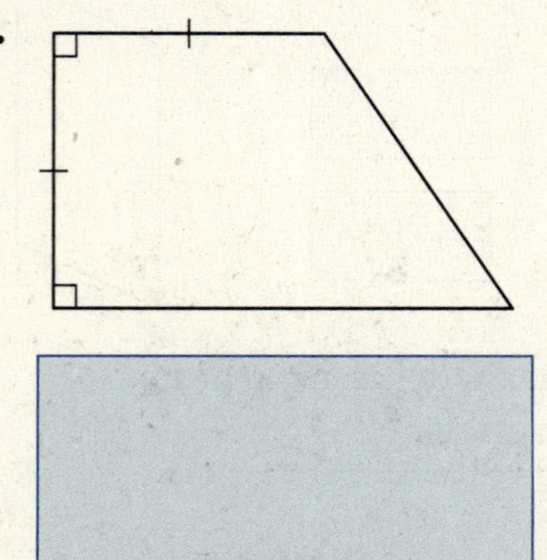

16.

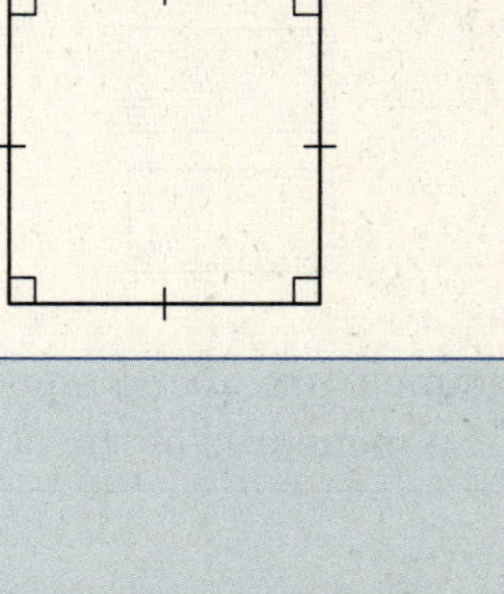

8-5 Trapezoids

Complete each statement.

17. The segment that joins the midpoints of each leg of a trapezoid is the ______.

18. A ______ is a quadrilateral with exactly one pair of parallel sides.

19. The nonparallel sides of a trapezoid are its ______.

20. The parallel sides of a trapezoid are its ______.

Refer to trapezoid *ABCD* with median $\overline{JK}$. Name each of the following.

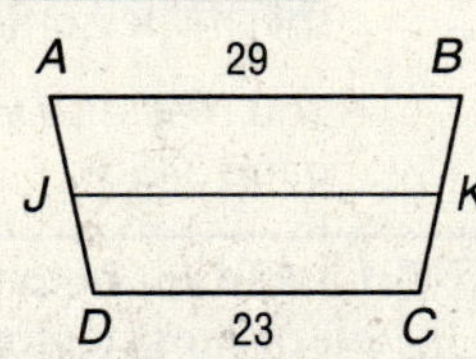

21. bases

22. legs

23. base angle pairs

24. If $AB = 29$ and $DC = 23$, what is JK?

25. If $AD = 18$, find JD.

26. If $WXYZ$ is an isosceles trapezoid and one base angle measures 66, what are the remaining angle measures?

ARE YOU READY FOR THE CHAPTER TEST?

Visit **geomconcepts.com** to access your textbook, more examples, self-check quizzes, and practice tests to help you study the concepts in Chapter 8.

Check the one that applies. Suggestions to help you study are given with each item.

☐ **I completed the review of all or most lessons without using my notes or asking for help.**

- You are probably ready for the Chapter Test.
- You may want to take the Chapter 8 Practice Test on page 345 of your textbook as a final check.

☐ **I used my Foldable or Study Notebook to complete the review of all or most lessons.**

- You should complete the Chapter 8 Study Guide and Review on pages 342–344 of your textbook.
- If you are unsure of any concepts or skills, refer back to the specific lesson(s).
- You may also want to take the Chapter 8 Practice Test on page 345.

☐ **I asked for help from someone else to complete the review of all or most lessons.**

- You should review the examples and concepts in your Study Notebook and Chapter 8 Foldable.
- Then complete the Chapter 8 Study Guide and Review on pages 342–344 of your textbook.
- If you are unsure of any concepts or skills, refer back to the specific lesson(s).
- You may also want to take the Chapter 8 Practice Test on page 345.

Student Signature

Parent/Guardian Signature

Teacher Signature

Proportions and Similarity

Use the instructions below to make a Foldable to help you organize your notes as you study the chapter. You will see Foldable reminders in the margin of this Interactive Study Notebook to help you in taking notes.

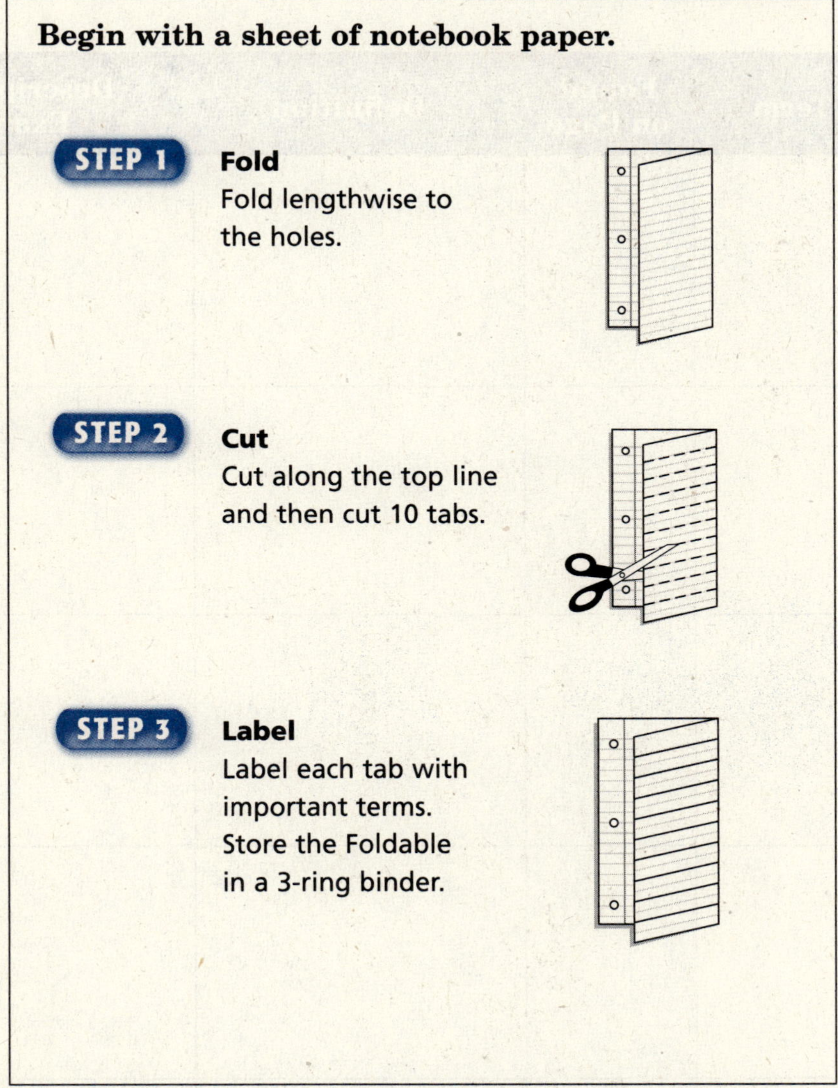

NOTE-TAKING TIP: You can design visuals such as graphs, diagrams, pictures, charts, and concept maps to help you organize information so that you can remember what you are learning.

BUILD YOUR VOCABULARY

This is an alphabetical list of new vocabulary terms you will learn in Chapter 9. As you complete the study notes for the chapter, you will see Build Your Vocabulary reminders to complete each term's definition or description on these pages. Remember to add the textbook page number in the second column for reference when you study.

Vocabulary Term	Found on Page	Definition	Description or Example
cross products			
extremes			
golden ratio			
means			
polygon [PA-lee-gon]			

Vocabulary Term	Found on Page	Definition	Description or Example
proportion [pro-POR-shun]			
ratio [RAY-she-oh]			
scale drawing			
scale factor			
sides			
similar polygons			

9–1 Using Ratios and Proportions

WHAT YOU'LL LEARN

- Use ratios and proportions to solve problems.

BUILD YOUR VOCABULARY (page 165)

A comparison of ______ numbers by division is called a **ratio**.

FOLDABLES™

ORGANIZE IT

Label the first tab *ratio*. Under the tab, write the definition and give an example.

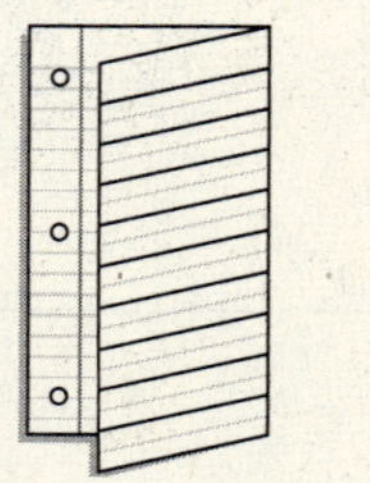

EXAMPLES

Write each ratio in simplest form.

1 $\frac{75}{400}$

$\frac{75 \div \square}{400 \div \square} = \square$ Divide the numerator and denominator by ______.

2 **24 inches to 3 feet**

The units of measure must be the same in a ratio. There are ______ inches in one foot, so 24 inches equals ______ feet.

The ratio is ______.

Your Turn **Write each ratio in simplest form.**

a. $\frac{169}{39}$

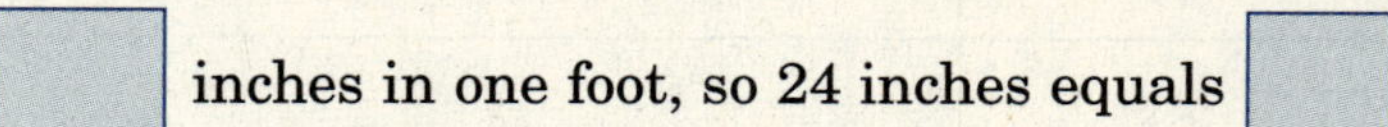

b. 30 minutes to $2\frac{1}{2}$ hours

BUILD YOUR VOCABULARY (pages 164–165)

An equation that shows two equivalent ratios is a **proportion**.

The **cross products** are the product of the ______ and the product of the ______.

In a proportion, the ______ of the first ratio and the ______ of the second ratio are the **extremes**.

In a proportion, the ______ of the first ratio and the ______ of the second ratio are the **means**.

FOLDABLES™

ORGANIZE IT

Label the next four tabs *proportion*, *cross products*, *extremes*, and *means*. Under each tab, write the definition and give an example.

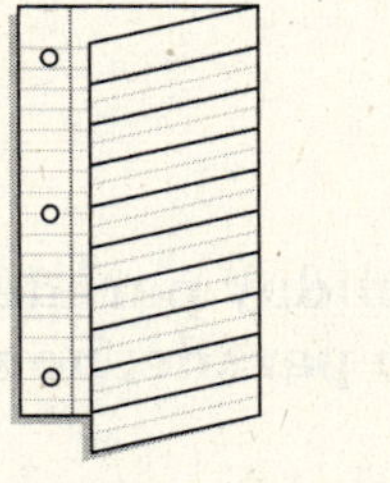

Theorem 9-1 Property of Proportions
For numbers a and c and any nonzero numbers b and d, if $\frac{a}{b} = \frac{c}{d}$, then $ad = bc$. Conversely, if $ad = bc$, then $\frac{a}{b} = \frac{c}{d}$.

EXAMPLE

3 **Solve $\frac{24}{30} = \frac{6x + 4}{35}$.**

$$\frac{24}{30} = \frac{6x + 4}{35}$$

$\_\_\_\_(35) = \_\_\_\_(6x + 4)$ ______

$840 = 180x + 120$ Distributive Property

$840 - \_\_\_\_ = 180x + 120 - \_\_\_\_$ Subtract ______ from each side.

$720 = 180x$

$\frac{720}{\_\_\_\_} = \frac{180x}{\_\_\_\_}$ Divide each side by .

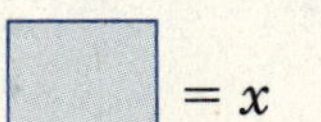 $= x$

REMEMBER IT

The denominator can never equal zero.

Your Turn Solve $\frac{15}{x-1} = \frac{4}{5}$.

EXAMPLE

4 The ratio of children to adults at a holiday parade is 2.5 to 1. If there are 1440 adults at the parade, how many children are there?

children → $\frac{2.5}{1} = \frac{x}{1440}$ ← adults

$2.5(1440) = \underline{\quad}$ Cross products

$\underline{\quad} = x$

Your Turn The ratio of Republicans to Democrats casting their votes in the local election was 73 to 27. If 135 Democrats voted, how many Republicans cast their votes?

HOMEWORK ASSIGNMENT

Page(s):

Exercises:

9–2 Similar Polygons

WHAT YOU'LL LEARN

- Identify similar polygons.

BUILD YOUR VOCABULARY (pages 164–165)

A **polygon** is a ________ figure in a plane formed by segments called **sides**.

Similar polygons are the same ________ but not necessarily the same ________.

KEY CONCEPT

Similar Polygons Two polygons are similar if and only if their corresponding angles are congruent and the measures of their corresponding sides are proportional.

FOLDABLES

Label the next three tabs *polygon*, *sides*, and *similar polygons*. Under each tab, write the definition and give an example.

EXAMPLE

1 Determine if the polygons are similar. Justify your answer.

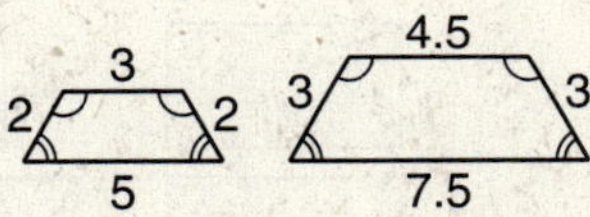

The polygons are ________. The corresponding angles are congruent and $\frac{2}{\square} = \frac{3}{\square} = \frac{2}{3} = \frac{\square}{7.5}$.

Your Turn Determine if the polygons are similar. Justify your answer.

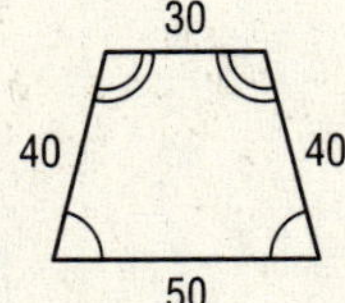

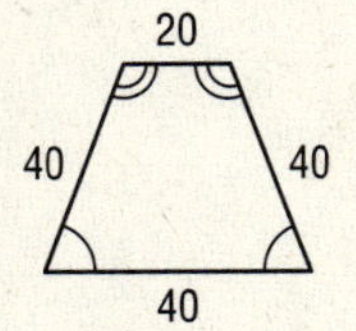

EXAMPLE

2 Find the values of x and y if $\triangle ABC \sim \triangle FED$.

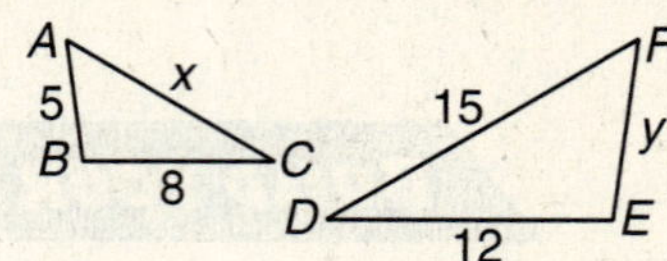

Use the corresponding order of the vertices to write proportions.

$$\frac{AB}{FE} = \frac{\square}{FD} = \frac{BC}{\square}$$ Definition of similar polygons

$$\frac{5}{y} = \frac{x}{\square} = \frac{8}{\square}$$ Substitution

Write the proportion to solve for x.

$$\frac{x}{15} = \frac{\square}{12}$$

$$\square = (8)\square$$ ______

$$\square = \square$$

$$\frac{12}{12}x = \frac{120}{12}$$ Divide each side by  ______.

$$x = \square$$

Now write the proportion that can be solved for y.

$$\frac{5}{y} = \frac{8}{\square}$$

$$\square(12) = y(8)$$ Cross products

$$60 = 8y$$

$$\frac{60}{8} = \frac{8y}{8}$$ Divide each side by ______.

$$\square = y$$

So, $x =$ ______ and $y =$ ______.

Your Turn The triangles are similar. Find the values of x and y.

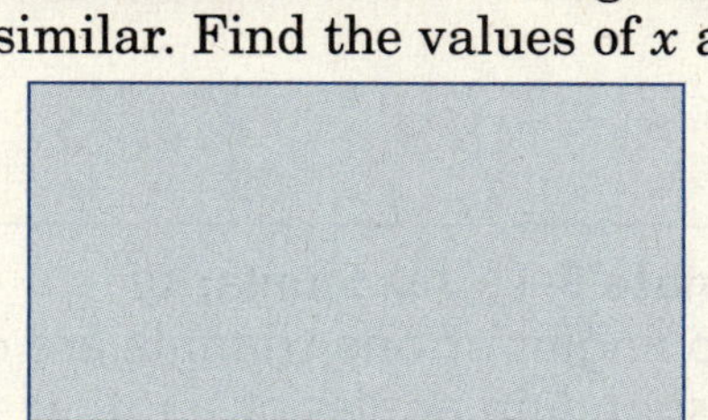

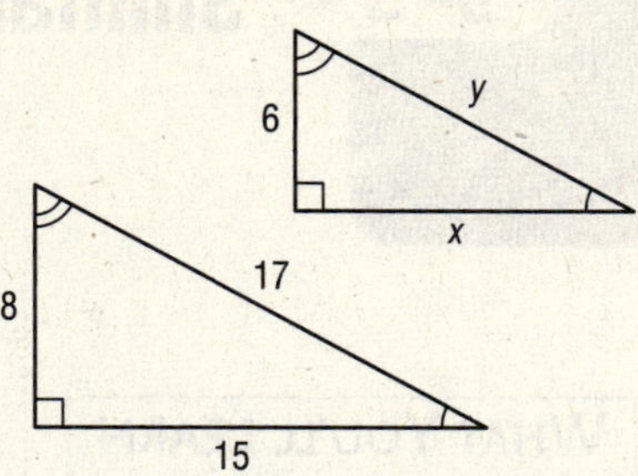

BUILD YOUR VOCABULARY (page 165)

Scale drawings are used to represent something either too

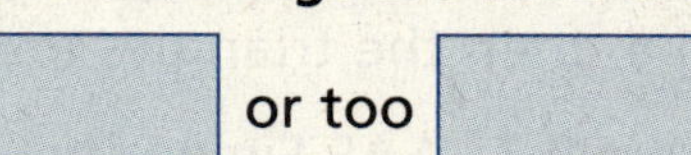

or too ______ to be drawn at its actual size.

EXAMPLE

FOLDABLES™

ORGANIZE IT

Label the next tab *scale drawings*. Under the tab, write the definition and give an example.

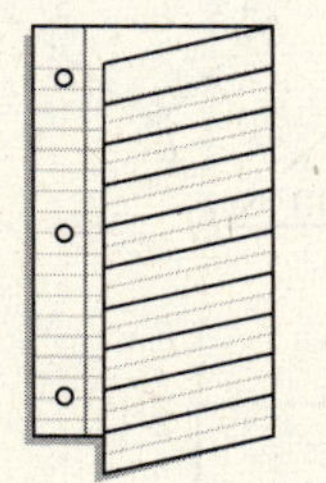

3 **In the blueprint, 1 inch represents an actual length of 16 feet. Use the blueprint to find the actual dimensions of the dining room.**

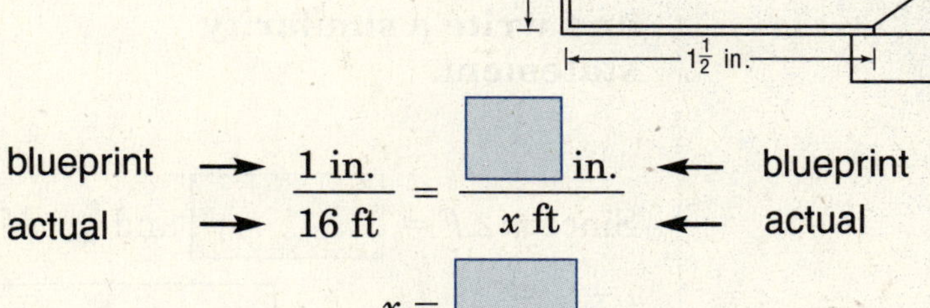

blueprint → $\frac{1 \text{ in.}}{16 \text{ ft}} = \frac{\square \text{ in.}}{x \text{ ft}}$ ← blueprint

actual → ← actual

$x = \square$

$\frac{1 \text{ in.}}{16 \text{ ft}} = \frac{\square \text{ in.}}{y \text{ ft}}$

$\square(y) = 16\left(\frac{3}{4}\right)$ Cross products

$y = \square$

The dimensions of the dining room are  ft by  ft.

Your Turn Refer to Example 3. Find the dimensions of the kitchen.

HOMEWORK ASSIGNMENT

Page(s):

Exercises:

9–3 Similar Triangles

What You'll Learn

- Use AA, SSS, and SAS similarity tests for triangles.

Postulate 9-1 AA Similarity
If two angles of one triangle are congruent to two corresponding angles of another triangle, then the two triangles are similar.

Theorem 9-2 SSS Similarity
If the measures of the sides of a triangle are proportional to the measures of the corresponding sides of another triangle, then the triangles are congruent.

Theorem 9-3 SAS Similarity
If the measures of two sides of a triangle are proportional to the measures of two corresponding sides of another triangle and their included angles are congruent, then the triangles are similar.

EXAMPLE

1 Determine whether the triangles are similar, If so, tell which similarity test is used and write a similarity statement.

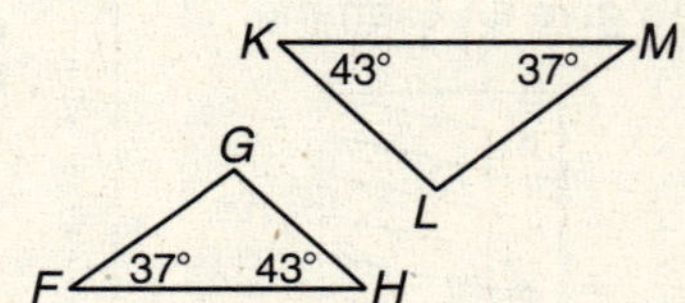

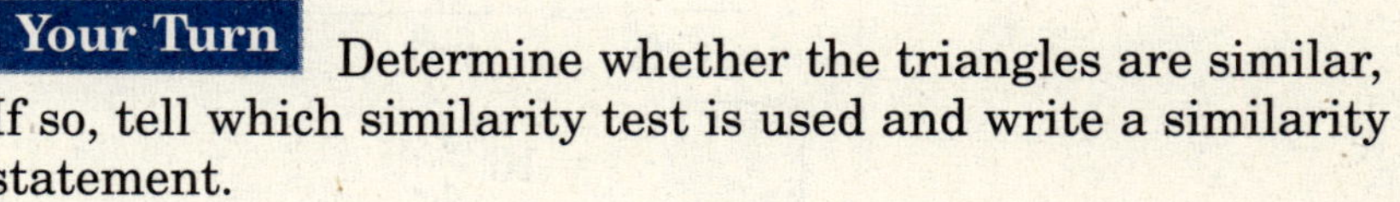

Since $m\angle F =$ ______ and $m\angle H =$ ______,

$\triangle FGH \sim \triangle MLK$ by ______.

Your Turn Determine whether the triangles are similar, If so, tell which similarity test is used and write a similarity statement.

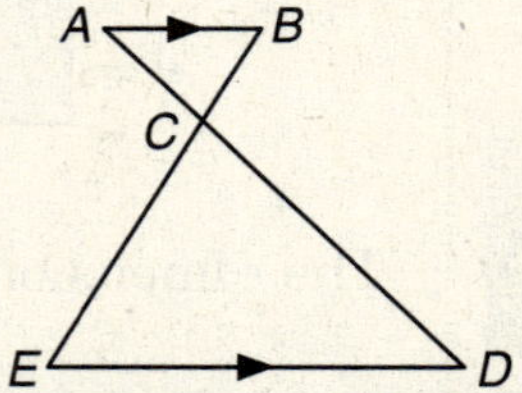

Write It

Why must only two pairs of corresponding angles be congruent for two triangles to be similar rather than three?

EXAMPLE

2 Find the value of x.

Since $\frac{8}{12} = \frac{12}{18}$, the triangles are similar by SAS similarity.

$\frac{8}{12} = \frac{14}{x}$ Definition of similar polygons

$\square x = (12)(14)$ Cross products

$8x = 168$

$\frac{8x}{\square} = \frac{168}{\square}$ Divide each side by .

$x = \square$

Your Turn Find the value of x.

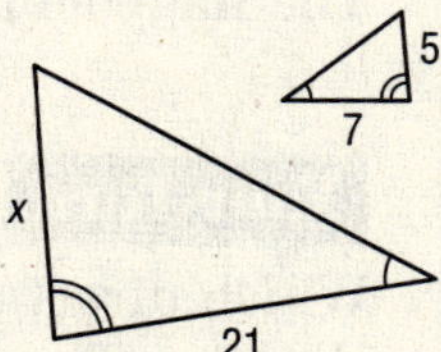

EXAMPLE

3 **The shadow of a flagpole is 2 meters long at the same time that a person's shadow is 0.4 meters long. If the person is 1.5 meters tall, how tall is the flagpole?**

flagpole's shadow → $\frac{2}{0.4} = \frac{x}{\square}$ ← flagpole's height

person's shadow → ← person's height

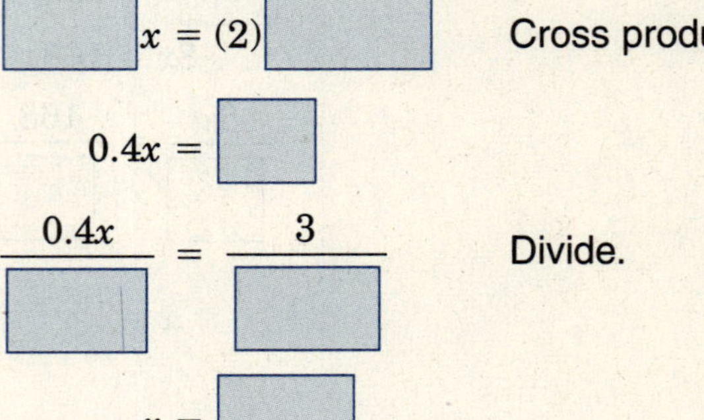

The flagpole is ☐ meters tall.

Your Turn A diseased tree must be cut down before it falls. Which direction the fall is directed depends on the height of the tree. The man who will cut the tree down is 74-in. tall and casts a shadow 60-in. long. If the tree's shadow measures 20 feet from its base, how tall is the tree?

HOMEWORK ASSIGNMENT

Page(s):

Exercises:

9–4 Proportional Parts and Triangles

WHAT YOU'LL LEARN

- Identify and use the relationships between proportional parts of triangles.

Theorem 9-4
If a line is parallel to one side of a triangle and intersects the other two sides, then the triangle formed is similar to the original triangle.

EXAMPLE

1 Using the figure, complete the proportion $\frac{?}{VW} = \frac{ST}{SW}$.

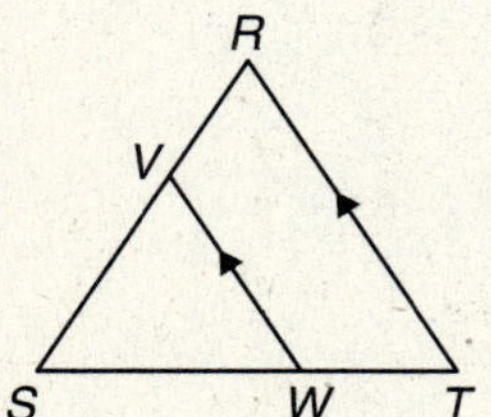

Since $\overline{VW} \parallel \overline{RT}$, $\triangle SVW \sim \triangle SRT$.

Therefore, $\frac{\square}{VW} = \frac{ST}{SW}$.

Your Turn Use the figure to complete the proportion $\frac{XY}{AY} = \frac{?}{BY}$.

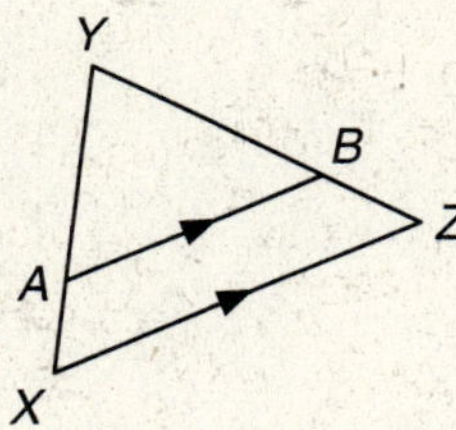

EXAMPLE

2 In the figure, $\overline{MN} \parallel \overline{KL}$. Find the value of x.

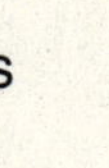

$\triangle JMN \sim \triangle JKL$

$\frac{MN}{KL} = \frac{JN}{JL}$ Definition of similar polygons

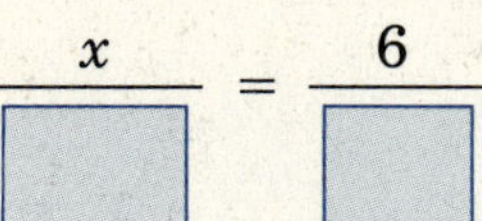

$\frac{x}{\square} = \frac{6}{\square}$ Substitution

$9x = (6)\square$ Cross products

$9x = \square$

$x = \square$ Divide each side by 9.

Your Turn Find the value of b.

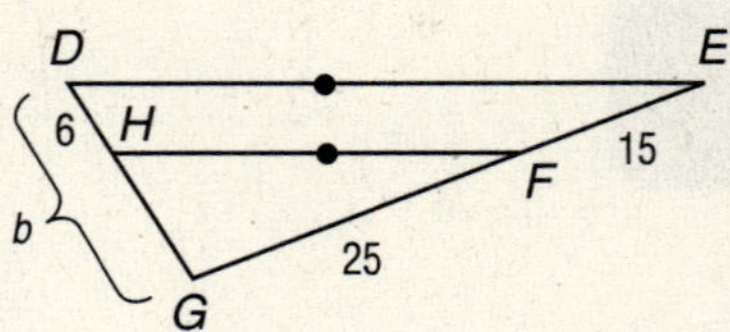

Theorem 9-5
If a line is parallel to one side of a triangle and intersects the other two sides, then it separates the sides into segments of proportional lengths.

EXAMPLE

3 **In the figure, $\overline{AB} \parallel \overline{DE}$. Find the value of x.**

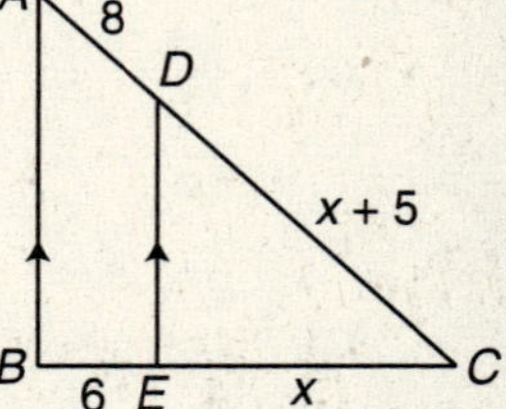

$$\frac{CE}{EB} = \frac{CD}{\square}$$ Theorem 9.5

$$\frac{x}{\square} = \frac{x + \square}{\square}$$ $CE = x$, $EB = 6$, $CD = x + 5$, $DA = 8$

$$x(8) = \square(x + 5)$$ Cross products

$$8x = 6x + \square$$ Distributive Property

$$8x - 6x = 6x + 30 - 6x$$ Subtract $6x$ from each side.

$$2x = 30$$

$$\frac{2x}{2} = \frac{30}{2}$$ Divide each side by 2.

$$x = \square$$

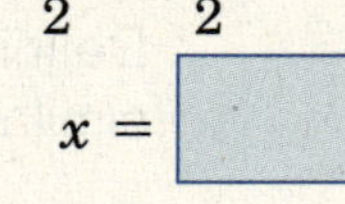

Your Turn Find the value of a.

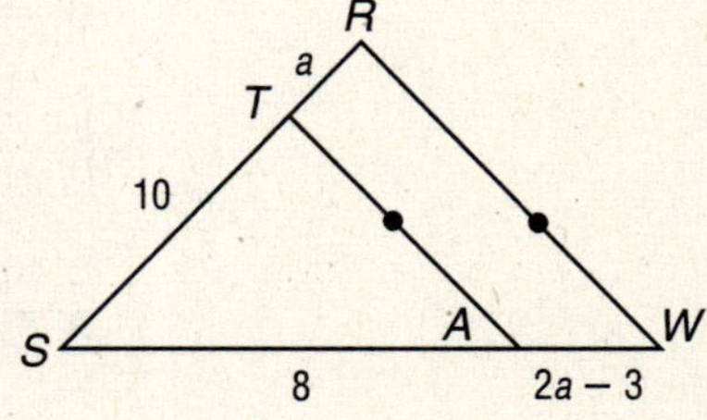

HOMEWORK ASSIGNMENT

Page(s):

Exercises:

9–5 Triangles and Parallel Lines

What You'll Learn

- Use proportions to determine whether lines are parallel to sides of triangles.

Theorem 9-6
If a line intersects two sides of a triangle and separates the sides into corresponding segments of proportional lengths, then the line is parallel to the third side.

EXAMPLE

1 Determine whether $\overline{DE} \parallel \overline{BC}$.

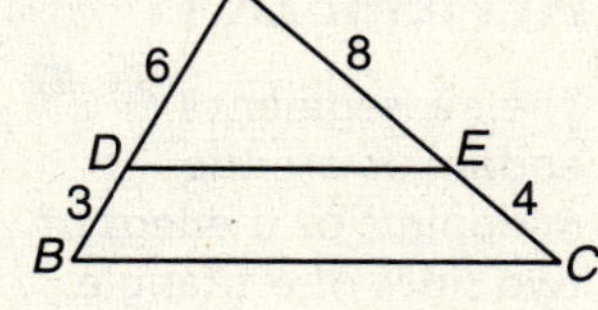

Determine whether $\frac{BD}{DA}$ and $\frac{CE}{EA}$ form a proportion.

$$\frac{BD}{\square} \stackrel{?}{=} \frac{CE}{EA}$$

$$\frac{\square}{6} \stackrel{?}{=} \frac{\square}{8}$$

$\square(8) \stackrel{?}{=} \square(4)$ Cross products

$24 = \square$ ✔

Therefore, $\overline{DE} \parallel \overline{BC}$ by Theorem 9-6.

Your Turn Determine whether $\overline{HJ} \parallel \overline{KM}$.

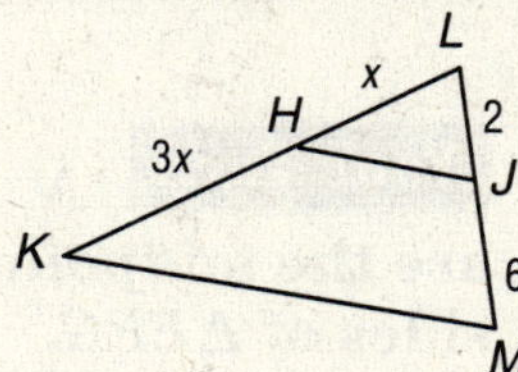

Theorem 9-7
If a segment joins the midpoint of two sides of a triangle, then it is parallel to the third side, and its measure equals one-half the measure of the third side.

EXAMPLES

For Examples 2 and 3, refer to the figure shown.

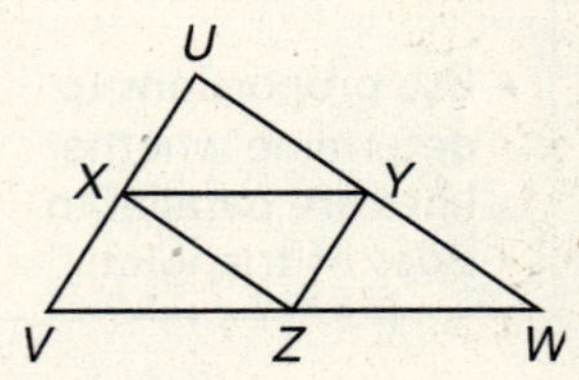
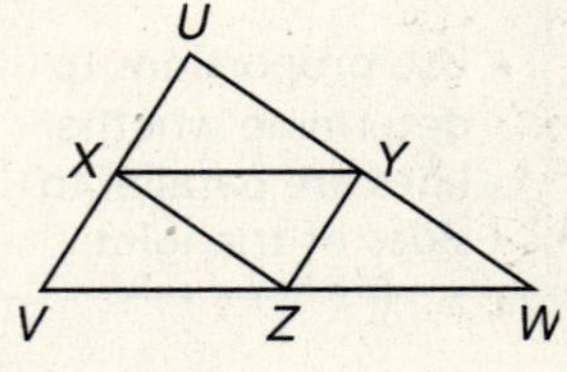

2 **In the figure, X, Y, and Z are midpoints of the sides of $\triangle UVW$. If $XZ = 7c$, then what does UW equal?**

$XZ = \frac{1}{2}$ ______ 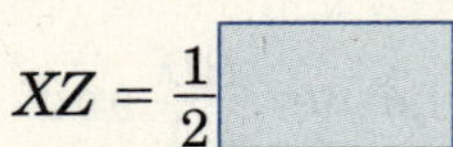 Theorem 9-7

______ $= \frac{1}{2}UW$ Replace XZ with ______.

______ $7c =$ ______ $\left(\frac{1}{2}UW\right)$ Multiply each side by ______.

______ $= UW$

REMEMBER IT
The midsegment's endpoints are the midpoints of the legs of two sides of a triangle.

3 **In the figure, if $m\angle UYX = d$, then what is $m\angle YWZ$?**

By Theorem 9-6, $\overline{XY} \parallel \overline{VW}$. Since $\overline{XY}$ and $\overline{VW}$ are parallel segments cut by transversal $\overline{UW}$, $\angle UYX$ and ______ are congruent ______ angles.

Therefore, $m\angle YWZ =$ ______.

Your Turn **N, O, and P are the midpoints of the sides of $\triangle EFG$.**

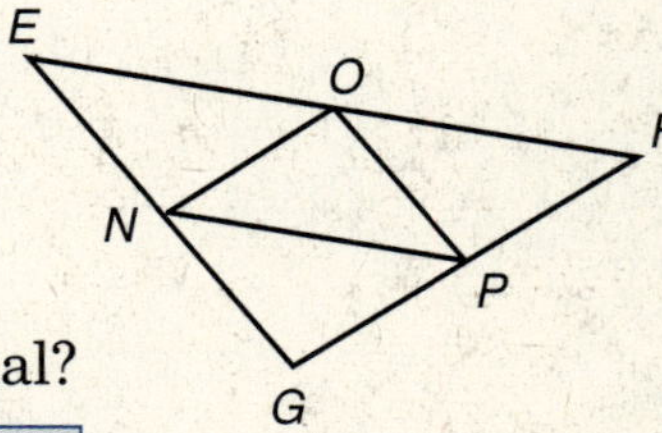

a. If $EF = 25$, then what does NP equal?

______

b. If $m\angle EGF = 85$, then what is $m\angle ENO$?

______

HOMEWORK ASSIGNMENT
Page(s):
Exercises:

9–6 Proportional Parts and Parallel Lines

WHAT YOU'LL LEARN

- Identify and use the relationships between parallel lines and proportional parts.

REVIEW IT

What is the definition of a *transversal*? (*Lesson 4-2*)

____

____

____

____

Theorem 9-8
If three or more parallel lines intersect two transversals, the lines divide the transversals proportionally.

EXAMPLES

1 **Complete the proportion $\frac{ST}{RT} = \frac{NP}{?}$.**

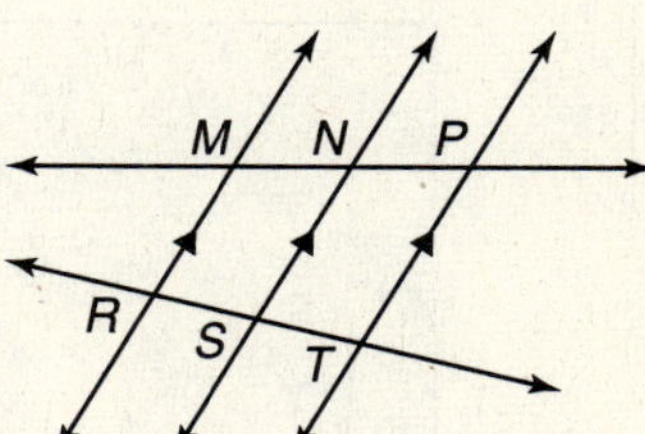

Since ______ $\parallel \overleftrightarrow{NS} \parallel \overleftrightarrow{PT}$, the transversals are divided

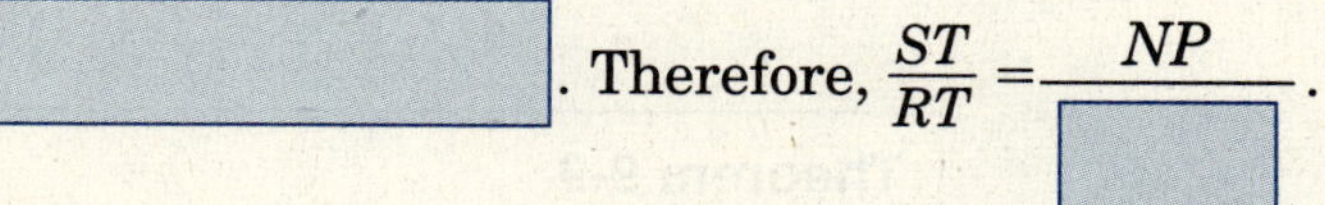

______________. Therefore, $\frac{ST}{RT} = \frac{NP}{\_\_\_\_}$.

2 **In the figure, $a \parallel b \parallel c$. Find the value of x.**

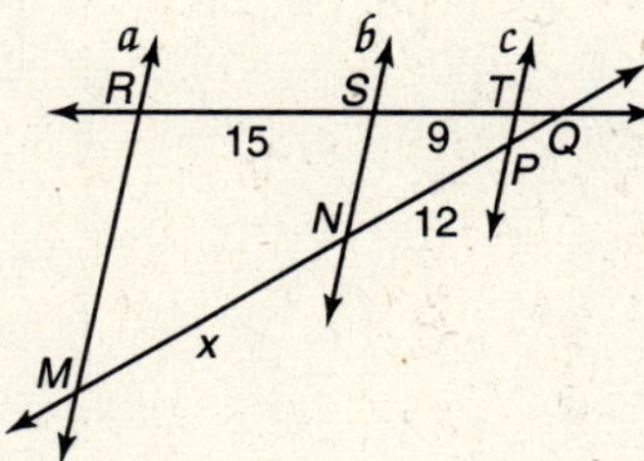

$\frac{TS}{SR} = \frac{\_\_\_\_}{NM}$

$\frac{9}{15} = \frac{\_\_\_\_}{x}$ $TS = 9$, $SR = 15$, $PN =$ ______, $NM = x$

$9(x) = 15(\_\_\_\_)$ Cross products

$9x =$ ______

$x =$ ______ Divide each side by ______.

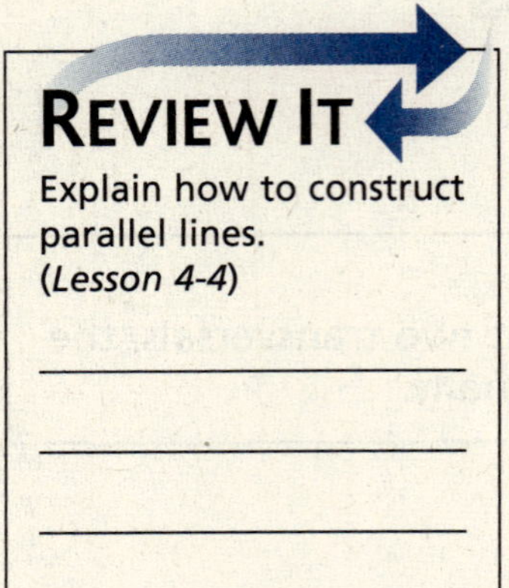

Explain how to construct parallel lines.
(*Lesson 4-4*)

Your Turn

a. Complete the proportion $\frac{ZP}{QP} = \frac{NB}{?}$.

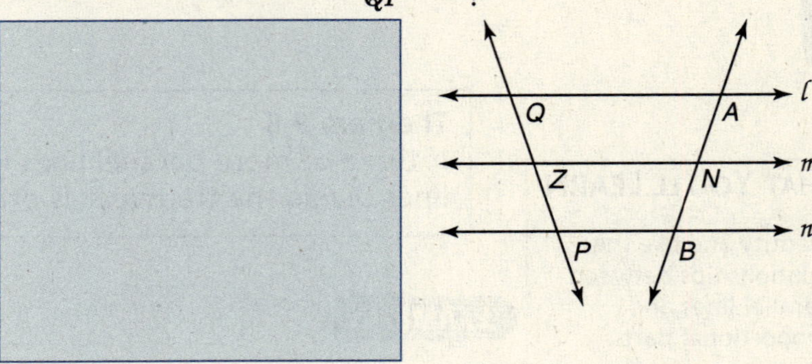

b. In the figure, $a \parallel b \parallel c$. Find the value of x.

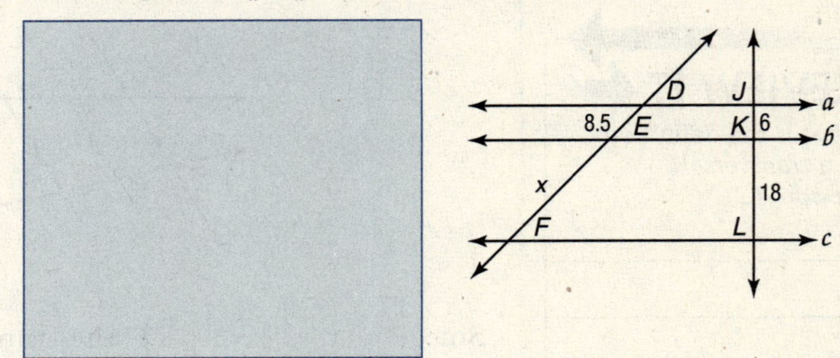

Theorem 9-9
If three or more parallel lines cut off congruent segments on one transversal, then they cut off congruent segments on every transversal.

HOMEWORK ASSIGNMENT

Page(s):

Exercises:

9–7 Perimeters and Similarity

WHAT YOU'LL LEARN

- Identify and use proportional relationships of similar triangles.

Theorem 9-10
If two triangles are similar, then the measures of the corresponding perimeters are proportional to the measures of the corresponding sides.

EXAMPLE

1 **The perimeter of $\triangle DEF$ is 90 units, and $\triangle ABC \sim \triangle DEF$. Find the value of each variable.**

$$\frac{DE}{AB} = \frac{\text{perimeter of } \triangle DEF}{\text{perimeter of } \triangle ABC}$$ Theorem 9-10

$$\frac{x}{26} = \frac{90}{60}$$ 26 + 10 + 24 = ____

$$\_\_\_\_(60) = \_\_\_\_(90)$$ Cross products

$$60x = 2340$$ Divide.

$$x = \_\_\_\_$$

Because the triangles are similar, find y and z.

$$\frac{DF}{DE} = \frac{AC}{AB}$$

$$\frac{y}{39} = \frac{\_\_\_\_}{26}$$

$$26y = 390$$

$$y = \_\_\_\_$$

$$\frac{EF}{DE} = \frac{BC}{AB}$$

$$\frac{z}{\_\_\_\_} = \frac{24}{26}$$

$$26z = \_\_\_\_$$

$$z = \_\_\_\_$$

FOLDABLES™

ORGANIZE IT

- Write each vocabulary word from the lesson.
- Explain its meaning in your own words.
- Use diagrams to clarify.

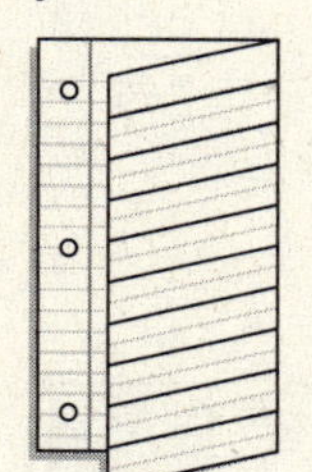

Your Turn The perimeter of $\triangle ABC$ is 20 units, and $\triangle ABC \sim \triangle XYZ$. Find the value of each variable.

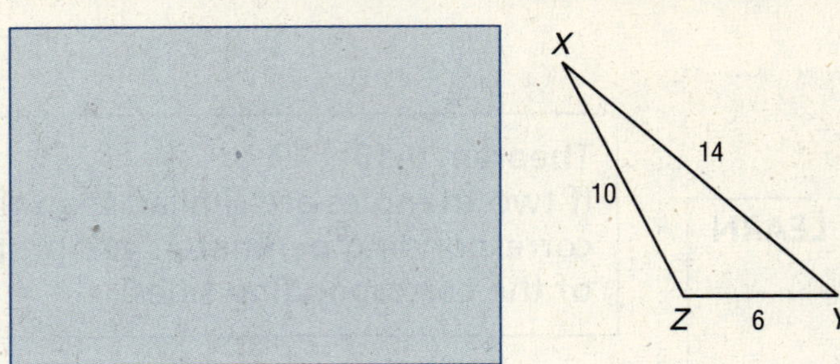

BUILD YOUR VOCABULARY (page 165)

The **scale factor**, also known as the constant of ______, is the ______ found by comparing the measures of corresponding sides of similar triangles.

EXAMPLE

FOLDABLES

ORGANIZE IT

Label the next tab *scale factor.* Under the tab, write the definition and give an example.

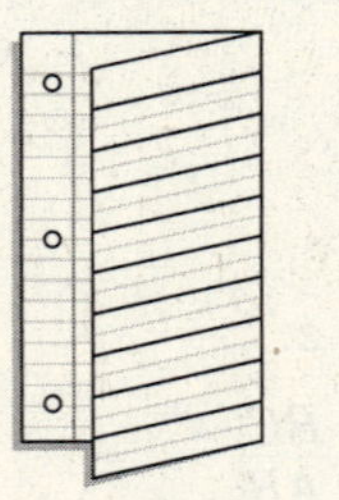

2 **Determine the scale factor of $\triangle ABC$ to $\triangle DEF$.**

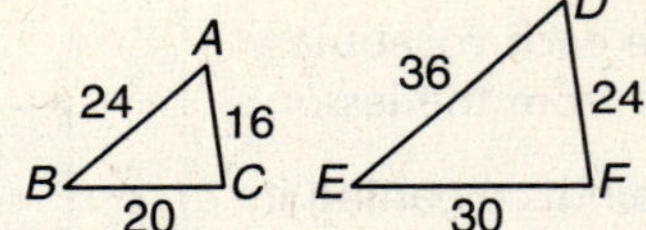

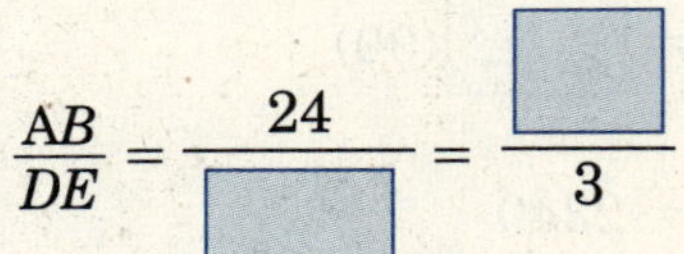

$\frac{AB}{DE} = \frac{24}{\square} = \frac{\square}{3}$

$\frac{BC}{EF} = \frac{\square}{30} = \frac{2}{\square}$

$\frac{AC}{DF} = \frac{\square}{24} = \frac{\square}{3}$

The scale factor is ______.

Your Turn Determine the scale factor of $\triangle RST$ to $\triangle XYZ$.

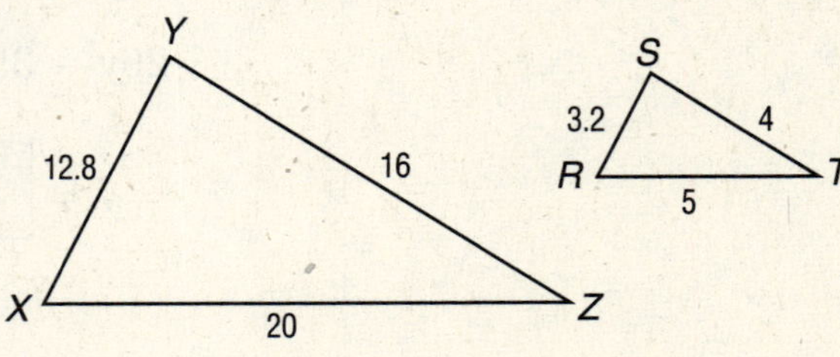

HOMEWORK ASSIGNMENT

Page(s):

Exercises:

CHAPTER 9

BRINGING IT ALL TOGETHER

STUDY GUIDE

FOLDABLES™	VOCABULARY PUZZLEMAKER	BUILD YOUR VOCABULARY
Use your **Chapter 9 Foldable** to help you study for your chapter test.	To make a crossword puzzle, word search, or jumble puzzle of the vocabulary words in Chapter 9, go to: www.glencoe.com/sec/math/t_resources/free/index.php	You can use your completed **Vocabulary Builder** (pages 164–165) to help you solve the puzzle.

9-1 Using Ratios and Proportions

Indicate whether the statement is *true* or *false*.

1. Every proportion has two cross products.
2. A ratio is a comparison of two numbers by division.
3. The two cross products of a ratio are the extremes and the means.
4. Cross products are always equal in a proportion.
5. Simplify $\frac{220}{70}$.
6. Solve: $\frac{84}{63} = \frac{12}{11 - x}$

9-2 Similar Polygons

Complete the sentence.

7. In ______ measures of corresponding sides are proportional, and corresponding angles are congruent.
8. ______ represent something either too large or too small to be drawn at actual size.

9. Given that the rectangles are similar, find the values of x and y to show similarity.

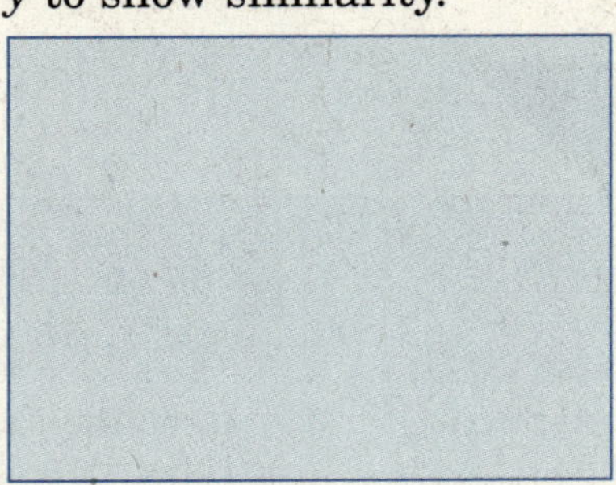

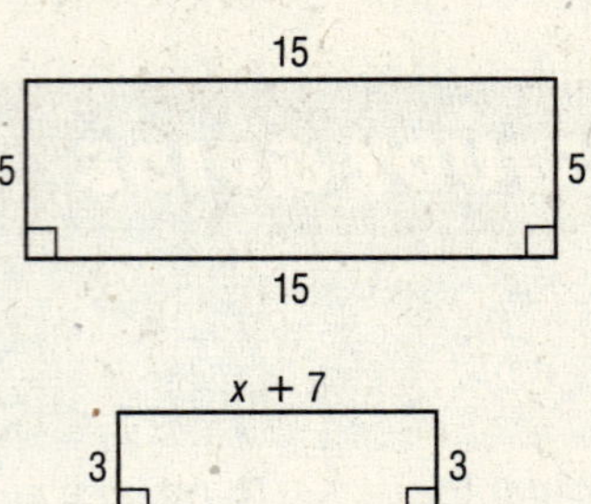

9-3 Similar Triangles

Determine whether the pair of triangles is similar. Justify your reasons.

10.

11.

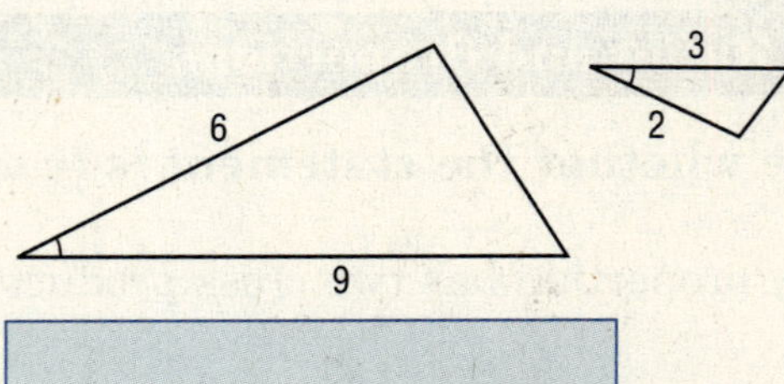

9-4 Proportional Parts and Triangles

Complete the proportions.

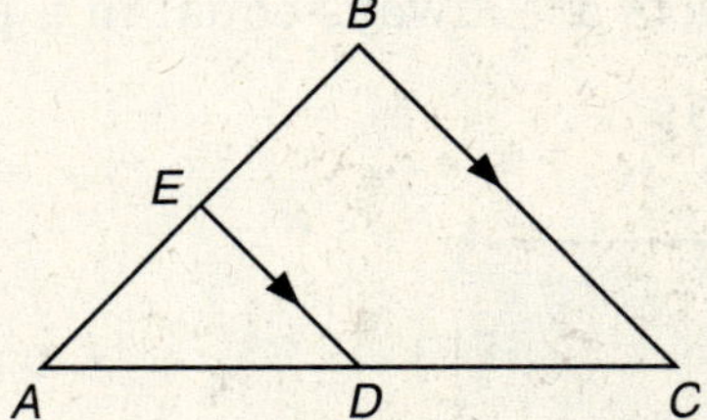

12. $\frac{AD}{DE} = \frac{\square}{CB}$

13. $\frac{AE}{EB} = \frac{AD}{\square}$

9-5 Triangles and Parallel Lines

Vertices *A*, *B*, and *C* are midpoints.

14. $\overline{AC} \parallel$ ______.

15. If $BC = 6$, then $RT =$ ______.

16. If $SB = 4$, $AC =$ ______.

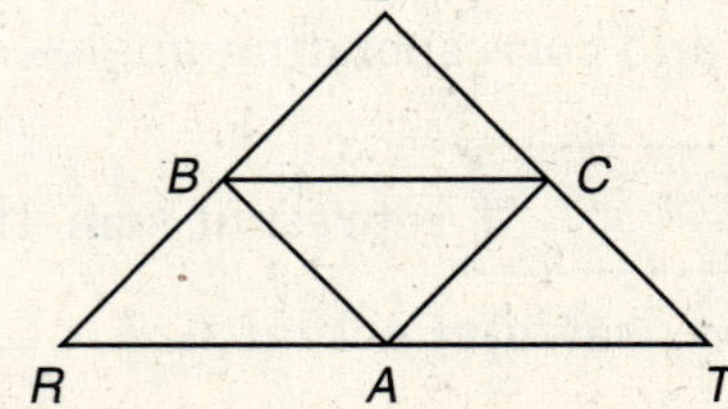

9-6

Proportional Parts and Parallel Lines

A tract of land bordering school property was divided into sections for five biology classes to plant gardens. The fences separating the plots are parallel, and the plots' front measures are shown. The entire back border measures 254 feet. What are the individual border lengths, to the nearest tenth of a foot?

22 ft. 20 ft. 25 ft. 28 ft. 16 ft. front

back

A B C D E

17. $A =$

18. $D =$

19. $E =$

9-7

Perimeters and Similarity

Complete the sentence.

20. The scale factor is also called the constant of

.

21. Find the scale factor.

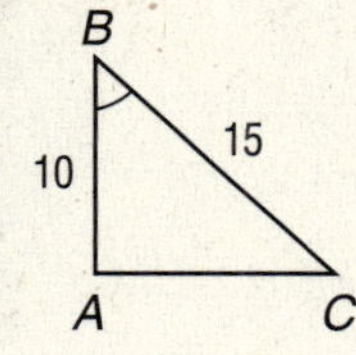

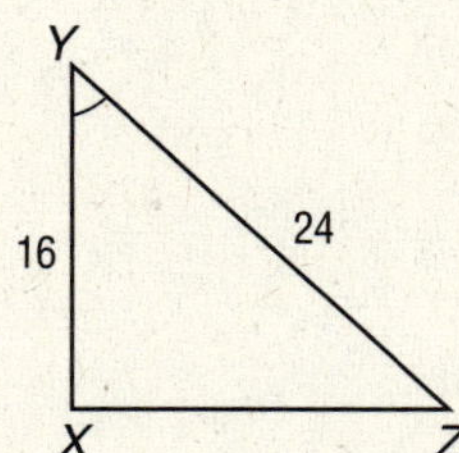

$\triangle JKL \sim \triangle MNO$. The perimeter of $\triangle JKL$ is 54. What are the values for the variables?

22. $a =$

23. $b =$

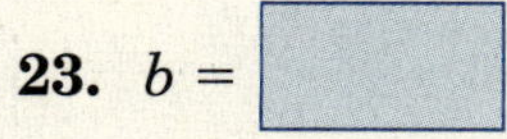

24. $c =$

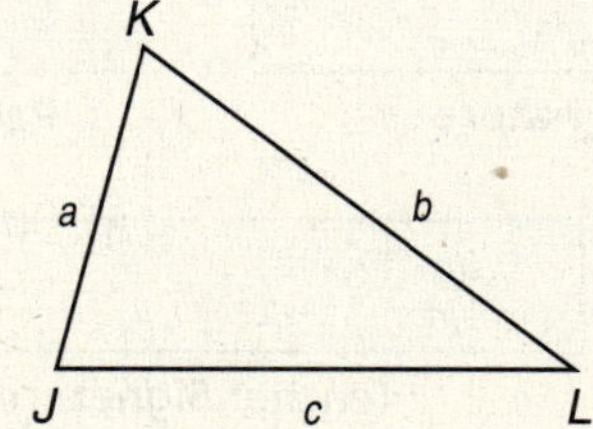

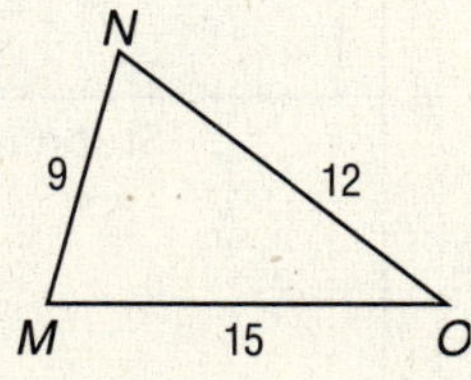

Checklist

ARE YOU READY FOR THE CHAPTER TEST?

Visit **geomconcepts.com** to access your textbook, more examples, self-check quizzes, and practice tests to help you study the concepts in Chapter 9.

Check the one that applies. Suggestions to help you study are given with each item.

☐ **I completed the review of all or most lessons without using my notes or asking for help.**

- You are probably ready for the Chapter Test.
- You may want to take the Chapter 9 Practice Test on page 397 of your textbook as a final check.

☐ **I used my Foldable or Study Notebook to complete the review of all or most lessons.**

- You should complete the Chapter 9 Study Guide and Review on pages 394–396 of your textbook.
- If you are unsure of any concepts or skills, refer back to the specific lesson(s).
- You may also want to take the Chapter 9 Practice Test on page 397.

☐ **I asked for help from someone else to complete the review of all or most lessons.**

- You should review the examples and concepts in your Study Notebook and Chapter 9 Foldable.
- Then complete the Chapter 9 Study Guide and Review on pages 394–396 of your textbook.
- If you are unsure of any concepts or skills, refer back to the specific lesson(s).
- You may also want to take the Chapter 9 Practice Test on page 397.

Student Signature

Parent/Guardian Signature

Teacher Signature

Polygons and Area

Use the instructions below to make a Foldable to help you organize your notes as you study the chapter. You will see Foldable reminders in the margin of this Interactive Study Notebook to help you in taking notes

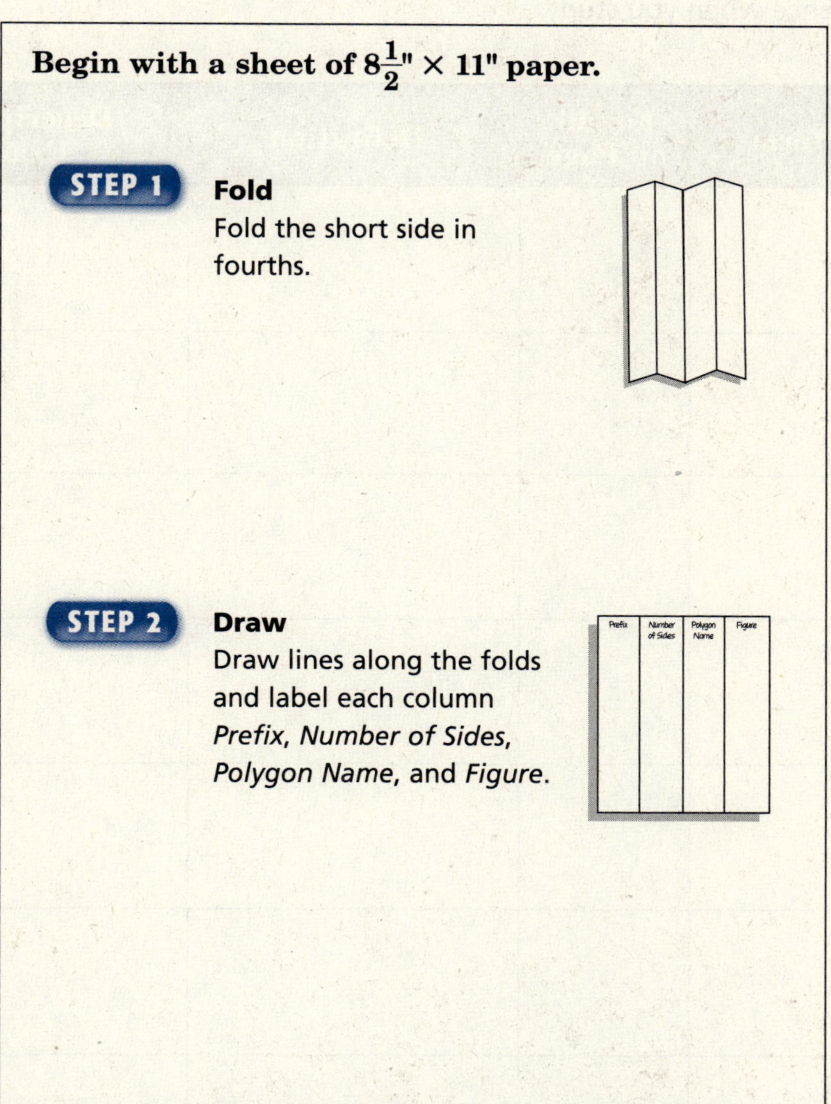

Begin with a sheet of $8\frac{1}{2}$" × 11" paper.

STEP 1 **Fold**
Fold the short side in fourths.

STEP 2 **Draw**
Draw lines along the folds and label each column *Prefix*, *Number of Sides*, *Polygon Name*, and *Figure*.

NOTE-TAKING TIP: When you take notes, it is important to record major concepts and ideas. Refer to your journal when reviewing for tests.

BUILD YOUR VOCABULARY

This is an alphabetical list of new vocabulary terms you will learn in Chapter 10. As you complete the study notes for the chapter, you will see Build Your Vocabulary reminders to complete each term's definition or description on these pages. Remember to add the textbook page number in the second column for reference when you study.

Vocabulary Term	Found on Page	Definition	Description or Example
altitude			
apothem [a-pa-thum]			
center			
composite figure [kahm-PA-sit]			
concave			
convex			
irregular figure			
line of symmetry [SIH-muh-tree]			

Vocabulary Term	Found on Page	Definition	Description or Example
line symmetry			
polygonal region			
regular polygon			
regular tessellation			
rotational symmetry			
semi-regular tessellation			
significant digits			
symmetry			
tessellation [tes-a-LAY-shun]			
turn symmetry			

10–1 Naming Polygons

WHAT YOU'LL LEARN

- Name polygons according to the number of sides and angles.

BUILD YOUR VOCABULARY (page 189)

A **regular polygon** has all ______ congruent and all ______ congruent.

FOLDABLES™

ORGANIZE IT

Under the tabs labeled *Prefix*, *Number of Sides*, and *Polygon Name*, write the information given in the table on page 402. Under the tab labeled *Figure*, draw a picture of each polygon. Include regular and irregular polygons, as well as convex and concave polygons.

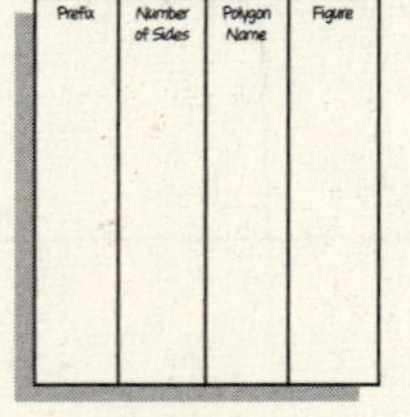

EXAMPLES

Refer to the figure for Examples 1–2.

W, X, Y, Z, V

1 a. Identify polygon *VWXYZ*.

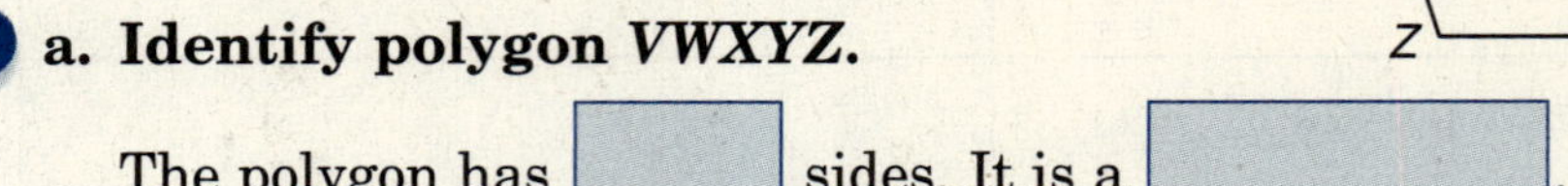

The polygon has ______ sides. It is a ______.

b. Determine whether the polygon *VWXYZ* appears to be *regular* or *not regular*. If not regular, explain why.

The ______ appear to be the same length, and the ______ appear to have the same measure. The polygon is regular.

2 Name two nonconsecutive vertices of polygon *VWXYZ*.

W and *Z*, *W* and *Y*, *V* and *X*, *V* and *Y*, *X* and *Z* are examples of ______ vertices.

Your Turn **Refer to the figure for parts a, b, and c.**

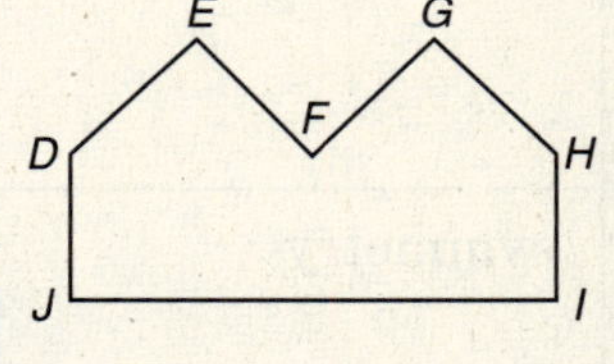

a. Identify polygon *DEFGHIJ* by its sides.

______

b. Determine whether the polygon *DEFGHIJ* appears to be *regular* or *not regular*. If not regular, explain why.

______

c. Name two nonconsecutive vertices of polygon *DEFGHIJ*.

______

BUILD YOUR VOCABULARY (page 188)

All of the diagonals of a **convex** polygon lie in the ______ of the polygon.

If any part of a diagonal lies ______ of the polygon, the polygon is **concave**.

REMEMBER IT

Most polygons have more than one diagonal. As the number of sides increases, so does the number of diagonals.

EXAMPLE

3 Classify each polygon as convex or concave.

a.

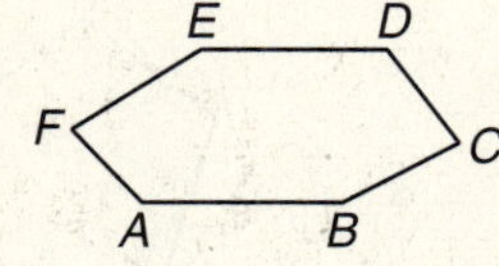

When all the diagonals are drawn, ______ points lie outside of the polygon. So polygon *ABCDEF* is

.

b.

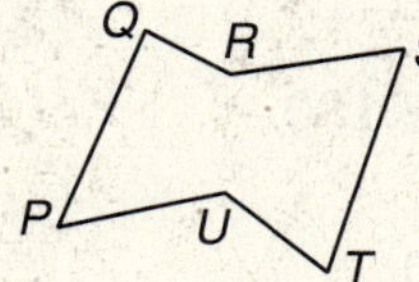

Diagonal $\overline{QS}$ lies outside the polygon, so *PQRSTU* is ______.

Your Turn **Classify each polygon as convex or concave.**

a.

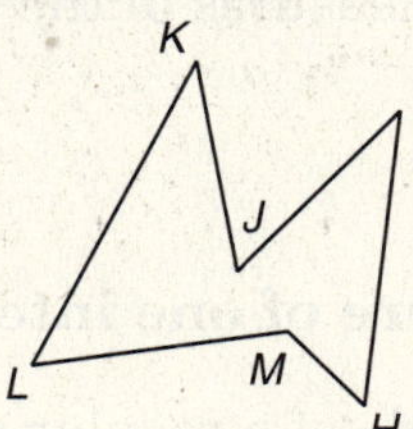

b.

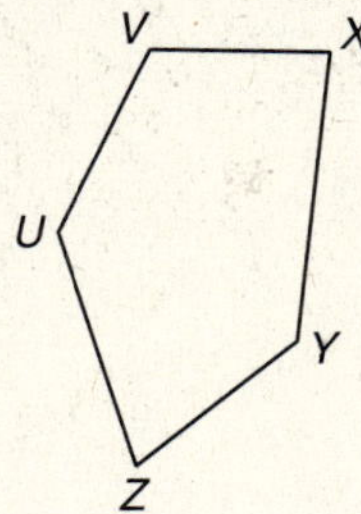

HOMEWORK ASSIGNMENT

Page(s):

Exercises:

10–2 Diagonals and Angle Measure

WHAT YOU'LL LEARN

- Find measures of interior and exterior angles of polygons.

Theorem 10-1
If a convex polygon has n sides, then the sum of the measures of the interior angles is $(n - 2)180$.

EXAMPLES

Refer to the regular pentagon for Examples 1–2.

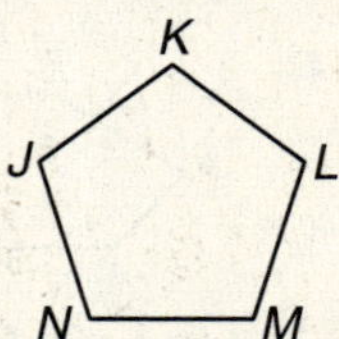

FOLDABLES™

ORGANIZE IT

On the back of your Foldable, you may wish to write the interior angle sum for each of the different polygons listed on your Foldable.

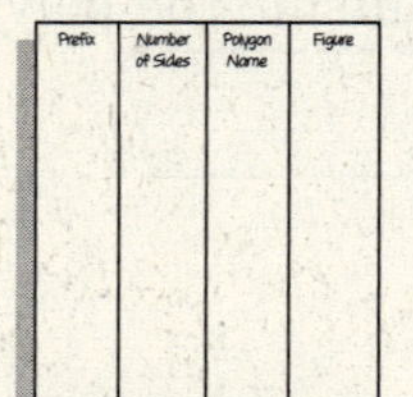

1 Find the sum of the measures of the interior angles.

Sum of measures of interior angles

$= (n - 2)180$	Theorem 10-1
$= (\_\_\_\_ - 2)180$	Substitution
$= (\_\_\_\_)180$	
$= \_\_\_\_$	

The sum of the measures of the interior angles of a pentagon is ______.

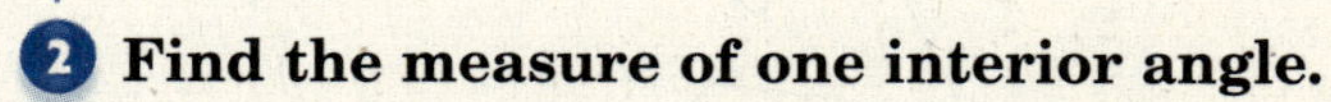

2 Find the measure of one interior angle.

Each interior angle of a regular polygon has the same measure. Divide the ______ of the measures by the ______ of angles.

measure of one interior angle $= \dfrac{\_\_\_\_}{\_\_\_\_}$ or ______

The measure of one interior angle of a regular pentagon is ______.

WRITE IT

How do you find the measure of an interior angle of an n-sided regular polygon?

______

______

______

______

REMEMBER IT

Theorems 10-1 and 10-2 only apply to *convex* polygons.

Your Turn

a. Find the sum of the measures of the interior angles of a regular 15-sided polygon.

b. Find the measure of one interior angle of a regular 15-sided polygon.

Theorem 10-2
In any convex polygon, the sum of the measures of the exterior angles, one at each vertex, is 360.

EXAMPLE

3 Find the measure of one exterior angle of a regular octagon.

By Theorem 10-2, the sum of the measures of exterior angles is ______. An octagon has ______ exterior angles.

measure of one exterior angle $= \frac{360}{8} =$ ______

Your Turn Find the measure of one exterior angle of a regular 15-sided polygon.

HOMEWORK ASSIGNMENT

Page(s):

Exercises:

10–3 Areas of Polygons

WHAT YOU'LL LEARN

- Estimate the areas of polygons.

Postulate 10-1 Area Postulate
For any polygon and a given unit of measure, there is a unique number *A* called the measure of the area of the polygon.

Postulate 10-2
Congruent polygons have equal areas.

Postulate 10-3 Area Addition Postulate
The area of a given polygon equals the sum of the areas of the nonoverlapping polygons that form the given polygon.

BUILD YOUR VOCABULARY (pages 188–189)

Any polygon and its ______ are called a **polygonal region.**

A **composite figure** is a figure made from ______ that have been placed together.

REVIEW IT
What formulas for area have you learned? *(Lesson 1-6)*

______

______

______

______

EXAMPLE

1 **Find the area of the polygon. Each square represents 1 square centimeter.**

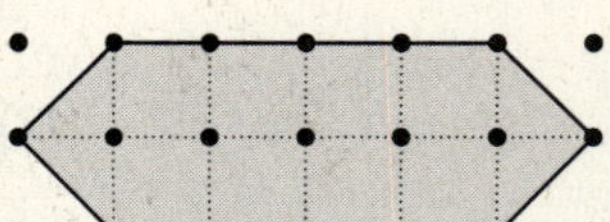

Since the area of each square represents one square centimeter, the area of each triangular half square represents 0.5 square centimeter. There are 8 squares and 4 half squares.

$A = 8(1)\text{ cm}^2 + 4(0.5)\text{ cm}^2$

$= \underline{\quad}\text{ cm}^2 + \underline{\quad}\text{ cm}^2$

$= \underline{\quad}\text{ cm}^2$

Your Turn Find the area of the polygon. Each square represents 1 square inch.

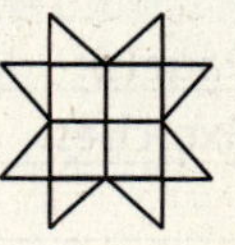

______

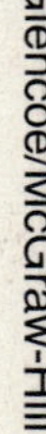

BUILD YOUR VOCABULARY (page 188)

Irregular figures are not polygons and cannot be made from combinations of polygons. Their areas can be approximated using combinations of polygons.

EXAMPLE

2 Estimate the area of the polygon. Each square represents 20 square miles.

Count each square as one unit and each partial square as a half unit regardless of size. There are ☐ whole squares and ☐ partial squares.

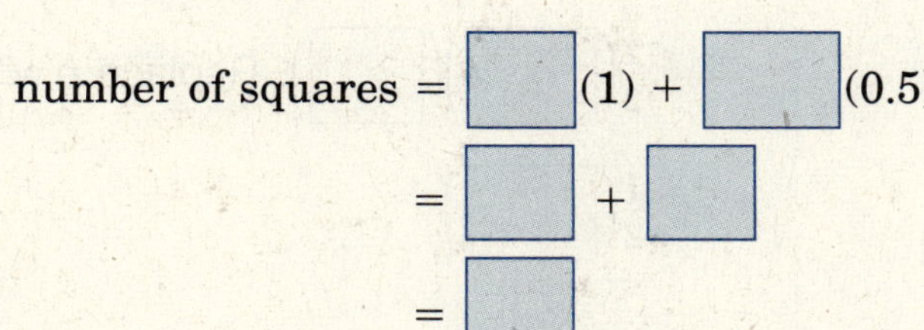

number of squares = ☐(1) + ☐(0.5)

= ☐ + ☐

= ☐

Area ≈ 20 × ☐

Each square represents 20 square miles.

= ☐

The area of the polygon is about ☐ square miles, or ☐.

WRITE IT

How can you determine the area of a polygon by dividing it into familiar shapes?

Your Turn A swimming pool at a resort is shaped as shown on the grid. Each square on the grid represents 16 square meters. Estimate the area of the pool.

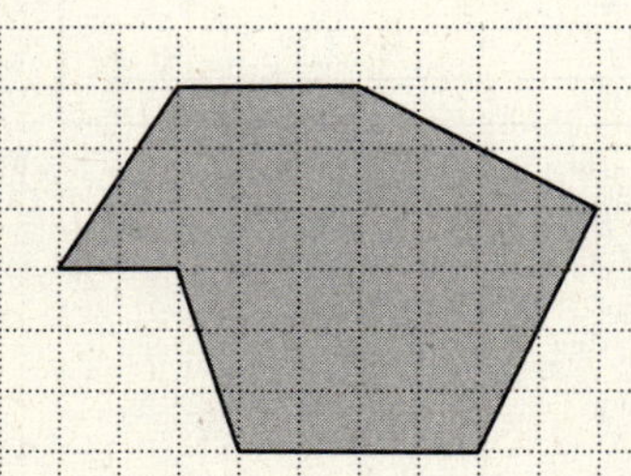

HOMEWORK ASSIGNMENT

Page(s):

Exercises:

10–4 Areas of Triangles and Trapezoids

What You'll Learn

- Find the areas of triangles and trapezoids.

Theorem 10-3 Area of a Triangle
If a triangle has an area of A square units, a base of b units, and a corresponding altitude of h units, then $A = \frac{1}{2}bh$.

EXAMPLES

Find the area of each triangle.

1

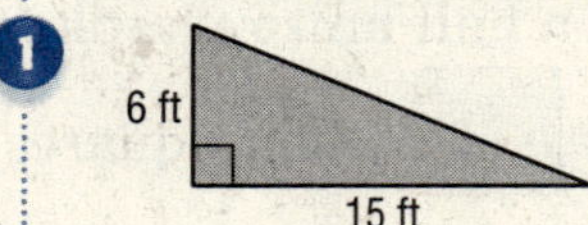

$A = \frac{1}{2}bh$ Theorem 10-3

$= \frac{1}{2}(\quad)(\quad)$ Replace b with ____ and h with ____.

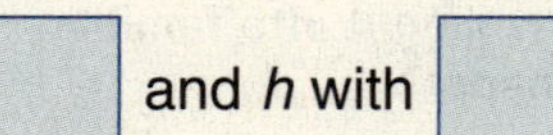

$= \frac{1}{2}(\quad)$

$=$ ____

2

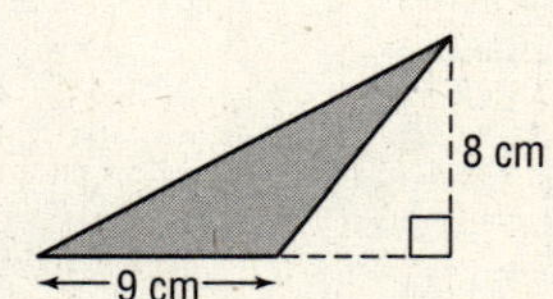

$A = \frac{1}{2}bh$ Theorem 10-3

$= \frac{1}{2}(\quad)(\quad)$ Replace b with ____ and h with ____.

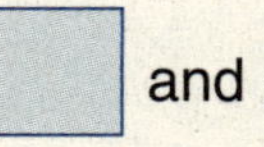

$= \frac{1}{2}(\quad)$

$=$ ____

Review It

What is an *altitude* of a triangle? *(Lesson 6-2)*

Your Turn **Find the area of each triangle.**

a.

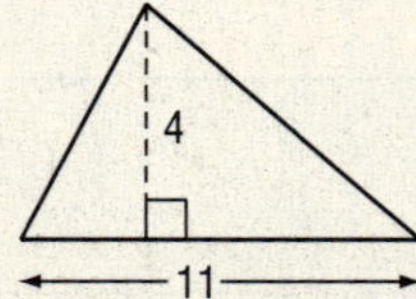

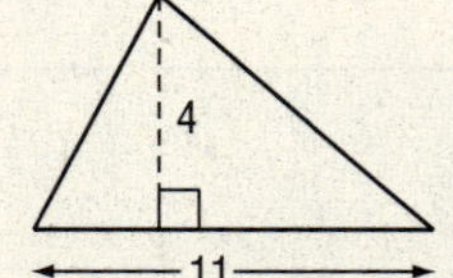

b.

c.

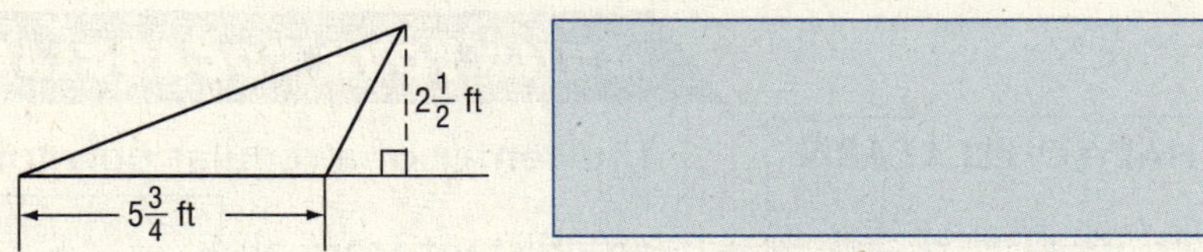

FOLDABLES™

ORGANIZE IT

Draw and label the base and altitude for each triangle on your Foldable. In addition, draw a trapezoid as an example of a quadrilateral. Draw and label the bases and altitude for the trapezoid.

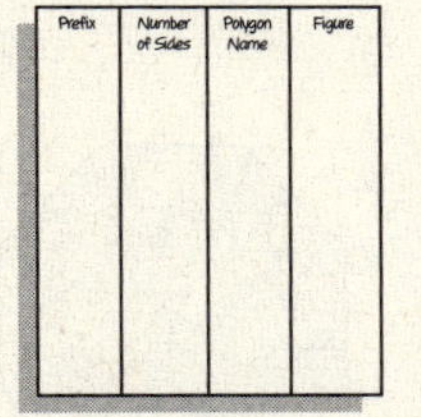

BUILD YOUR VOCABULARY (page 188)

The **altitude** of a trapezoid is a segment perpendicular to each ______.

Theorem 10-4 Area of a Trapezoid
If a trapezoid has an area of A square units, bases of b_1 and b_2 units, and an altitude of h units, then $A = \frac{1}{2}h(b_1 + b_2)$.

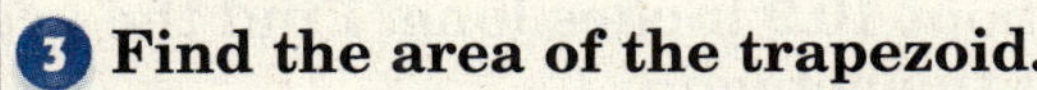

EXAMPLE

3 **Find the area of the trapezoid.**

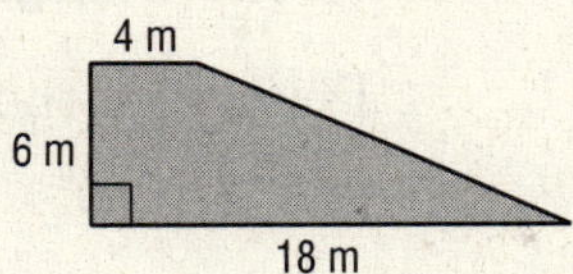

$A = \frac{1}{2}h(b_1 + b_2)$ Theorem 10-4

$= \frac{1}{2}(\ \ )(\ \ + \ \ )$ Replace h with 6, b_1 with 4, and b_2 with 18.

$= \frac{1}{2}(\ \ )(\ \ )$

$= (\ \ )(\ \ )$ or ______

HOMEWORK ASSIGNMENT

Page(s):

Exercises:

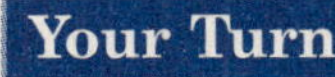

Your Turn Find the area of the trapezoid.

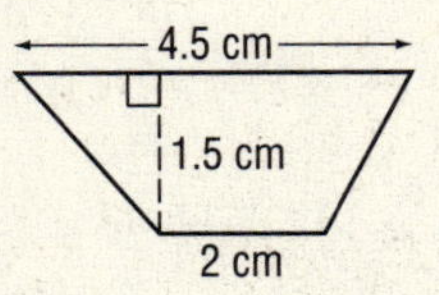

10–5 Areas of Regular Polygons

What You'll Learn

- Find the areas of regular polygons.

Build Your Vocabulary (page 188)

The **center** of a regular polygon is an interior point that is equidistant from all ________.

The segment drawn from the center and ________ to a side of a regular polygon is an **apothem**.

Foldables

Organize It

Draw an apothem for each of the regular polygons drawn on your Foldable.

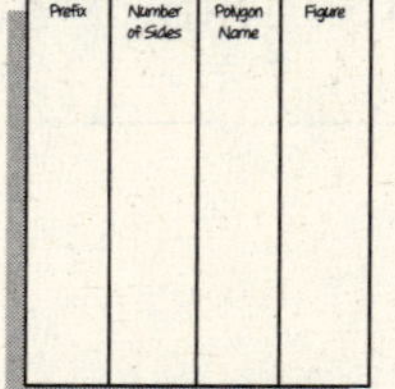

Theorem 10-5 Area of a Regular Polygon
If a regular polygon has an area of A square units, an apothem of a units, and a perimeter of P units, then $A = \frac{1}{2}aP$.

Example

1 A regular octagon has a side length of 9 inches and an apothem that is about 10.9 inches long. Find the area of the octagon.

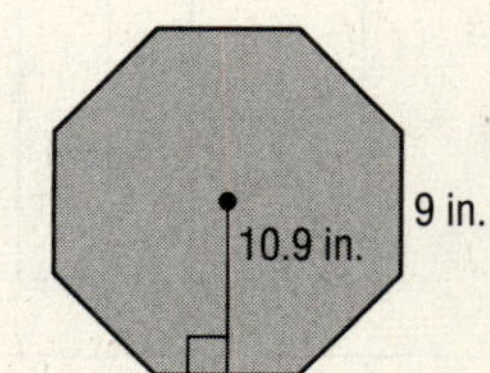

First, find the perimeter of the octagon.

$P = 8s$ — All sides of a regular octagon are congruent.

$= 8(9)$ or 72 — Replace s with 9.

Now find the area.

$A = \frac{1}{2}aP$ — Theorem 10-5

$= \frac{1}{2}(10.9)(72)$ or ________ — Replace a with 10.9 and P with 72.

The area of the octagon is about ________ in^2.

Remember It

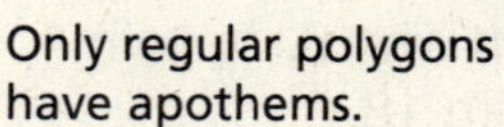

Only regular polygons have apothems.

Your Turn A regular pentagon has a side length of 8 inches and an apothem that is about 5.5 inches long. Find the area of the pentagon.

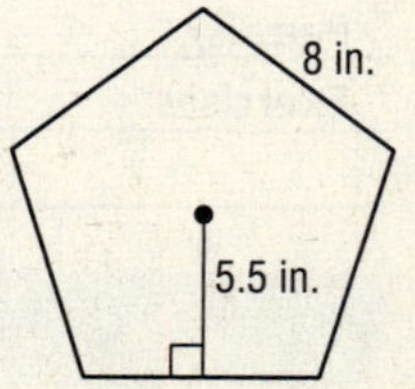

____

EXAMPLE

2 A regular octagon has a side length of 12 inches and an apothem that is about 14.5 inches long. Find the area of the shaded region of the octagon.

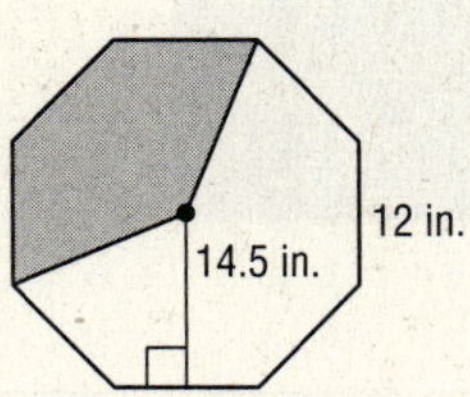

Find the area of the octagon minus the area of the unshaded region.

Area of an octagon:

$A = \frac{1}{2}aP$ Theorem 10-5

$= \frac{1}{2}(\quad)(\quad)$ Replace *a* with 14.5 and *P* with 96.

$= \quad$ in^2

Area of a Triangle:

$A = \frac{1}{2}bh$ Theorem 10-3

$= \frac{1}{2}(12)(14.5)$ Replace *b* with 12 and *h* with 14.5.

$= \quad$ in^2

The area of one triangular section is 87 in^2. There are 5 triangular sections in the unshaded region.

The area of the unshaded region is $5(\quad) = \quad$ in^2.

Subtract the area of the unshaded region from the area of the octagon.

Area of shaded region = 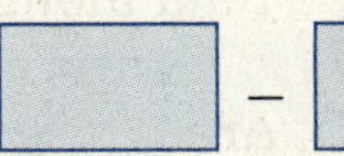 –  or  in^2

Your Turn Find the area of the shaded region of the regular hexagon.

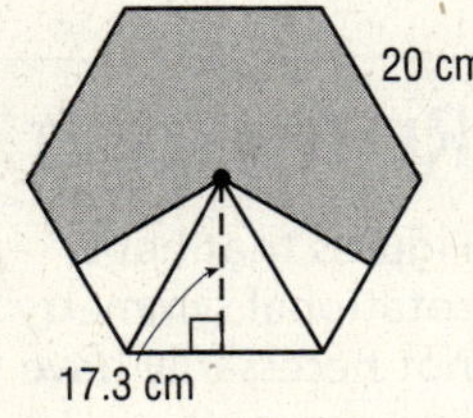

HOMEWORK ASSIGNMENT

Page(s):

Exercises:

BUILD YOUR VOCABULARY (page 189)

Significant digits represent the precision of a ______.

10–6 Symmetry

What You'll Learn

- Identify figures with line symmetry and rotational symmetry.

BUILD YOUR VOCABULARY (pages 188–189)

Symmetry is when a figure has balanced proportions across a reference ______, line, or plane.

When a line is drawn through the ______ of a figure and one half is the ______ image of the other, the figure is said to have **line symmetry**.

The reference line is known as the **line of symmetry.**

EXAMPLE

1 **Find all lines of symmetry for equilateral triangle *ABC*.**

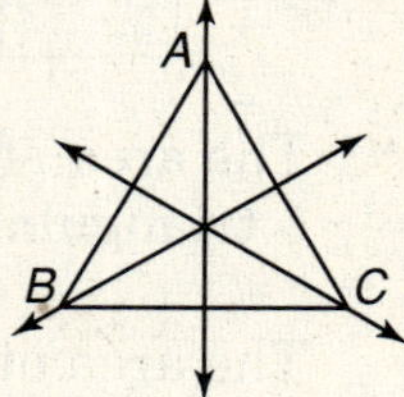

Fold along all possible lines to see if the sides match. There are ______ lines of symmetry along the lines shown in the figure.

Remember It 

Figures that have rotational symmetry do not necessarily have line symmetry.

Your Turn Draw all lines of symmetry for regular pentagon *JKXYZ*.

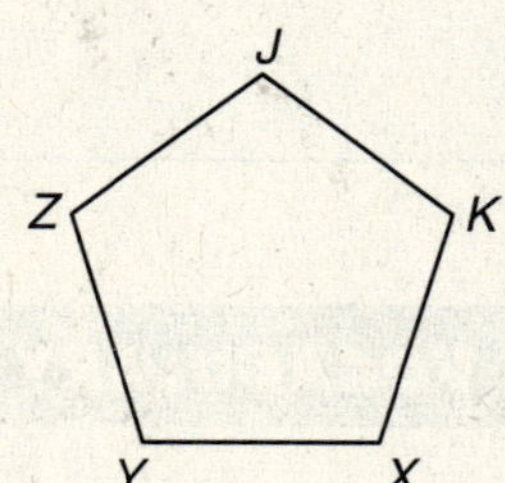

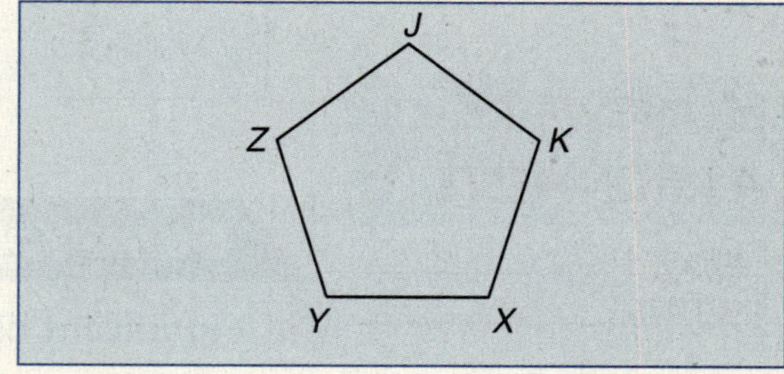

BUILD YOUR VOCABULARY (page 189)

A figure that can be turned or rotated less than 360° about a fixed point and that looks exactly as it does in the ______ is said to have **turn symmetry** or **rotational symmetry**.

WRITE IT

Draw a polygon that has line symmetry but does not have rotational symmetry. Do you think it is possible to draw a figure with more than 1 line of symmetry, but that does not have rotational symmetry? Explain.

______

______

______

______

EXAMPLE

2 Which of the figures have rotational symmetry?

a.

The figure can be turned 120° and 240° to look like the original. The figure has ______ symmetry.

b.

The figure must be turned 360° about its center to look like the original. Therefore, it ______ have rotational symmetry.

Your Turn **Which of the figures has rotational symmetry?**

a. ______

b. ______

HOMEWORK ASSIGNMENT

Page(s):

Exercises:

10–7 Tessellations

What You'll Learn

- Identify tessellations and create them by using transformations.

Remember It

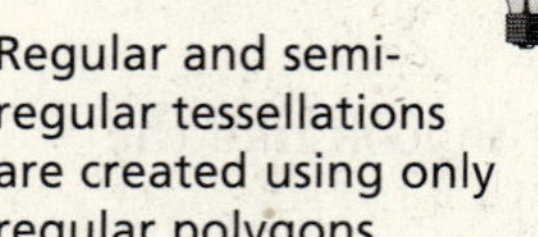

Regular and semi-regular tessellations are created using only regular polygons.

Build Your Vocabulary (page 189)

Tessellations are tiled patterns created by ________ figures to fill a plane without gaps or overlaps. They can be made by translating, rotating, or reflecting polygons.

A pattern is a **regular tessellation** when only ________ type of regular polygon is used to form the pattern.

When two or more regular polygons are used in the same order at every vertex to form a pattern, it is a **semi-regular tessellation**.

Examples

Identify the figures used to create each tessellation. Then identify the tessellation as *regular*, *semi-regular*, or *neither*.

1 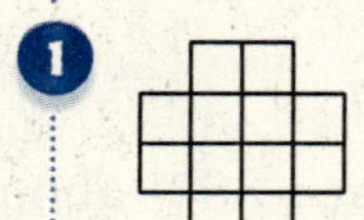

Only squares are used. A square is a regular polygon. The tessellation is ________.

2 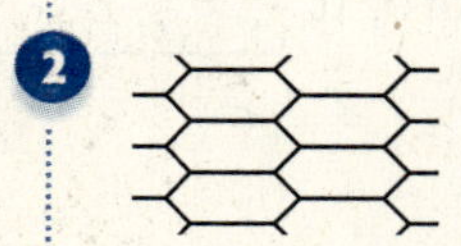

Hexagons are used and there are no gaps in the pattern, but the hexagons are not ________. The tessellation is ________ a regular nor a semi-regular tessellation.

Your Turn **Identify the tessellation as *regular*, *semi-regular*, or *neither*.**

a.

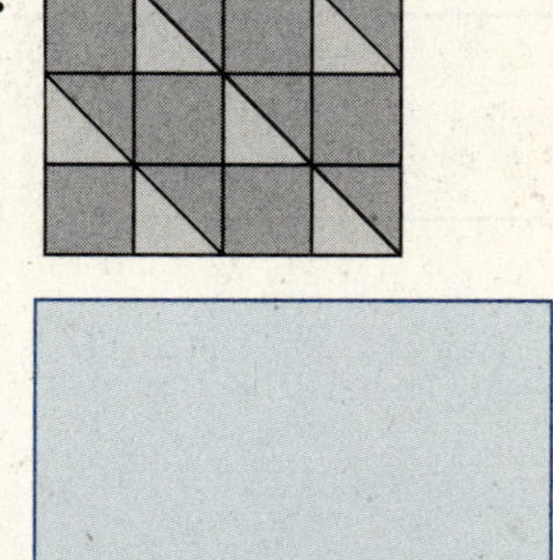

b.

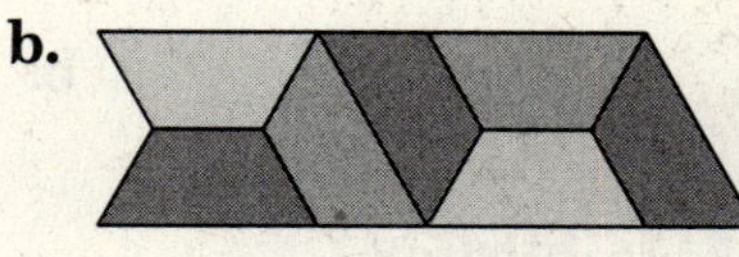

Homework Assignment

Page(s):

Exercises:

CHAPTER 10

BRINGING IT ALL TOGETHER

STUDY GUIDE

FOLDABLES™	VOCABULARY PUZZLEMAKER	BUILD YOUR VOCABULARY
Use your **Chapter 10 Foldable** to help you study for your chapter test.	To make a crossword puzzle, word search, or jumble puzzle of the vocabulary words in Chapter 10, go to: www.glencoe.com/sec/math/t_resources/free/index.php.	You can use your completed **Vocabulary Builder** (pages 188–189) to help you solve the puzzle.

10-1 Naming Polygons

Indicate whether the statement is *true* or *false*.

1. All the diagonals of a concave polygon lie on the interior.
2. A regular polygon is both equilateral and equiangular.

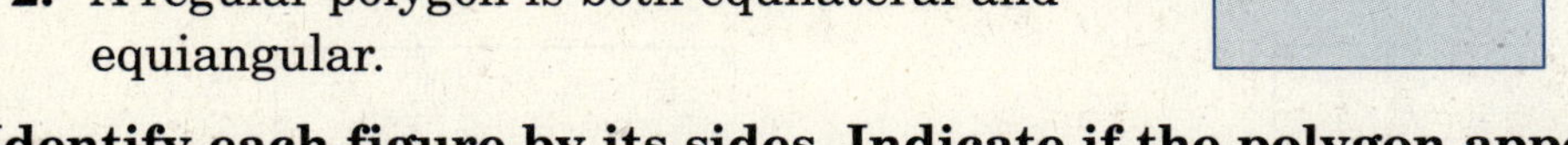

Identify each figure by its sides. Indicate if the polygon appears to be regular or not regular. If not regular, justify your reason.

3. 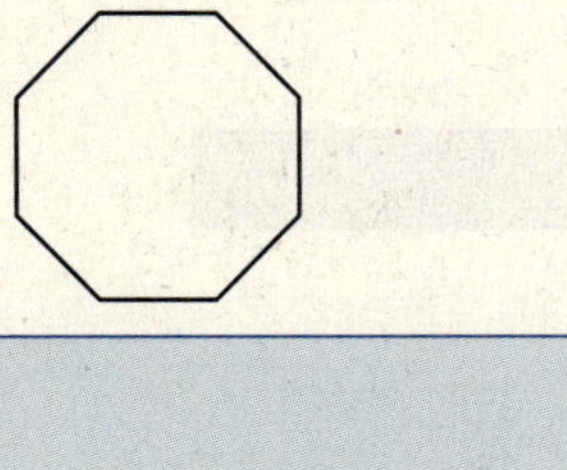

4. 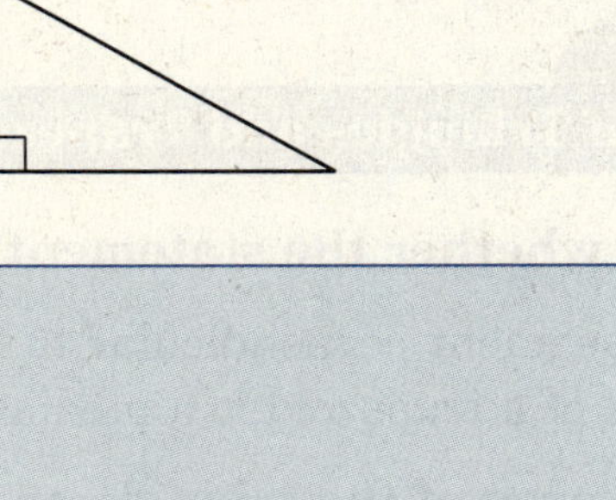

10-2 Diagonals and Angle Measure

Find the sum of the measures of the interior angles.

5. 

6. 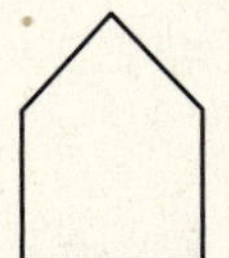

Find the measure of one interior angle and one exterior angle of the regular polygon.

7. dodecagon

8. decagon

9. The sum of the measures of four exterior angles of a pentagon is 280. What is the measure of the fifth exterior angle?

10-3 Areas of Polygons

Indicate whether the statement is *true* or *false*.

10. A polygon and its interior are known as a polygonal region.

Find the area of the polygon in square units.

11.

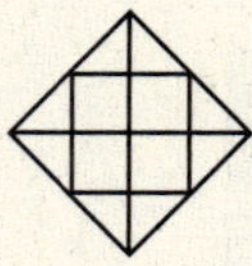

12.

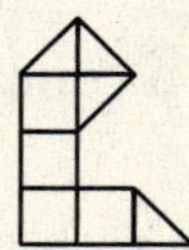

10-4 Areas of Triangles and Trapezoids

Indicate whether the statement is *true* or *false*.

13. The segment perpendicular to the parallel bases of a trapezoid is a median.

Find the area of the triangle or trapezoid.

14.

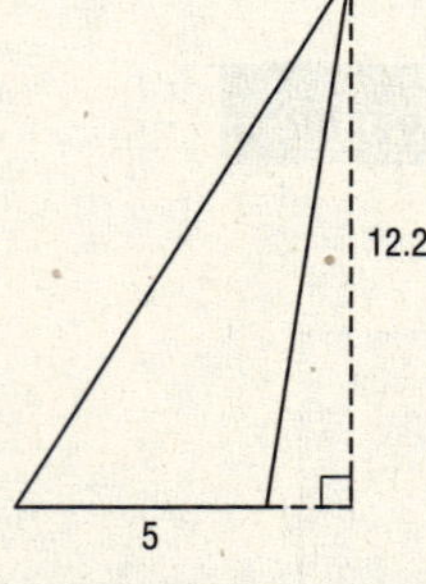

15.

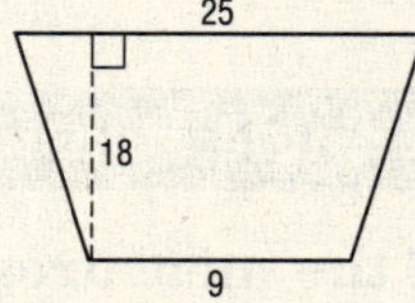

16. Find the area of a trapezoid whose altitude measures 4.5 cm and has bases measuring 6.2 and 8.8 cm.

17. What is the area of a triangle with base length $6\frac{1}{3}$ in. and height 2 in.?

10-5 Areas of Regular Polygons

18. Find the area of a regular 11-sided polygon with each side measuring 7 cm and an apothem length of 11.9 cm.

19. Find the area of the shaded region.

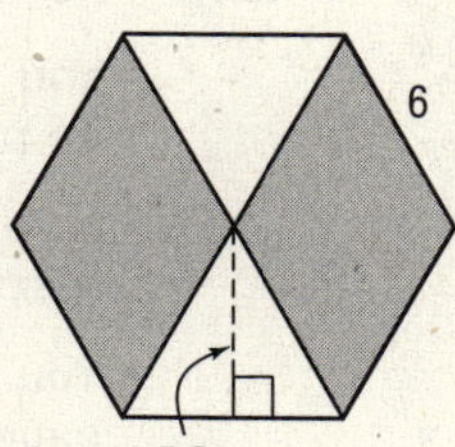

10-6 Symmetry

Underline the best term to make the statement true.

20. When a line is drawn through a figure and makes each half a mirror image of the other, the figure has [line/rotational] symmetry.

21. When a figure looks exactly as it does in its original position after being turned less than 360° around a fixed point, it has [line/rotational] symmetry.

Determine whether the figure has *line symmetry*, *rotational symmetry*, *both*, or *neither*.

22.

23.

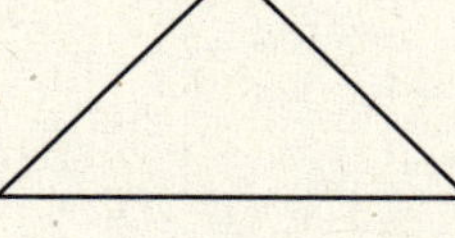

10-7 Tessellations

Identify the tessellation as *regular*, *semi-regular*, or *neither*.

24.

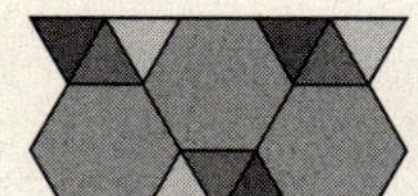

25.

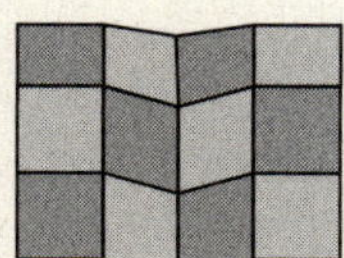

ARE YOU READY FOR THE CHAPTER TEST?

Visit **geomconcepts.com** to access your textbook, more examples, self-check quizzes, and practice tests to help you study the concepts in Chapter 10.

Check the one that applies. Suggestions to help you study are given with each item.

☐ **I completed the review of all or most lessons without using my notes or asking for help.**

- You are probably ready for the Chapter Test.
- You may want to take the Chapter 10 Practice Test on page 449 of your textbook as a final check.

☐ **I used my Foldable or Study Notebook to complete the review of all or most lessons.**

- You should complete the Chapter 10 Study Guide and Review on pages 446–448 of your textbook.
- If you are unsure of any concepts or skills, refer back to the specific lesson(s).
- You may also want to take the Chapter 10 Practice Test on page 449 of your textbook.

☐ **I asked for help from someone else to complete the review of all or most lessons.**

- You should review the examples and concepts in your Study Notebook and Chapter 10 Foldable.
- Then complete the Chapter 10 Study Guide and Review on pages 446–448 of your textbook.
- If you are unsure of any concepts or skills, refer back to the specific lesson(s).
- You may also want to take the Chapter 10 Practice Test on page 449 of your textbook.

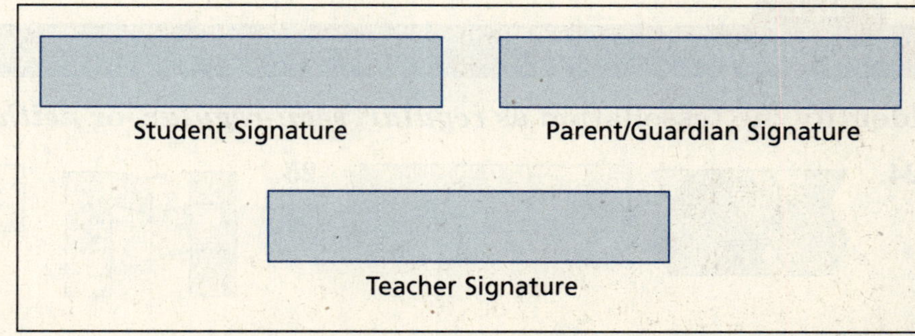

Circles

Use the instructions below to make a Foldable to help you organize your notes as you study the chapter. You will see Foldable reminders in the margin of this Interactive Study Notebook to help you in taking notes.

Begin with seven sheets of plain paper.

STEP 1 **Draw**
Draw and cut a circle from each sheet. Use a small plate or a CD to outline the circle.

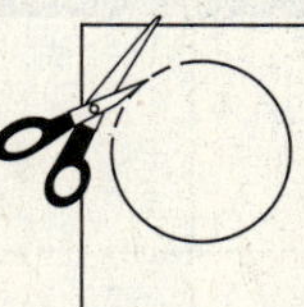

STEP 2 **Staple**
Staple the circles together to form a booklet.

STEP 3 **Label**
Label the chapter name on the front. Label the inside six pages with the lesson titles.

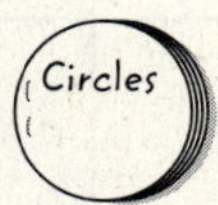

NOTE-TAKING TIP: When you take notes, write concise definitions in your own words. Add examples that illustrate the concepts.

Chapter 11

BUILD YOUR VOCABULARY

This is an alphabetical list of new vocabulary terms you will learn in Chapter 11. As you complete the study notes for the chapter, you will see Build Your Vocabulary reminders to complete each term's definition or description on these pages. Remember to add the textbook page number in the second column for reference when you study.

Vocabulary Term	Found on Page	Definition	Description or Example
adjacent arcs			
arcs			
center			
central angle			
chord			
circle			
circumference [sir-KUM-fur-ents]			
circumscribed			
concentric			
diameter			

Vocabulary Term	Found on Page	Definition	Description or Example
experimental probability [ek-speer-uh-MEN-tul]			
inscribed			
loci			
locus			
major arc			
minor arc			
pi (π)			
radius [RAY-dee-us]			
sector			
semicircle			
theoretical probability [thee-uh-RET-i-kul]			

11–1 Parts of a Circle

BUILD YOUR VOCABULARY (pages 208–209)

A **circle** is the set of all points in a plane that are a given distance from a given point in the plane, called the ______ of the circle.

In a circle, all points are ______ from the **center**.

A **radius** is a segment whose endpoints are the ______ of the circle and a ______ on the circle.

A **chord** is a segment whose ______ are on the circle.

A **diameter** is a ______ that contains the ______ of the circle.

Two circles are **concentric** if they lie in the same plane, have the same ______, and have ______ of different lengths.

WHAT YOU'LL LEARN

- Identify and use parts of circles.

FOLDABLES™ ORGANIZE IT

Under the tab for Lesson 11-1, draw a circle with a radius, a chord and a diameter. Label each special segment.

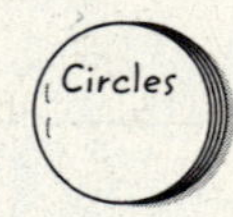

EXAMPLES

Use circle P to determine whether each statement is true or false.

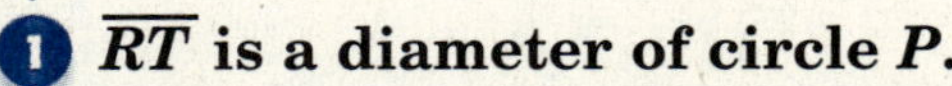

1 $\overline{RT}$ **is a diameter of circle** P**.**

______; $\overline{RT}$ ______ go through the center P. Therefore, $\overline{RT}$ is not a diameter.

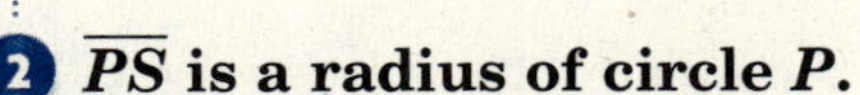

2 $\overline{PS}$ **is a radius of circle** P**.**

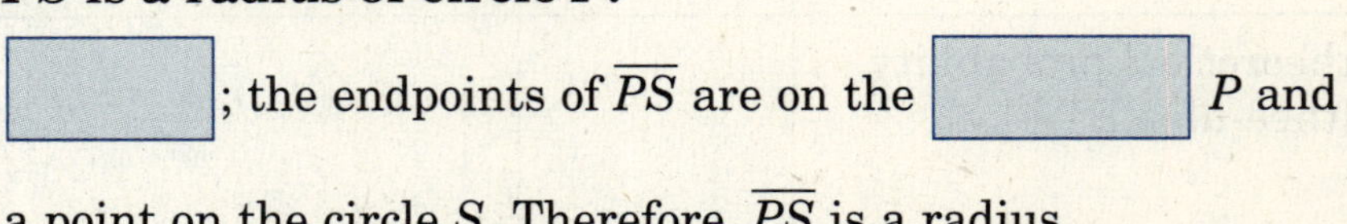

______; the endpoints of $\overline{PS}$ are on the ______ P and a point on the circle S. Therefore, $\overline{PS}$ is a radius.

Your Turn **Use circle T to determine whether each statement is *true* or *false*.**

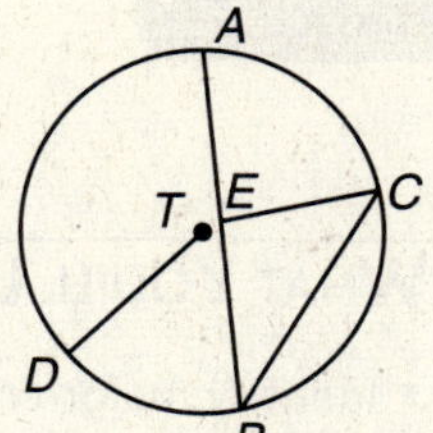

a. $\overline{AB}$ is not a diameter.

b. $\overline{TD}$ is not a radius.

WRITE IT

Describe the differences between a radius, a diameter, and a chord.

Theorem 11-1
All radii of a circle are congruent.

Theorem 11-2
The measure of the diameter d of a circle is twice the measure of the radius r of the circle.

EXAMPLE

3 In circle R, $\overline{QT}$ is a diameter. If $QR = 7$, find QT.

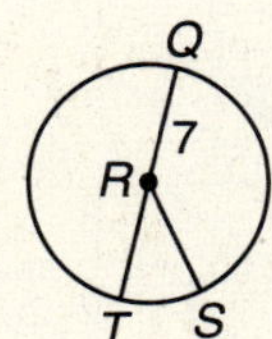

$\overline{QR}$ is a radius, and $d = 2r$.

$QT = 2(QR)$ Replace d and r.

$QT = 2(\quad)$ Replace QR with .

$QT = $

Your Turn In circle A, $\overline{FC}$ is a diameter. If $FC = 25$, find AB.

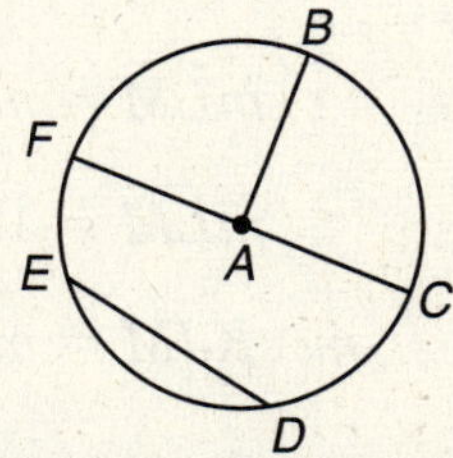

HOMEWORK ASSIGNMENT

Page(s):
Exercises:

11–2 Arcs and Central Angles

BUILD YOUR VOCABULARY (pages 208–209)

When two sides of an angle meet at the center of a circle, a **central angle** is formed.

Each side of the central angle intersects a point on the circle, dividing it into ______ lines called **arcs**.

A **minor arc** is formed by the intersection of the circle and sides of a central angle with interior degree measure less than 180.

A **major arc** is the part of the circle in the ______ of the central angle that measures greater than 180.

Semicircles are ______ arcs whose endpoints lie on the diameter of the circle.

Adjacent arcs are arcs of a circle with exactly one point in common.

WHAT YOU'LL LEARN

- Identify major arcs, minor arcs, and semicircles and find the measures of arcs and central angles.

KEY CONCEPTS

The degree measure of a minor arc is the degree measure of its central angle.

The degree measure of a major arc is 360 minus the degree measure of its central angle.

The degree measure of a semicircle is 180.

FOLDABLES Under the tab for Lesson 11-2, draw a circle with a central angle. Label the central angle, the major and minor arcs and give examples of degree measurements for each.

EXAMPLE

1 **In circle *J*, find $m\widehat{LM}$, $m\angle KJM$, and $m\widehat{LK}$.**

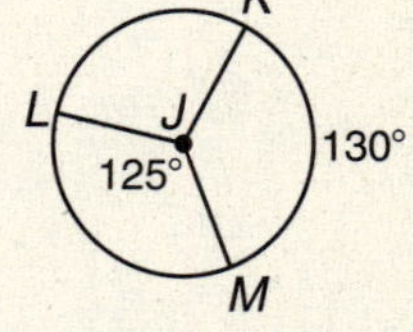

$m\widehat{LM} = m\angle LJM$ — Measure of minor arc

$m\widehat{LM} = 125$

$m\angle KJM = m\widehat{KM}$ — Measure of central angle

$m\angle KJM =$ ______

$m\widehat{LK} = 360 - m\angle LJM - m\angle KJM$ — Measure of major arc

$m\widehat{LK} = 360 - 125 - 130$ — Substitution

$m\widehat{LK} =$ ______

Postulate 11-1 Arc Addition Postulate
The sum of the measures of two adjacent arcs is the measure of the arc formed by the adjacent arcs.

EXAMPLE

2 **In circle A, $\overline{CE}$ is a diameter. Find $m\widehat{BC}$, $m\widehat{BE}$, and $m\widehat{BDE}$.**

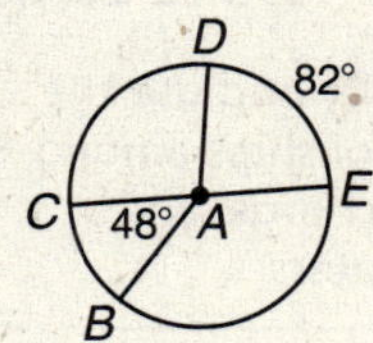

REMEMBER IT
A circle contains 360°.

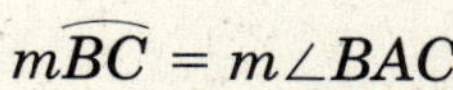

$m\widehat{BC} = m\angle BAC$	Measure of minor arc
$m\widehat{BC} = \square$	Substitution
$\square + m\widehat{BC} = m\widehat{EBC}$	Arc Addition Postulate
$m\widehat{BE} + \square = 180$	Substitution
$m\widehat{BE} = 132$	Subtract.
$m\widehat{BDE} = \square - m\widehat{BE}$	Measure major arc
$m\widehat{BDE} = \square - 132$	Substitution
$m\widehat{BDE} = \square$	

Your Turn **In circle X, $m\angle AXB = 70$, $m\widehat{DC} = 45$, and $\overline{BE}$ and $\overline{AD}$ are diameters.**

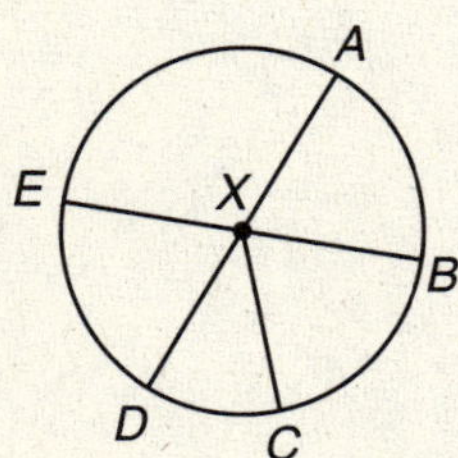

a. Find $m\widehat{EA}$, $m\angle BXC$, and $m\widehat{ED}$.

b. Find $m\widehat{AC}$, $m\widehat{DAE}$, and $m\widehat{ABE}$.

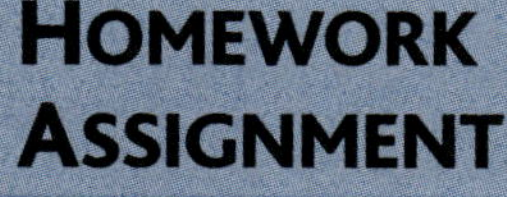

HOMEWORK ASSIGNMENT

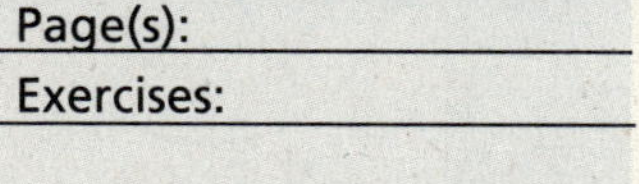

Page(s):
Exercises:

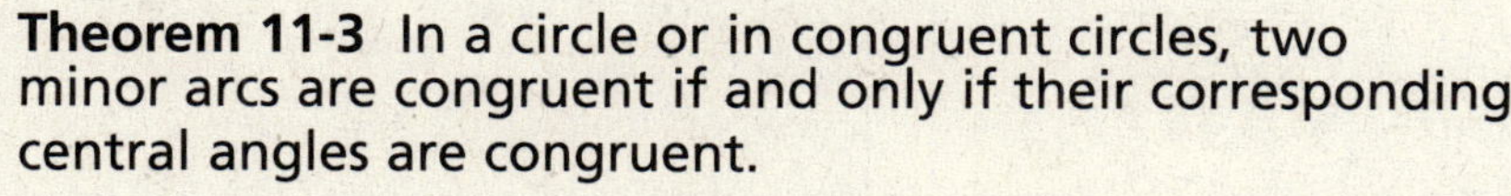

Theorem 11-3 In a circle or in congruent circles, two minor arcs are congruent if and only if their corresponding central angles are congruent.

11–3 Arcs and Chords

What You'll Learn

- Identify and use the relationships among arcs, chords, and diameters.

Theorem 11-4
In a circle or in congruent circles, two minor arcs are congruent if and only if their corresponding chords are congruent.

Theorem 11-5
In a circle, a diameter bisects a chord and its arc if and only if it is perpendicular to the chord.

Foldables™

Organize It

Under the tab for Lesson 11-3, draw diagrams and give descriptions to summarize Theorems 11-4 and 11-5.

EXAMPLE

1 **In circle R, if $\overline{PR} \perp \overline{QT}$, find PQ.**

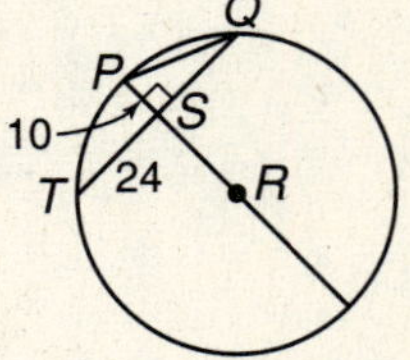

$\angle PSQ$ is a right angle.	Definition of perpendicular
$\triangle PSQ$ is a ______ triangle.	Definition of right triangle
$(\_\_\_\_)^2 + (SQ)^2 = (PQ)^2$	Pythagorean Theorem
$SQ = 24$	Theorem 11-5
$\_\_\_\_^2 + 24^2 = (PQ)^2$	Replace PS with ______ and SQ with 24.
$100 + 576 = (PQ)^2$	
$\sqrt{676} = \sqrt{(PQ)^2}$	Take the square root of each side.
______ $= PQ$	

Your Turn In circle P, $AB = 8$ and $PD = 3$. Find PC.

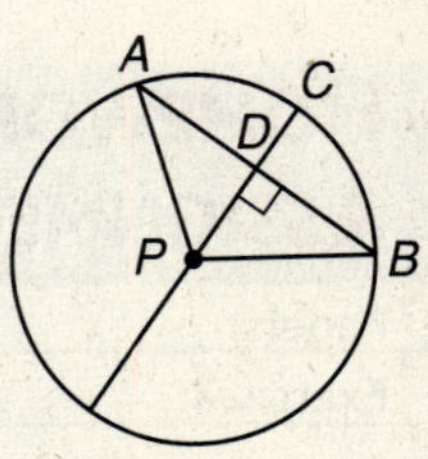

EXAMPLE

2 **In circle W, find XV if $\overline{UW} \perp \overline{XV}$, $VW = 35$, and $WY = 21$.**

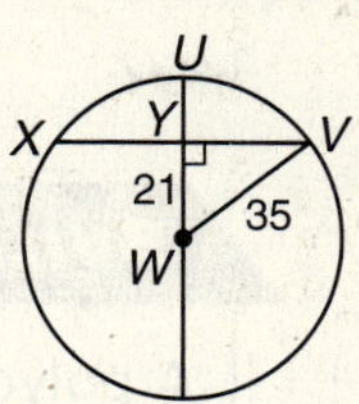

$\angle VYW$ is a ______ angle.	Definition of perpendicular
$\triangle VYW$ is a right triangle.	Definition of right triangle
$(WY)^2 + (YV)^2 = (\ \_\_\_\_\ )^2$	Pythagorean Theorem
$21^2 + (YV)^2 = 35^2$	Replace *WY* and *VW*.
______ $+ (YV)^2 = 1225$	
$(YV)^2 =$ ______	Subtract.
$\sqrt{(YV)^2} = \sqrt{\_\_\_\_}$	Take the square root of each side.
$YV =$ ______ $= XY$	Theorem 11-5
$XV = YV + XY$	Segment addition
$XV =$ ______ $+ 28$	Substitution
$XV =$ ______	

Your Turn In circle G, if $\overline{CG} \perp \overline{AE}$, $EG = 20$, $CG = 12$, find AE.

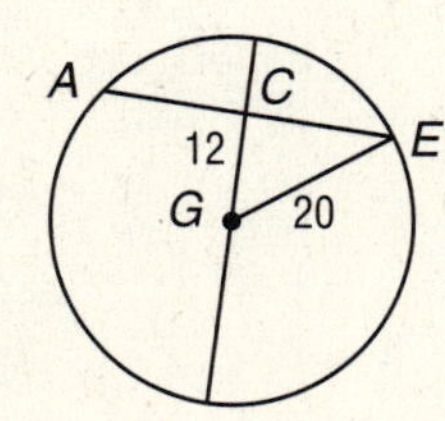

HOMEWORK ASSIGNMENT

Page(s):

Exercises:

11–4 Inscribed Polygons

WHAT YOU'LL LEARN

- Inscribe regular polygons in circles and explore the relationship between the length of a chord and its distance from the center of the circle.

BUILD YOUR VOCABULARY (pages 208–209)

A polygon is **inscribed** in a circle if and only if every of the polygon lies on the circle.

A **circumscribed** polygon is a polygon with each side to a .

FOLDABLES™ ORGANIZE IT

Under the tab for Lesson 11-4, draw a polygon inscribed in a circle and another circumscribed about the circle. Label each drawing appropriately.

EXAMPLE

1 Construct a regular octagon.

Construct a quadrilateral by connecting the consecutive of two diameters.

Bisect adjacent . Extend the bisectors through the of the circle to the edges of the circle. The other four are where the other two perpendicular intersect the circle. Connect all of the consecutive to form the regular .

Your Turn Construct a regular hexagon.

Theorem 11-6
In a circle or in congruent circles, two chords are congruent if and only if they are equidistant from the center.

EXAMPLE

2 **In circle O, point O is the midpoint of $\overline{AB}$. If $CR = 2x - 1$ and $ST = x + 10$, find x.**

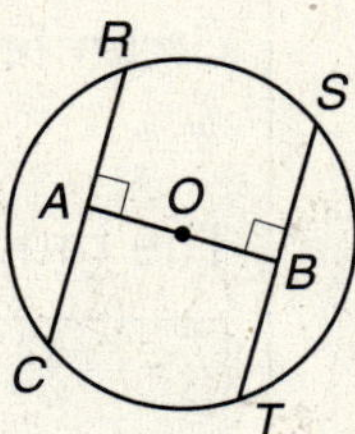

$OA =$ ____	Definition of midpoint
____ $= TS$	Theorem 11-6
$2x - 1 = x +$ ____	Substitution
$2x = x +$ ____	Add ____ to each side.
$x =$ ____	Subtract ____ from each side.

Your Turn In circle Y, $NY = YO$. If $AX = 2x + 15$ and $BZ = 3x + 6$, what is the value of x?

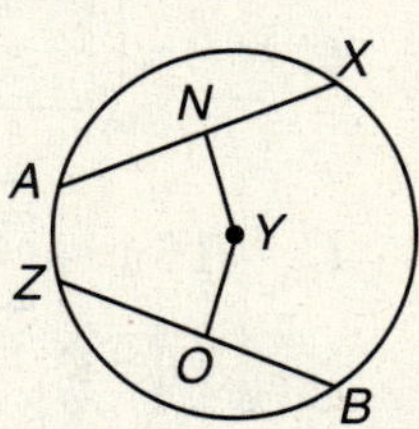

WRITE IT

Explain Theorem 11-6 in your own words.

HOMEWORK ASSIGNMENT

Page(s):
Exercises:

11–5 Circumference of a Circle

BUILD YOUR VOCABULARY (pages 208–209)

The perimeter of a ______ is known as the **circumference**. It is the ______ around the circle.

The ratio of the ______ of a circle to its ______ is always equal to the irrational number called **pi**.

WHAT YOU'LL LEARN

- Solve problems involving circumference of circles.

FOLDABLES

ORGANIZE IT

Under the tab for Lesson 11-5, give the formulas for finding the circumference of a circle and give an example of how they are used.

Theorem 11-7 Circumference of a Circle
If a circle has a circumference of *C* units and a radius of *r* units, then $C = 2\pi r$ or $C = \pi d$.

EXAMPLES

1 The radius of a circle is 8 feet. Find the circumference of the circle to the nearest tenth.

$C = 2\pi$ ______ Theorem 11-7

$C = 2\pi($ ______ $)$ Replace *r* with ______.

$C = 16\pi \approx$ ______ feet

2 The diameter of a plastic pipe is 5 cm. Find the circumference of the pipe to the nearest centimeter.

$C = \pi$ ______ Theorem 11-7

$C = \pi($ ______ $)$ Substitution

$C = 5\pi \approx$ ______ cm

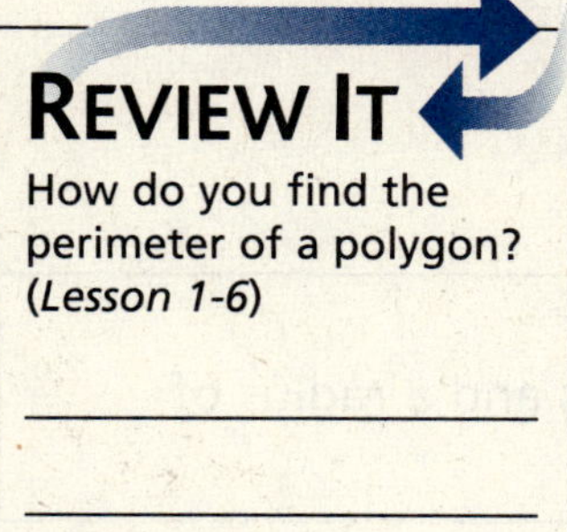

How do you find the perimeter of a polygon? (*Lesson 1-6*)

______

______

______

Your Turn

a. Find the circumference of circle A to the nearest tenth.

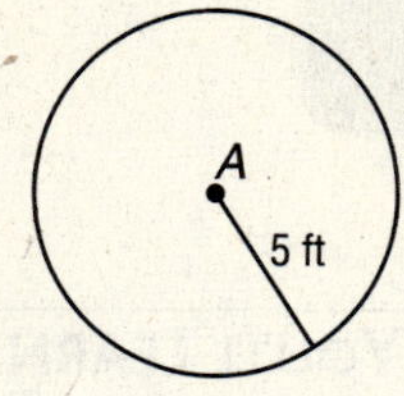

b. The diameter of a *CD* is 4.5 inches. Find its circumference to the nearest tenth.

REMEMBER IT

Pi (π) is an exact constant. The decimal approximation 3.14... is only an estimate.

EXAMPLE

3 **A circular garden has a radius of 20 feet. There is a path around the garden that is 3 feet wide. Jasmine stands on the inside edge of the path, and Hitesh stands on the outside edge. They each walk around the garden exactly once while staying along their edge of the path. To the nearest foot, how much farther does Hitesh walk than Jasmine?**

Jasmine:		Hitesh:
$C = 2\pi r$	Theorem 11-7	$C = 2\pi r$
$C = 2\pi(\quad)$	Substitution	$C = 2\pi(\quad)$
$C =$ ______		$C =$ ______

So, Hitesh walked ______ – ______ or approximately ______ feet more than Jasmine.

HOMEWORK ASSIGNMENT

Page(s):

Exercises:

Your Turn A circle has a circumference of 20.5 meters. Find the radius of the circle to the nearest tenth.

11–6 Area of a Circle

WHAT YOU'LL LEARN

- Solve problems involving areas and sectors of circles.

FOLDABLES™

ORGANIZE IT

Under the tab for Lesson 11-6, give the formula for finding the area of a circle and give an example of how it is used.

Theorem 11-8 Area of a Circle
If a circle has an area of A square units and a radius of r units, then $A = \pi r^2$.

EXAMPLE

1 Find the area of circle *G*.

G
10 cm

$A = \pi r^2$	Theorem 11-8
$A = \pi(\_\_\_\_)^2$	Replace r.
$A = 100\pi \approx \_\_\_\_$ cm^2	

Your Turn Find the area of a circle to the nearest tenth whose diameter is 10 cm.

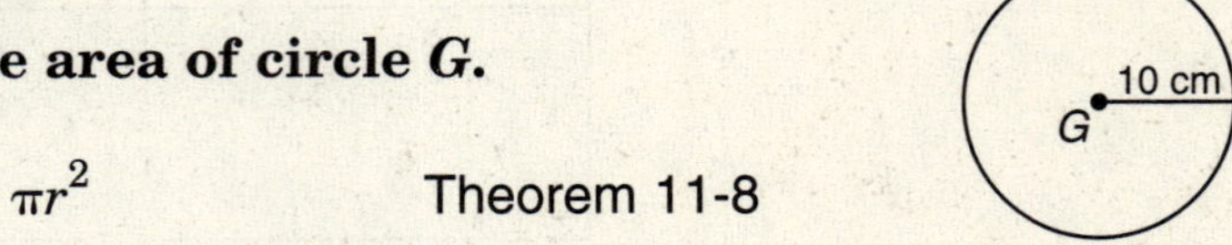

EXAMPLE

2 If circle *S* has a circumference of 16π inches, find the area of the circle to the nearest hundredth.

$C = 2\pi r$	Theorem 11-7
$\_\_\_\_ = 2\pi r$	Replace C with ____.
$\frac{16\pi}{2\pi} = \frac{2\pi r}{2\pi}$	Divide each side by ____.
$\_\_\_\_ = r$	
$A = \pi r^2$	Theorem 11-8
$A = \pi(\_\_\_\_)^2$	Replace r with ____.
$A = 64\pi \approx \_\_\_\_$ in^2	

Your Turn Find the area of the circle to the nearest hundredth whose circumference is 84π cm.

______

BUILD YOUR VOCABULARY (pages 208–209)

Theoretical probability is the chance for a successful outcome based on ______.

Experimental probability is calculated from actual observations and recording ______. It is the chance for a successful outcome based on observing patterns of occurrences.

EXAMPLE

3 **A pond has a radius of 10 meters. In the center of the pond is a square island with a side length of 5 meters. The seeds of a nearby maple tree float down randomly over the pond. What is the probability that a randomly-chosen seed will land in the water rather than on the island? Assume that the seed will land somewhere within the circular edge of the pond.**

A of pond $= \pi \square^2$

$= \square \approx \square$ m^2

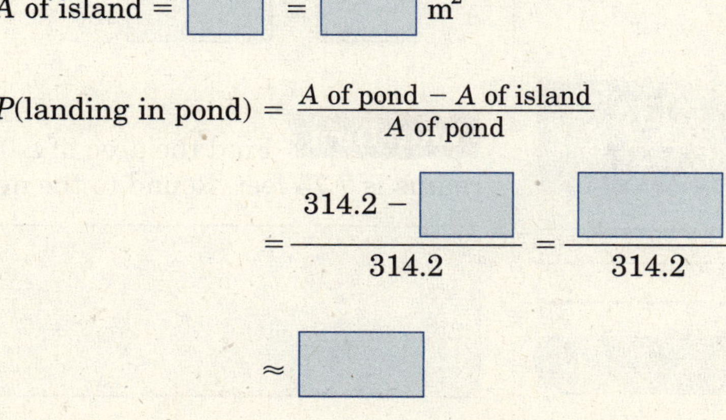

A of island $= \square^2 = \square$ m^2

$P(\text{landing in pond}) = \dfrac{A \text{ of pond} - A \text{ of island}}{A \text{ of pond}}$

$= \dfrac{314.2 - \square}{314.2} = \dfrac{\square}{314.2}$

$\approx \square$

Your Turn Assume that all darts will land on the dartboard. Find the probability that a randomly-thrown dart will land in the shaded region.

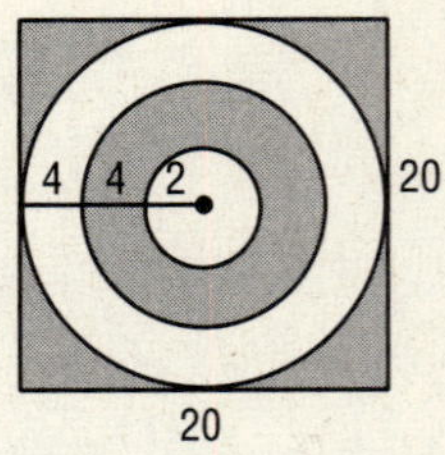

BUILD YOUR VOCABULARY (page 209)

A **sector** of a circle is a region bounded by a central ______ and its corresponding ______.

Theorem 11-9 Area of a Sector of a Circle
If a sector of a circle has an area of A square units, a central angle measurement of N degrees, and a radius of r units, then $A = \left(\frac{N}{360}\right)\pi r^2$.

EXAMPLE

4 **Find the area of a 45° sector of a circle whose radius is 8 in. Round to the nearest hundredth.**

$A = \left(\frac{N}{360}\right)\pi r^2$ Theorem 11-9

$A = \left(\frac{45}{360}\right)\pi 8^2$ Substitution

$A = (0.125)(64)\pi$

$A =$ ______ $\approx$ ______ in^2

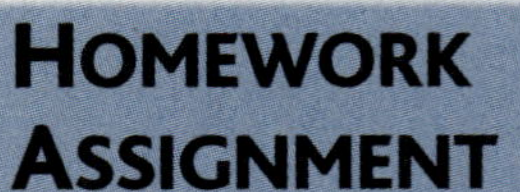

Page(s):
Exercises:

Your Turn Find the area of a 30° sector of a circle whose radius is 7.75 feet. Round to the nearest hundredth.

CHAPTER 11

BRINGING IT ALL TOGETHER

STUDY GUIDE

FOLDABLES™	VOCABULARY PUZZLEMAKER	BUILD YOUR VOCABULARY
Use your **Chapter 11 Foldable** to help you study for your chapter test.	To make a crossword puzzle, word search, or jumble puzzle of the vocabulary words in Chapter 11, go to: www.glencoe.com/sec/math/t_resources/free/index.php	You can use your completed **Vocabulary Builder** (pages 208–209) to help you solve the puzzle.

11-1 Parts of a Circle

Underline the term that best completes the statement.

1. A chord that contains the center of the circle is the [diameter/radius].
2. A [chord/radius] is a segment with endpoints of the circle.
3. Two circles are [circumscribed/concentric] if they lie on the same plane, have the same center, and have radii of different lengths.

11-2 Arcs and Central Angles

In circle C, $\overline{BD}$ is a diameter and $m\angle GCF = 63$. Find each measure.

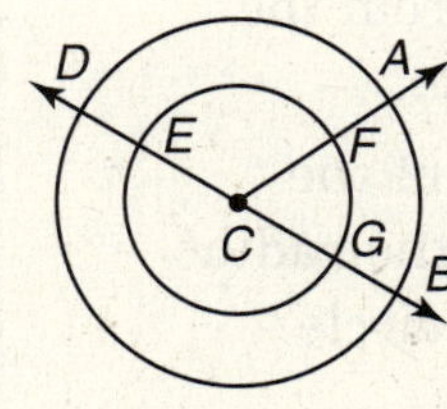

4. $m\widehat{FG}$

5. $m\widehat{AD}$

6. $m\widehat{AB}$

7. $m\widehat{GEF}$

11-3 Arcs and Chords

Complete each statement.

8. If two chords are congruent in the same circle, the intercepted ______ are also congruent.

9. When the diameter of the circle bisects a chord of the circle, then it is ______ to the chord and ______ the corresponding arc.

10. In a circle, if two arcs are ______, their ______ are congruent.

11-4 Inscribed Polygons

11. Construct an equilateral triangle inscribed in a circle with radius 1 inch.

12. Draw a circle inscribed in the triangle from the previous problem.

Which segment of the triangle equals the radius of the inscribed circle?

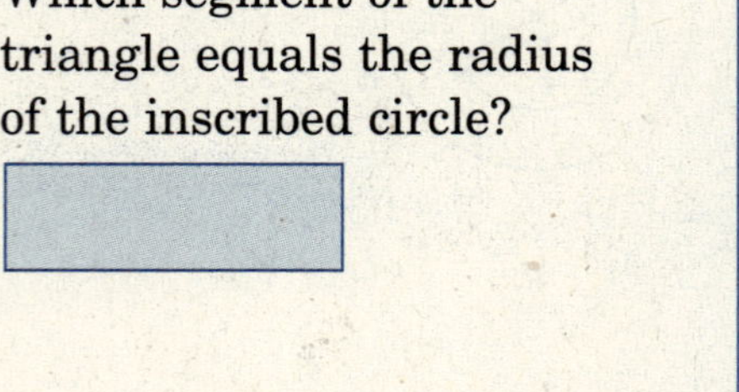

13. What is the approximate length of the segment in Exercise 12?

11-5

Circumference of a Circle

Find the circumference of each circle.

14. $r = \frac{1}{2}$ yd

15. $d = 4.2$ in.

Find the radius of the circle whose circumference is given.

16. 47 ft

17. 22.7 in.

11-6

Area of a Circle

Underline the term that best completes the statement.

18. A region of a circle bounded by a central angle and its corresponding arc is a(n) [arc/sector].

19. The segment with endpoints at the center and on the circle is a [sector/radius].

20. Find the area of the shaded region in circle B to the nearest hundredth.

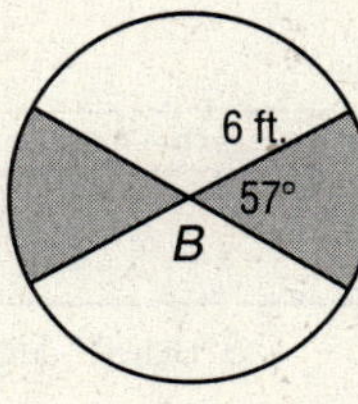

ARE YOU READY FOR THE CHAPTER TEST?

Visit **geomconcepts.com** to access your textbook, more examples, self-check quizzes, and practice tests to help you study the concepts in Chapter 11.

Check the one that applies. Suggestions to help you study are given with each item.

☐ **I completed the review of all or most lessons without using my notes or asking for help.**

- You are probably ready for the Chapter Test.
- You may want to take the Chapter 11 Practice Test on page 491 of your textbook as a final check.

☐ **I used my Foldable or Study Notebook to complete the review of all or most lessons.**

- You should complete the Chapter 11 Study Guide and Review on pages 488–490 of your textbook.
- If you are unsure of any concepts or skills, refer back to the specific lesson(s).
- You may also want to take the Chapter 11 Practice Test on page 491.

☐ **I asked for help from someone else to complete the review of all or most lessons.**

- You should review the examples and concepts in your Study Notebook and Chapter 11 Foldable.
- Then complete the Chapter 11 Study Guide and Review on pages 488–490 of your textbook.
- If you are unsure of any concepts or skills, refer back to the specific lesson(s).
- You may also want to take the Chapter 11 Practice Test on page 491.

Student Signature

Parent/Guardian Signature

Teacher Signature

Surface Area and Volume

Use the instructions below to make a Foldable to help you organize your notes as you study the chapter. You will see Foldable reminders in the margin of this Interactive Study Notebook to help you in taking notes.

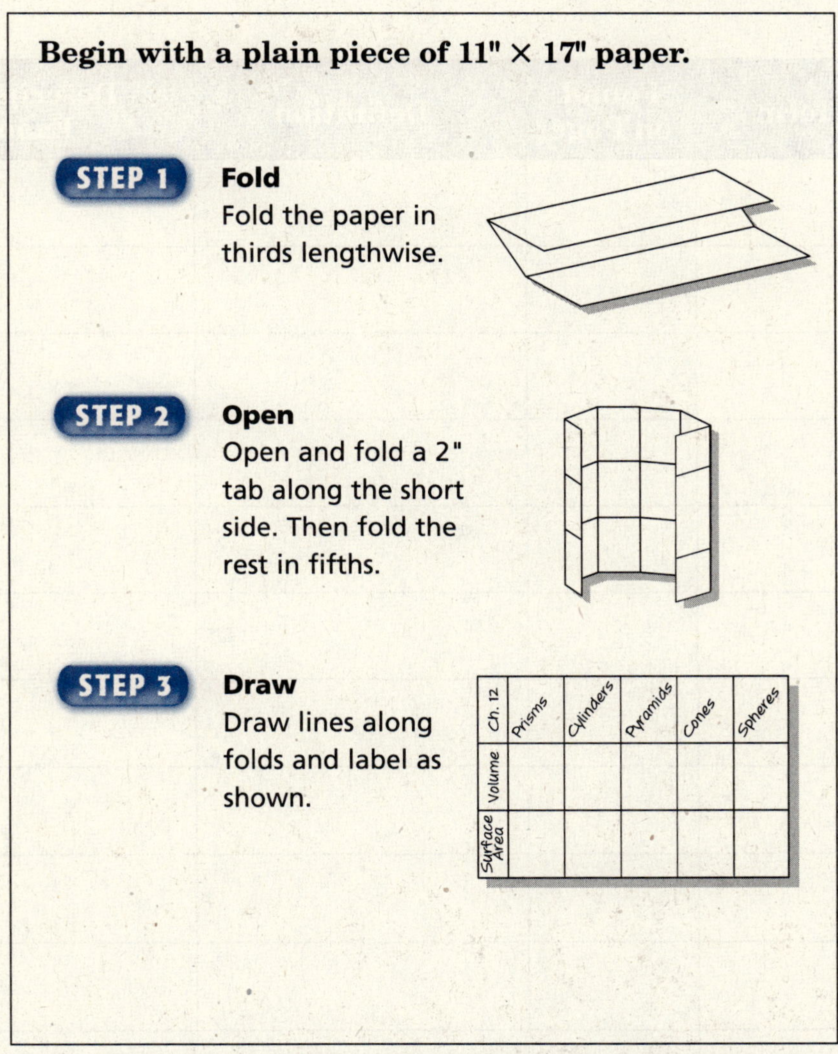

Begin with a plain piece of 11" × 17" paper.

STEP 1 **Fold**
Fold the paper in thirds lengthwise.

STEP 2 **Open**
Open and fold a 2" tab along the short side. Then fold the rest in fifths.

STEP 3 **Draw**
Draw lines along folds and label as shown.

NOTE-TAKING TIP: When taking notes, explain each new idea or concept in words and give one or more examples.

BUILD YOUR VOCABULARY

This is an alphabetical list of new vocabulary terms you will learn in Chapter 12. As you complete the study notes for the chapter, you will see Build Your Vocabulary reminders to complete each term's definition or description on these pages. Remember to add the textbook page number in the second column for reference when you study.

Vocabulary Term	Found on Page	Definition	Description or Example
axis			
composite solid			
cone			
cube			
cylinder [SIL-in-dur]			
edge			
face			
lateral area [LAT-er-ul]			
lateral edge			
lateral face			
net			
oblique cone [oh-BLEEK]			
oblique cylinder			
oblique prism			

Vocabulary Term	Found on Page	Definition	Description or Example
oblique pyramid			
Platonic solid			
polyhedron [pa-lee-HEE-drun]			
prism [PRIZ-um]			
pyramid [PEER-a-MID]			
regular pyramid			
right cone			
right cylinder			
right prism			
right pyramid			
similar solids			
slant height			
solid figures			
sphere [SFEER]			
surface area			
tetrahedron			
volume			

12–1 Solid Figures

What You'll Learn

- Identify solid figures.

Build Your Vocabulary (pages 228–229)

Solid figures enclose a part of space.

Solids with flat surfaces that are ______ are known as **polyhedrons.**

The two-dimensional polygonal surfaces of a polyhedron are its **faces.**

Two faces of a polyhedron ______ in a segment called an **edge**.

A **prism** is a ______ with two faces, called bases, which are formed by congruent polygons that lie in parallel planes.

Faces in a prism that are not bases are parallelograms and are called **lateral faces.**

The intersection of two ______ lateral faces in a prism are called **lateral edges** and are parallel segments.

A **pyramid** is a solid with all faces but one intersecting at a common point called the vertex. The face not intersecting at the vertex is the base. The base of a pyramid is a polygon. The faces meeting at the vertex are lateral faces and are triangles.

Remember It

Euclidean solids are also called solid figures.

Example

1 Name the faces, edges, and vertices of the polyhedron.

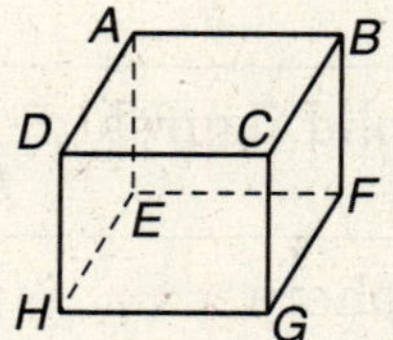

The faces are quadrilaterals *ABCD*, ______, *DCGH*, *ADHE*, *ABFE*, ______.

The edges are ______, $\overline{BC}$, $\overline{CD}$, ______, $\overline{BF}$, $\overline{AE}$, $\overline{DH}$, $\overline{CG}$, $\overline{EF}$, $\overline{FG}$, $\overline{GH}$, $\overline{EH}$.

The vertices are *A*, *B*, ______, *D*, *E*, *F*, ______, *H*.

Your Turn Name the faces, edges, and vertices of the polyhedron.

Y
X
Z
W

WRITE IT

Give three real-world examples of polyhedrons.

______

______

______

______

______

______

BUILD YOUR VOCABULARY (pages 228–229)

A **Platonic solid** is a ______ polyhedron.

A **cube** is a special rectangular prism where all the faces are ______.

A triangular pyramid is known as a **tetrahedron** because all of its faces are ______.

A **cylinder** is a solid that is not a ______. Its bases are two congruent ______ in parallel planes, and its lateral surface is curved.

A **cone** is a solid that is not a ______. Its base is a ______, and the lateral surface is curved.

A **composite solid** is a solid made by ______ two or more solids.

EXAMPLE

2 Is the pyramid in the figure a tetrahedron or a rectangular pyramid?

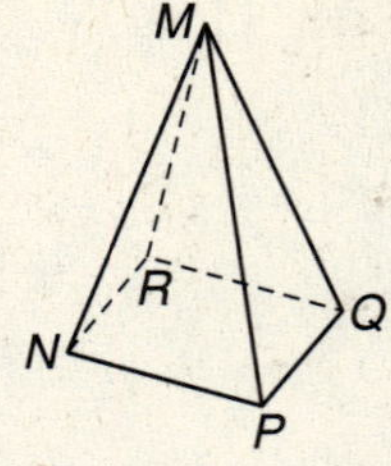

The pyramid has a ________ base and ________ lateral faces. It is a ________ pyramid.

Your Turn Describe the Washington Monument in terms of solid figures.

____

REMEMBER IT

Cylinders and *cones* are terms referring to circular cylinders and circular cones.

HOMEWORK ASSIGNMENT

Page(s):

Exercises:

12–2 Surface Areas of Prisms and Cylinders

What You'll Learn

- Find the lateral areas and surface areas of prisms and cylinders.

BUILD YOUR VOCABULARY (pages 228–229)

In a **right prism**, a lateral edge is also an altitude.

In an **oblique prism**, a lateral edge is *not* an altitude.

The **lateral area** of a solid figure is the ________ of all the areas of its lateral faces.

The **surface area** of a solid figure is the ________ of the areas of all its surfaces.

A **net** is a two-dimensional figure that ________ to form a solid.

Theorem 12-1 Lateral Area of a Prism
If a prism has a lateral area of L square units and a height of h units and each base has a perimeter of P units, then $L = Ph$.

Theorem 12-2 Surface Area of a Prism
If a prism has a surface area of S square units and a height of h units and each base has a perimeter of P units and an area of B square units, then $S = Ph + 2B$.

FOLDABLES™

ORGANIZE IT

In the box for *Surface Area of Prisms*, make a sketch of a prism. Then write the formula for finding the surface area of a prism.

Ch. 12	Prisms	Cylinders	Pyramids	Cones	Spheres
Volume					
Surface Area					

EXAMPLE

1 Find the lateral area and total surface area of a cube with side length 6 inches.

Perimeter of Base

$P = 4s$

$= 4(6)$ or ________

Area of Base

$B = s^2$

$= 6^2$ or ________

Lateral Area

$L = Ph$

$= (24)(6)$ or ________

Surface Area

$S = L + 2B$

$= 144 + 2(36)$

$= 144 + 72$ or ________

The lateral area of the cube is ________ in^2, and the surface area is ________ in^2

WRITE IT

What is the difference between lateral area and surface area?

______

______

______

______

Your Turn Find the lateral area and the surface area of the rectangular prism.

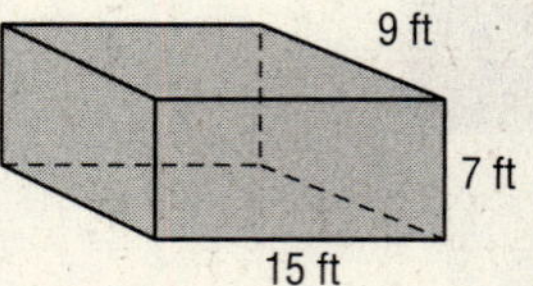

______

EXAMPLE

2 Find the lateral area and the surface area of the triangular prism.

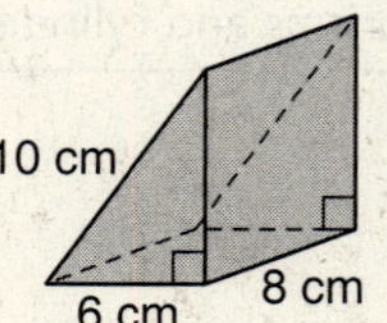

Use the Pythagorean Theorem to find the length of side b.

$c^2 = a^2 + b^2$

$10^2 = 6^2 + b^2$

$100 = 36 + b^2$

$\_\_\_\_ = b^2$

$\sqrt{64} = \sqrt{b^2}$

$\_\_\_\_ = b$

Perimeter of Base

$P = 10 + 6 + b$

$= 10 + 6 + 8$

$= \_\_\_\_$

Area of Base

$B = \frac{1}{2}bh$

$= \frac{1}{2}(6)(8)$

$= \_\_\_\_$

REVIEW IT

What is the length of the hypotenuse of a right triangle with legs 5 cm and 12 cm long? *(Lesson 6-6)*

______

______

______

______

Find the lateral and surface areas.

$L = Ph$

$= (24)(8)$

$= \_\_\_\_ \text{ cm}^2$

$S = L + 2B$

$= 192 + 2(24)$

$= 192 + 48$

$= \_\_\_\_ \text{ cm}^2$

Your Turn Find the lateral area and the surface area of the triangular prism.

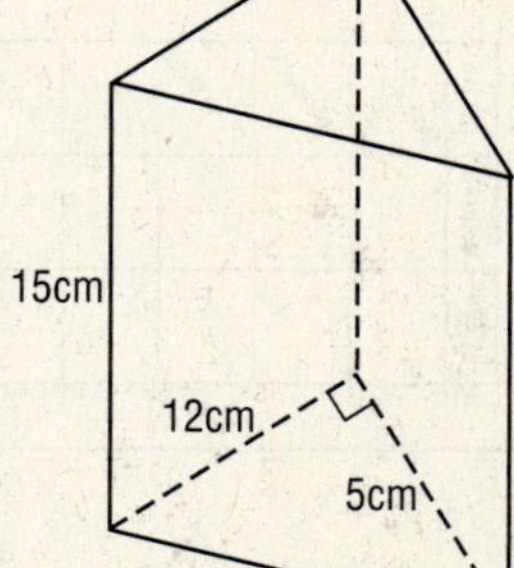

______

BUILD YOUR VOCABULARY (pages 228–229)

The **axis** of a cylinder is the segment whose ______ are centers of the circular bases.

In a **right cylinder**, the axis is also an ______.

In an **oblique cylinder**, the axis is *not* an altitude.

FOLDABLES™

ORGANIZE IT

In the box for *Surface Area of Cylinders*, make a sketch of a cylinder. Then write the formula for finding the surface area of a cylinder.

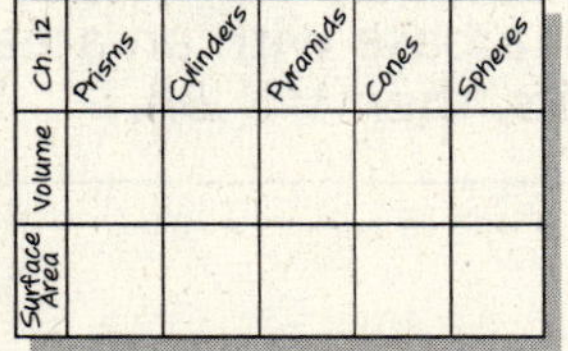

Theorem 12-3 Lateral Area of a Cylinder
If a cylinder has a lateral area of L square units and a height of h units and the bases have radii of r units, then $L = 2\pi rh$.

Theorem 12-4 Surface Area of a Cylinder
If a cylinder has a surface area of S square units and a height of h units and the bases have radii of r units, then $S = 2\pi rh + 2\pi r^2$.

EXAMPLE

3 **Find the lateral area and surface area of the cylinder to the nearest hundredth.**

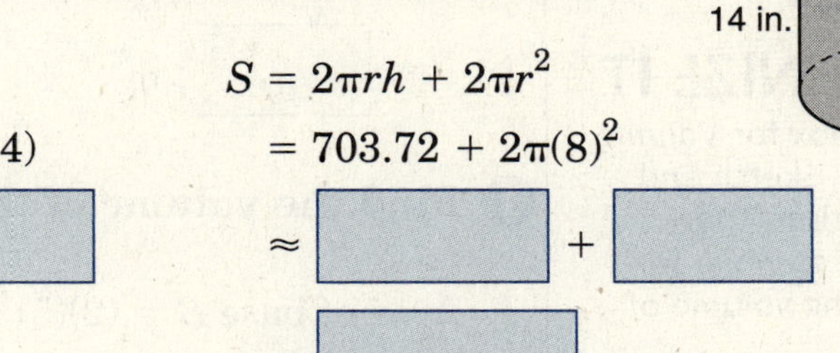

$L = 2\pi rh$
$= 2\pi(8)(14)$
$\approx$ ______

$S = 2\pi rh + 2\pi r^2$
$= 703.72 + 2\pi(8)^2$
$\approx$ ______ + ______
$\approx$ ______

The lateral area is about ______ in^2, and the surface area is about ______ in^2.

Your Turn Find the lateral area and surface area of the cylinder to the nearest hundredth.

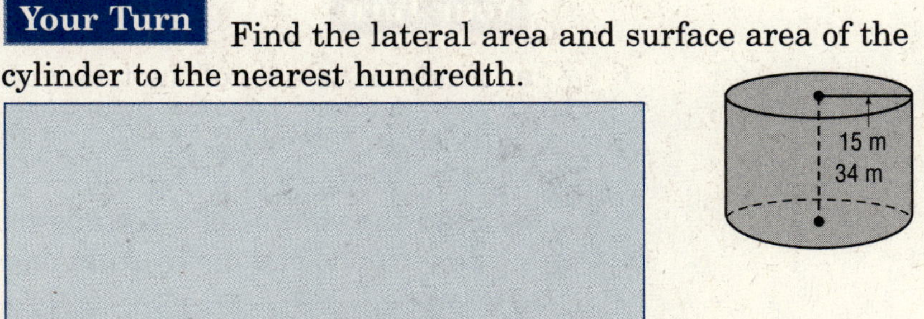

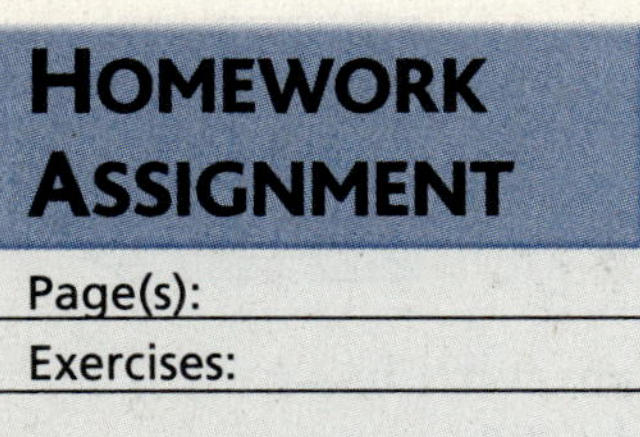

12–3 Volumes of Prisms and Cylinders

What You'll Learn

- Find the volumes of prisms and cylinders.

Build Your Vocabulary (page 229)

Volume measures the space contained within a solid.

Theorem 12-5 Volume of a Prism
If a prism has a volume of V cubic units, a base with an area of B square units, and a height of h units, then $V = Bh$.

EXAMPLES

1 Find the volume of the triangular prism.

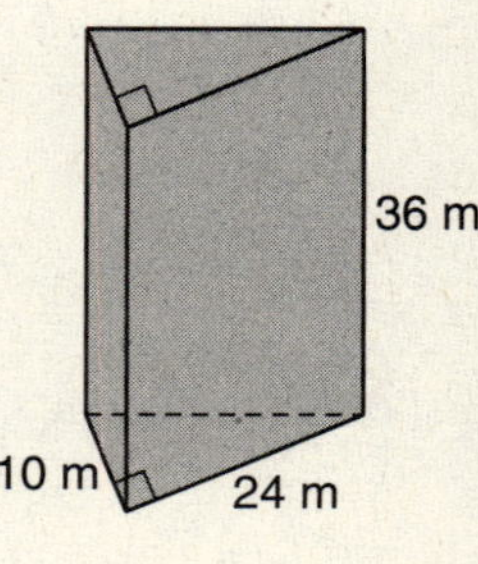

Area of triangular base

$B = \frac{1}{2}(10)(24)$ or ______.

$V = Bh$ Theorem 12-5

$= (\_\_\_\_)(\_\_\_\_)$ 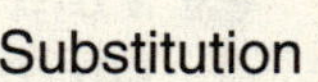

$=$ ______ m^3

2 Find the volume of the rectangular prism.

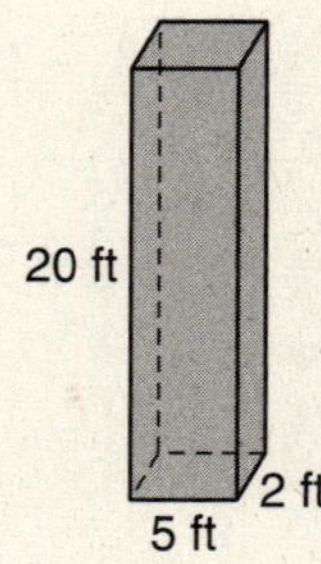

Area of base $B = (2)(5)$ or ______.

$V = Bh$ Theorem 12-5

$= (\_\_\_\_)(\_\_\_\_)$ 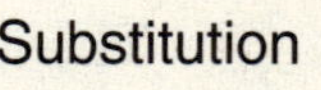

$=$ ______ ft^3

Foldables

Organize It

Use the box for *Volume of Prisms*. Sketch and label a prism. Then write the formula for finding the volume of a prism.

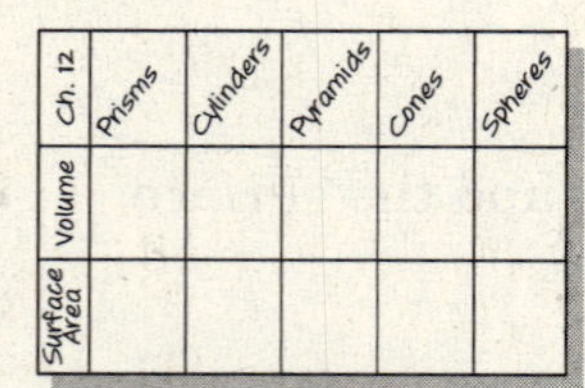

Your Turn

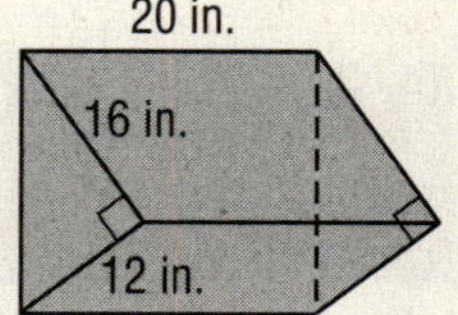

a. Find the volume of the triangular prism.

______

b. Find the volume of a rectangular prism with base dimensions of 8 cm by 9 cm and height 4.1 cm.

______

Theorem 12-6 Volume of a Cylinder
If a cylinder has a volume of V cubic units, a radius of r units, and a height of h units, then $V = \pi r^2 h$.

EXAMPLE

FOLDABLES

ORGANIZE IT

Use the box for *Volume of Cylinders*. Sketch and label a cylinder. Then write the formula for finding the volume of a cylinder.

Ch. 12	Prisms	Cylinders	Pyramids	Cones	Spheres
Volume					
Surface Area					

3 Find the volume of the cylinder to the nearest hundredth.

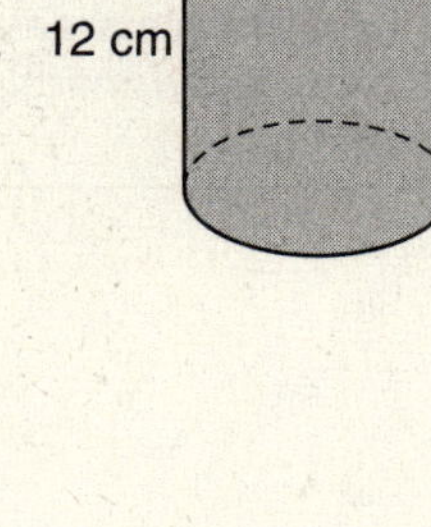

$V = \pi r^2 h$ Theorem 12-6

$= \pi(5)^2(12)$ Substitution

$= 300\pi$

$\approx$ ______ cm^3

Your Turn Find the volume of the cylinder to the nearest hundredth.

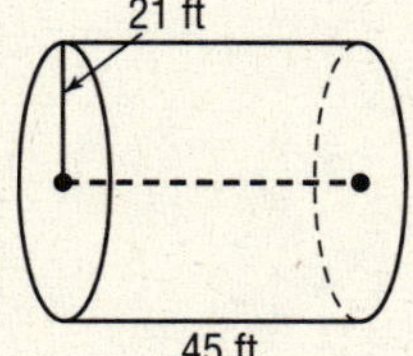

EXAMPLE

4 Leticia is making a sand sculpture by filling a glass tube with layers of different-colored sand. The tube is 24 inches high and 1 inch in diameter. How many cubic inches of sand will Leticia use to fill the tube?

$V = \pi r^2 h$ Theorem 12-6

$= \pi(0.5)^2(24)$ Substitution

$= (0.25)(24)\pi$

$= 6\pi$

$\approx$ ______

Leticia will need about ______ in^3 of sand.

HOMEWORK ASSIGNMENT

Page(s):

Exercises:

Your Turn Sam fills the cylindrical coffee grind containers. One bag has 32π cubic inches of grinds. How many cylindrical containers can Sam fill with two bags of grinds if each cylinder is 4 inches wide and 4 inches high?

12–4 Surface Areas of Pyramids and Cones

WHAT YOU'LL LEARN

- Find the lateral areas and surface areas of regular pyramids and cones.

BUILD YOUR VOCABULARY (page 229)

In a **right pyramid** or a **right cone**, the ______ is perpendicular to the base at its center.

In a **oblique pyramid** or a **oblique cone**, the altitude is ______ to the base at a point other than its center.

A pyramid is a **regular pyramid** if and only if it is a

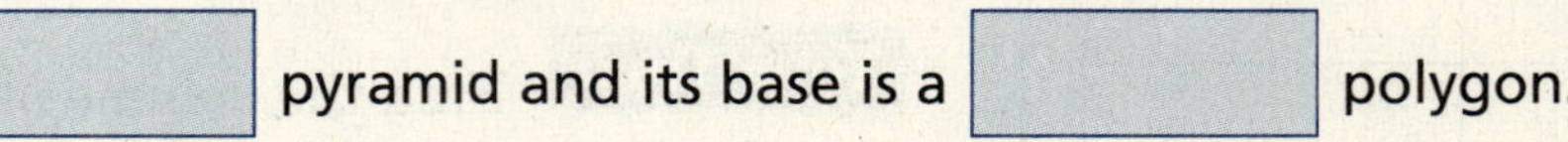

______ pyramid and its base is a ______ polygon.

The height of each ______ face of a regular pyramid is called the **slant height** of the pyramid.

FOLDABLES™

ORGANIZE IT

Use the box for *Surface Area of Pyramids*. Sketch and label a pyramid. Then write the formula for finding the surface area of a pyramid.

Ch. 12	Prisms	Cylinders	Pyramids	Cones	Spheres
Volume					
Surface Area					

Theorem 12-7 Lateral Area of a Regular Pyramid
If a regular pyramid has a lateral area of L square units, a base with a perimeter of P units, and a slant height of ℓ units, then $L = \frac{1}{2}P\ell$.

Theorem 12-8 Surface Area of a Regular Pyramid
If a regular pyramid has a surface area of S square units, a slant height of ℓ units, and a base with perimeter of P units and an area of B square units, then $S = \frac{1}{2}P\ell + B$.

EXAMPLE

1 Find the lateral area and the surface area of the square pyramid.

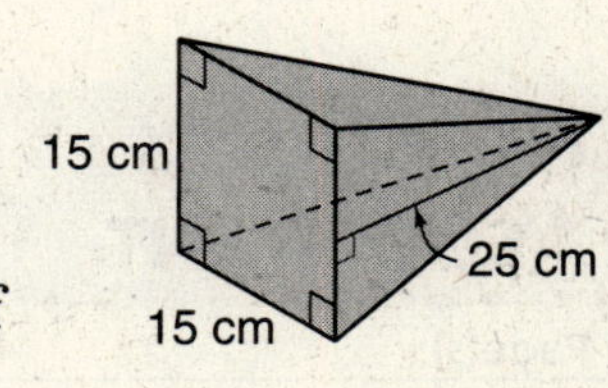

First, find the perimeter and the area of the base.

$P = 4s$
$= 4(15)$ or ______

$B = s^2$
$= 15^2$ or ______

Lateral Area	Surface Area
$L = \frac{1}{2}P\ell$	$S = L + B$
$= \frac{1}{2}(60)(25)$	$= 750 + 225$
$=$ ☐ cm^2	$=$ ☐ cm^2

Your Turn Find the lateral area and surface area of the square pyramid.

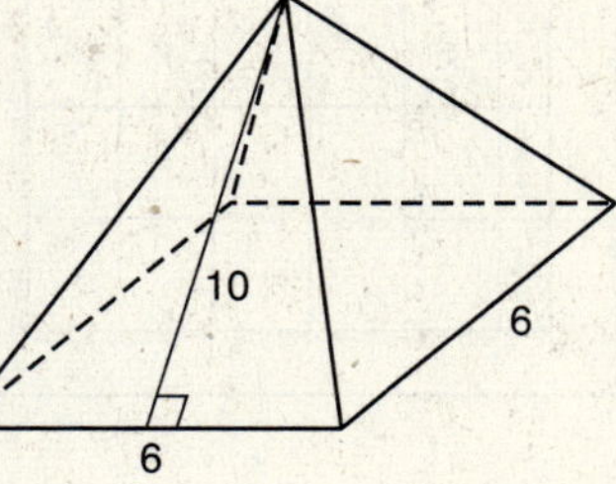

EXAMPLE

2 **Find the lateral area and the surface area of a regular triangular pyramid with a base perimeter of 24 inches, a base area of 27.7 square inches, and a slant height of 8 inches.**

$L = \frac{1}{2}P\ell$	Theorem 12-7
$= \frac{1}{2}(24)(8)$	Substitution
$=$ ☐ in^2	
$S = L + B$	Theorem 12-7
$= 96 + 27.7$	Substitution
$=$ ☐ in^2	

Your Turn Find the lateral area and the surface area of a regular triangular pyramid with a base perimeter of 18 inches, a base area of 15.6 square inches, and a slant height of 11 inches.

FOLDABLES™

ORGANIZE IT

Use the box for *Surface Area of Cones*. Sketch and label a cone. Then write the formula for finding the surface area of a cone.

Ch. 12	Prisms	Cylinders	Pyramids	Cones	Spheres
Volume					
Surface Area					

Theorem 12-9 Lateral Area of a Cone
If a cone has a lateral area of L square units, a slant height of ℓ units, and a base with a radius of r units, then $L = \frac{1}{2}(2\pi r\ell)$ or $\pi r\ell$.

Theorem 12-10 Surface Area of a Cone
If a cone has a surface area of S square units, a slant height of ℓ units, and a base with a radius of r units, then $S = \pi r\ell + \pi r^2$.

EXAMPLE

3 **Find the lateral area and the surface area of the cone to the nearest hundredth.**

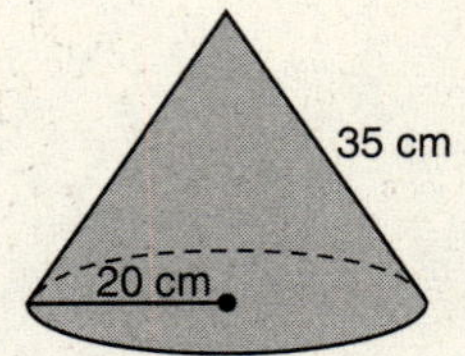

$L = \pi r\ell$	Theorem 12-9
$= \pi(20)(35)$	Substitution
$= 700\pi$	
$\approx$  cm^2	

$S = \pi r\ell + \pi r^2$	Theorem 12-10
$= \pi(\;\;\;)(35) + \pi(20)^2$	Substitution
$= 2199.11 + 400\pi$	
$\approx 2199.11 + 1256.64$	
$\approx$ 	

The lateral area is about ______ cm^2, and the surface area is about ______ cm^2.

Your Turn Find the lateral area and the surface area of the cone to the nearest hundredth.

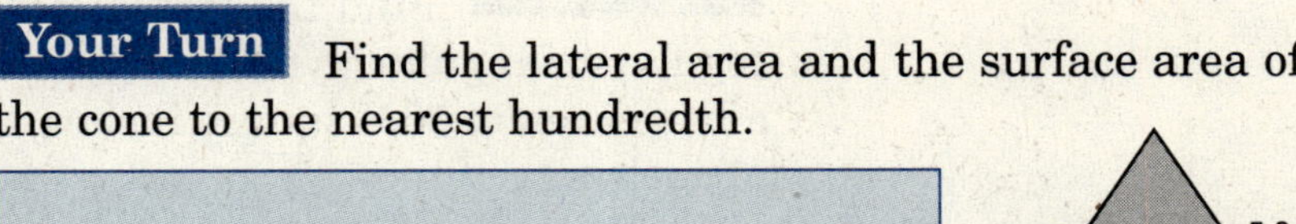

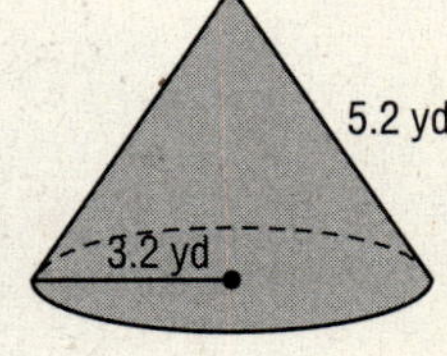

HOMEWORK ASSIGNMENT

Page(s):

Exercises:

12–5 Volumes of Pyramids and Cones

WHAT YOU'LL LEARN

- Find the volumes of pyramids and cones.

Theorem 12-11 Volume of a Pyramid
If a pyramid has a volume of V cubic units and a height of h units and the area of the base is B square units, then $V = \frac{1}{3}Bh$.

EXAMPLES

1 Find the volume of the rectangular pyramid.

12 cm
4 cm
10 cm

$B = \ell w$

$= (10)(4)$ or ______

$V = \frac{1}{3}Bh$ Theorem 12-11

$= \frac{1}{3}(40)(12)$ Substitution

$=$  cm^3

FOLDABLES™

ORGANIZE IT

Use the boxes for *Volume of Pyramids* and *Cones*. Sketch and label a pyramid and a cone. Then write the formula for finding the volumes of a pyramid and a cone.

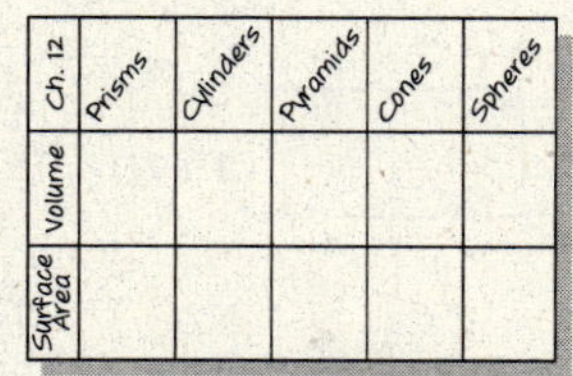

2 Find the volume of the cone to the nearest hundredth.

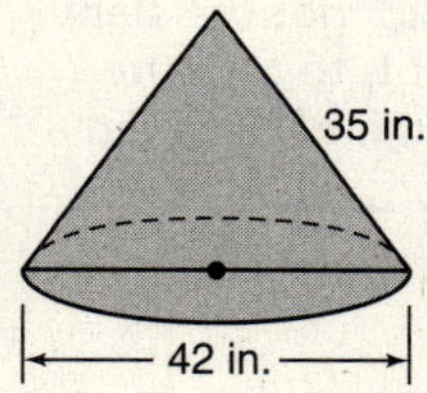

Find the height h

$h^2 + 21^2 = 35^2$

$h^2 + 441 = 1225$

$h^2 = 784$

$\sqrt{h^2} = \sqrt{784}$

$h =$ ______

$V = \frac{1}{3}\pi r^2 h$ Theorem 12-11

$= \frac{1}{3}\pi(21)^2(28)$ Substitution

$\approx$ ______ in^3

Theorem 12-12 Volume of a Cone
If a cone has a volume of V cubic units, a radius of r units, and a height of h units, then $V = \frac{1}{3}\pi r^2 h$.

Your Turn

a. Find the volume of the triangular pyramid.

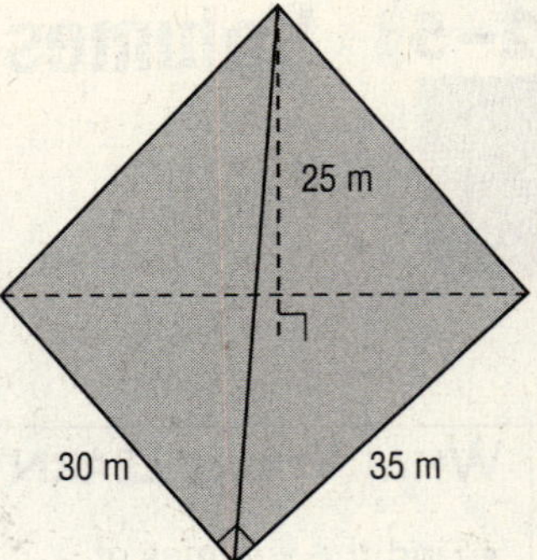

b. Find the volume of the cone to the nearest hundredth.

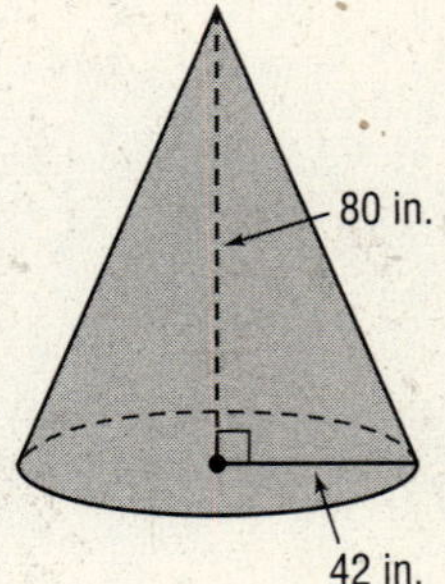

EXAMPLE

3 **The sand in a cone with radius 3 cm and height 10 cm is poured into a square prism with height of 29.5 cm and base area of 4 cm². How far up the side of the prism will the sand reach when leveled?**

REMEMBER IT
Use the altitude of a solid, not the slant height, to find the volume of the solid.

Volume of Cone

$V = \frac{1}{3}\pi r^2 h$

$= \frac{1}{3}\pi(3)^2(10)$

$= 30\pi$

$\approx$ ______

Volume of Prism

$V = Bh$

$94.25 = 4h$

$h \approx$ ______

The sand will level off at a height of about ______ cm in the prism.

Your Turn The salt in a cone with radius 6 cm and height 8 cm is poured into a square prism with height of 20 cm and base area of 12 cm². Will the prism be able to hold all of the salt?

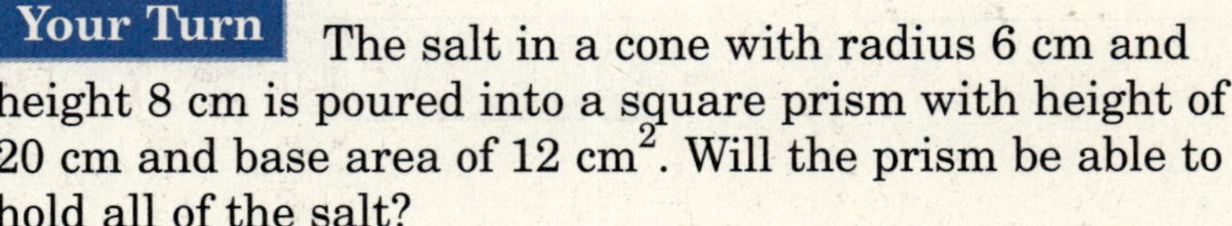

HOMEWORK ASSIGNMENT
Page(s):
Exercises:

12–6 Spheres

WHAT YOU'LL LEARN

- Find the surface areas and volumes of spheres.

BUILD YOUR VOCABULARY (page 229)

A **sphere** is the set of all points that are a fixed ______ from a given ______ called the center.

Theorem 12-13 Surface Area of a Sphere
If a sphere has a surface area of S square units and a radius of r units, then $S = 4\pi r^2$.

Theorem 12-14 Volume of a Sphere
If a sphere has a volume of V cubic units and a radius of r units, then $V = \left(\frac{4}{3}\right)\pi r^3$.

FOLDABLES

ORGANIZE IT

Use the boxes for *Volume* and *Surface Area of Spheres.* Sketch and label a sphere. Then write the formulas for finding the surface area and volume of a sphere.

Ch. 12	Prisms	Cylinders	Pyramids	Cones	Spheres
Volume					
Surface Area					

EXAMPLE

1 Find the surface area and volume of a sphere with radius 5 cm.

<u>Surface Area</u>

$S = 4\pi r^2$

$= 4\pi(5)^2$

$= 100\pi$

$\approx$ ______ cm^2

<u>Volume</u>

$V = \left(\frac{4}{3}\right)\pi r^3$

$= \left(\frac{4}{3}\right)\pi(5)^3$

$= \left(\frac{500}{3}\right)\pi$

$\approx$ ______ cm^3

Your Turn Find the surface area and volume of a sphere with diameter 15 in.

EXAMPLE

2 **Some students build a snow sculpture from a cylinder and a sphere of snow. Both the sphere and the cylinder have a radius of 1 ft. and the height of the cylinder is 4 ft. Find the volume of the snow used to build the sculpture.**

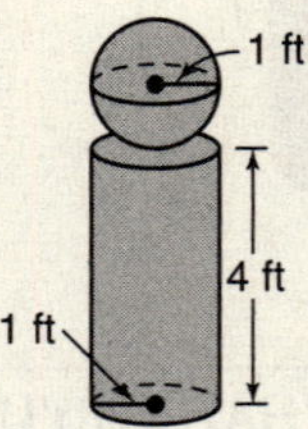

Volume of Cylinder

$V = \pi r^2 h$

$= \pi(1)^2(4)$

$= 4\pi$

$\approx$ _____

Volume of Sphere

$V = \left(\frac{4}{3}\right)\pi r^3$

$= \left(\frac{4}{3}\right)\pi(1)^3$

$= \left(\frac{4}{3}\right)\pi$

$\approx$ _____

The volume of the snow used for the sculpture is about

12.57 + 4.19, or _____ ft^3.

Your Turn Felix and Brenda want to share an ice cream cone. Brenda wants half the scoop of ice cream on top, while Felix wants the ice cream inside the cone. Assuming the half scoop of ice cream on top is a perfect sphere, who will have more ice cream? The cone and scoop both have radii of 1.5 inch; the cone is 3.25 inches long.

HOMEWORK ASSIGNMENT

Page(s):

Exercises:

12–7 Similarity of Solid Figures

What You'll Learn

- Identify and use the relationship between similar solid figures.

BUILD YOUR VOCABULARY (page 229)

Similar solids are solids that have the same ______ but not necessarily the same ______.

Key Concept

Characteristics of Similar Solids For similar solids, the corresponding lengths are proportional, and the corresponding faces are similar.

EXAMPLE

1 **Determine whether the pair of solids is similar.**

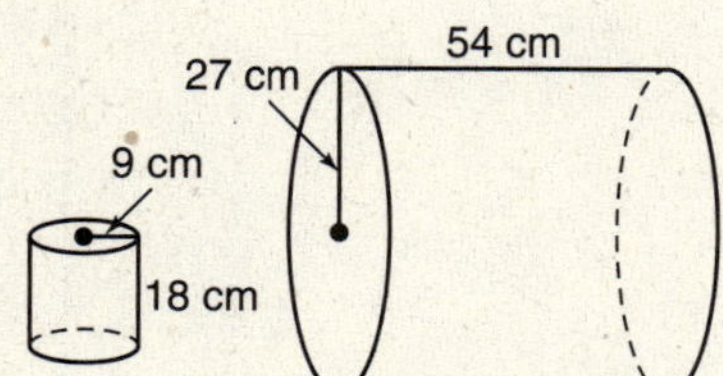

$\frac{9}{27} \stackrel{?}{=} \frac{18}{54}$ Definition of similarity

$(9)(54) \stackrel{?}{=} (27)(18)$ Cross products

$486 = 486$

The corresponding lengths are in , so the solids  similar.

Your Turn **Determine whether each pair of solids is similar.**

a.

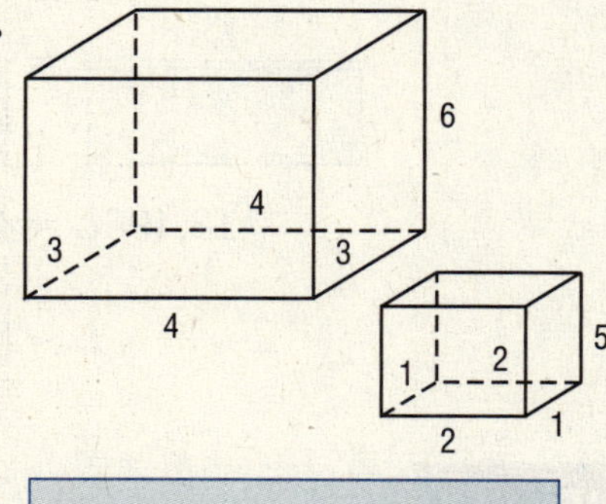

b.

8, 12, 12

12, 18, 18

Remember It

A scale factor is a one-dimensional measure. Surface area is a two-dimensional measure. Volume is a three-dimensional measure.

Theorem 12-15
If two solids are similar with a scale factor of $a{:}b$, then the surface areas have a ratio of $a^2{:}b^2$ and the volumes have a ratio of $a^3{:}b^3$.

EXAMPLE

2 **For the similar prisms, find the scale factor of the prism on the left to the prism on the right. Then find the ratios of the surface areas and the volumes.**

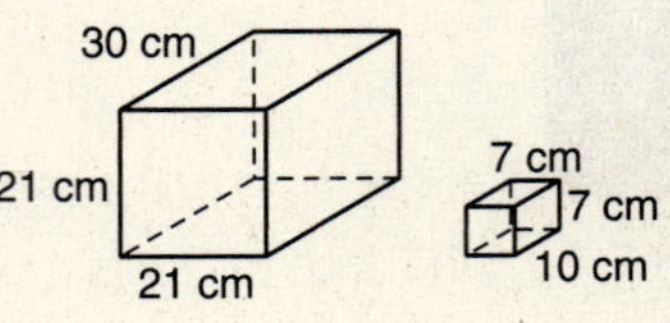

The scale factor is $\frac{21}{7} = \frac{30}{10}$ or ______.

The ratio of the surface areas is $\frac{3^2}{1^2}$ or ______.

The ratio of the volumes is $\frac{3^3}{1^3}$ or ______.

Your Turn Find the scale factor of the prism on the left to the prism on the right. Then find the ratios of the surface areas and the volumes.

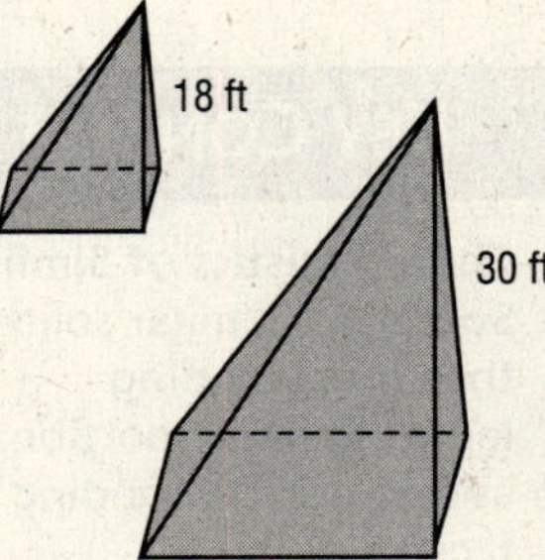

______

EXAMPLE

3 **Sara made a scale model of the Great American Pyramid in Memphis, Tennessee, which has a base side length of 544 ft and a lateral area of 456,960 ft^2. If the scale factor of the model to the original is 1:136, what will be the lateral area of the model?**

$$\frac{\text{surface area of the model}}{\text{surface area of Great Amer. Pyr.}} = \frac{1^2}{136^2}$$

$$\frac{L}{\_\_\_\_} = \frac{1}{\_\_\_\_}$$

$$18{,}496L = 456{,}960$$

$$L \approx \_\_\_\_ \text{ ft}^2$$

Your Turn A scale model of a house is made using a scale factor of $\frac{1}{112}$. What fraction of the actual house material would would the dollhouse need to cover all of its floors?

______

HOMEWORK ASSIGNMENT

Page(s):

Exercises:

CHAPTER 12

BRINGING IT ALL TOGETHER

STUDY GUIDE

FOLDABLES™	VOCABULARY PUZZLEMAKER	BUILD YOUR VOCABULARY
Use your **Chapter 12 Foldable** to help you study for your chapter test.	To make a crossword puzzle, word search, or jumble puzzle of the vocabulary words in Chapter 12, go to: www.glencoe.com/sec/math/t_resources/free/index.php	You can use your completed **Vocabulary Builder** (pages 228–229) to help you solve the puzzle.

12-1 Solid Figures

Complete each sentence.

1. Two faces of a polyhedron intersect at a(n) ______.

2. A triangular pyramid is called a ______.

3. A ______ is a figure that encloses a part of space.

4. Three faces of a polyhedron intersect at a point called a(n)

.

12-2 Surface Areas of Prisms and Cylinders

Find the lateral area and surface area of each solid to the nearest hundredth.

5. a regular pentagonal prism with apothem $a = 4$, side length $s = 6$, and height $h = 12$

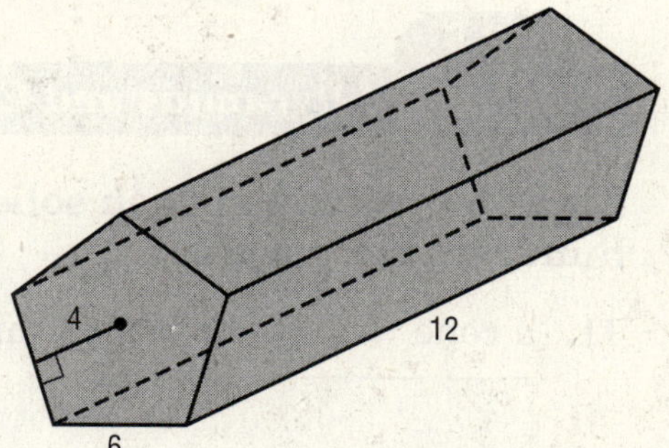

a. $L =$ ______

b. $S =$ ______

6. a cylinder with radius $r = 42$ and height $h = 10$

a. $L \approx$ ______ **b.** $S \approx$ ______

12-3

Volumes of Prisms and Cylinders

Find the volume of each solid and round to the nearest hundredth.

7. the regular pentagonal prism from Exercise #5

8. How much water will a 24 in. by 15 in. by 10 in. fish tank hold?

12-4

Surface Areas of Pyramids and Cones

Find the lateral and surface areas of each solid. Round to the nearest hundredth if necessary.

9. a rectangular pyramid with base dimensions 2 ft by 3 ft and lateral height $h = 1$ ft

a. $L \approx$

b. $S \approx$

10. a cone with diameter 3.6 cm and lateral height 2.4 cm

a. $L \approx$

b. $S \approx$

12-5

Volumes of Pyramids and Cones

Find the volume of each solid rounded to the nearest hundredth, if necessary.

11. a cone with its height as three times the radius

12. the cone in Exercise #10

12-6 Spheres

Complete the sentence.

13. The set of all points a given distance from the center is a ______.

A beach ball will have a diameter of 30 in.

14. How much material will be used to make the beach ball?

15. How much air will be needed to fill it?

12-7 Similarity of Solid Figures

16. Solids having the same shape but not always the same size are ______.

If the radius of a sphere is doubled:

17. How does the surface area change?

18. How does the volume change?

The diameter of the moon is about 2160 miles. The diameter of the Earth is about 7900 miles.

19. Assuming both are spheres, what is the scale factor of the Earth to the moon?

20. Are they similar solid figures?

ARE YOU READY FOR THE CHAPTER TEST?

Checklist

Visit **geomconcepts.com** to access your textbook, more examples, self-check quizzes, and practice tests to help you study the concepts in Chapter 12.

Check the one that applies. Suggestions to help you study are given with each item.

☐ **I completed the review of all or most lessons without using my notes or asking for help.**

- You are probably ready for the Chapter Test.
- You may want to take the Chapter 12 Practice Test on page 543 of your textbook as a final check.

☐ **I used my Foldable or Study Notebook to complete the review of all or most lessons.**

- You should complete the Chapter 12 Study Guide and Review on pages 540–542 of your textbook.
- If you are unsure of any concepts or skills, refer back to the specific lesson(s).
- You may also want to take the Chapter 12 Practice Test on page 543.

☐ **I asked for help from someone else to complete the review of all or most lessons.**

- You should review the examples and concepts in your Study Notebook and Chapter 12 Foldable.
- Then complete the Chapter 12 Study Guide and Review on pages 540–542 of your textbook.
- If you are unsure of any concepts or skills, refer back to the specific lesson(s).
- You may also want to take the Chapter 12 Practice Test on page 543.

Student Signature

Parent/Guardian Signature

Teacher Signature

Right Triangles and Trigonometry

Use the instructions below to make a Foldable to help you organize your notes as you study the chapter. You will see Foldable reminders in the margin of this Interactive Study Notebook to help you in taking notes.

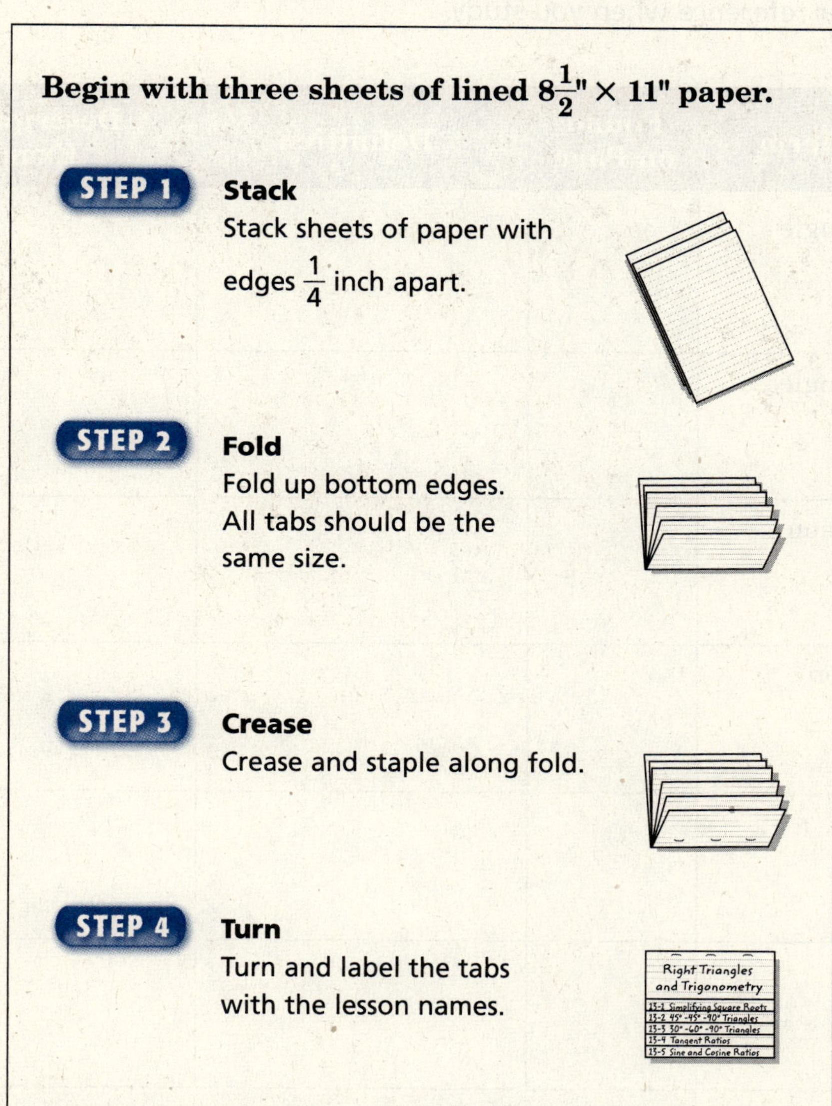

Begin with three sheets of lined $8\frac{1}{2}" \times 11"$ paper.

STEP 1 **Stack**
Stack sheets of paper with edges $\frac{1}{4}$ inch apart.

STEP 2 **Fold**
Fold up bottom edges. All tabs should be the same size.

STEP 3 **Crease**
Crease and staple along fold.

STEP 4 **Turn**
Turn and label the tabs with the lesson names.

NOTE-TAKING TIP: When taking notes, it is often helpful to remember what you've learned if you can paraphrase or summarize key terms and concepts in your own words.

BUILD YOUR VOCABULARY

This is an alphabetical list of new vocabulary terms you will learn in Chapter 13. As you complete the study notes for the chapter, you will see Build Your Vocabulary reminders to complete each term's definition or description on these pages. Remember to add the textbook page number in the second column for reference when you study.

Vocabulary Term	Found on Page	Definition	Description or Example
30°-60°-90° triangle			
45°-45°-90° triangle			
angle of depression			
angle of elevation			
cosine			
hypsometer			
perfect square			
radical expression [RAD-ik-ul]			

Vocabulary Term	Found on Page	Definition	Description or Example
radical sign			
radicand [RAD-i-KAND]			
simplest form			
sine			
square root			
tangent [TAN-junt]			
trigonometric identity [TRIG-guh-no-MET-rik]			
trigonometric ratio			
trigonometry			

13–1 Simplifying Square Roots

What You'll Learn

- Multiply, divide, and simplify radical expressions.

Build Your Vocabulary (pages 252–253)

Perfect squares are products of two ________ factors, or when a number multiplies itself.

The **square root**, therefore, is one of ________ equal factors.

A ________ number has both positive (+) and negative (−) square roots, indicated by the **radical sign** $\sqrt{}$.

A **radical expression** is an expression that contains a ________.

The number under the radical sign $\sqrt{}$ is the **radicand**.

Write It

What are the next three perfect squares after 16?

____

____

____

Examples

Simplify each expression.

1 $\sqrt{36}$

$\sqrt{36} =$ ________, because $6^2 = 36$.

2 $\sqrt{81}$

$\sqrt{81} =$ ________, because $9^2 = 81$.

3 $\sqrt{24}$

$\sqrt{24} = \sqrt{2 \cdot 2 \cdot 2 \cdot 3}$ Prime factorization

$= \sqrt{2 \cdot 2} \cdot \sqrt{2 \cdot 3}$ Product Property of Square Roots

$= 2 \cdot \sqrt{6}$ $\sqrt{2 \cdot 2} = 2$

$=$ ________

Key Concept

Product Property of Square Roots The square root of a product is equal to the product of each square root.

KEY CONCEPT

Quotient Property of Square Roots The square root of a quotient is equal to the quotient of each square root.

FOLDABLES On the tab for Lesson 13-1, write the names of the two properties introduced in this lesson. Then write your own example of each property on the back of the tab.

4 $\sqrt{6} \cdot \sqrt{30}$

$\sqrt{6} \cdot \sqrt{30} = \sqrt{6} \cdot$ ______ Prime factorization

$= \sqrt{6 \cdot 6 \cdot 5}$ Product Property of Square Roots

$=$ ______ $\cdot \sqrt{5}$ Product Property of Square Roots

$=$ ______ $\sqrt{6 \cdot 6} = 6$

Your Turn **Simplify each expression.**

a. $\sqrt{25}$

b. $\sqrt{121}$

c. $\sqrt{18}$

d. $\sqrt{3} \cdot \sqrt{12}$

EXAMPLES

Simplify each expression.

5 $\dfrac{\sqrt{16}}{\sqrt{8}}$

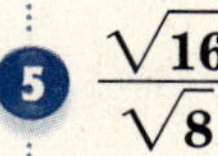

$\dfrac{\sqrt{16}}{\sqrt{8}} = \sqrt{\dfrac{16}{8}}$ Quotient Property

$=$ ______

6 $\sqrt{\dfrac{121}{49}}$

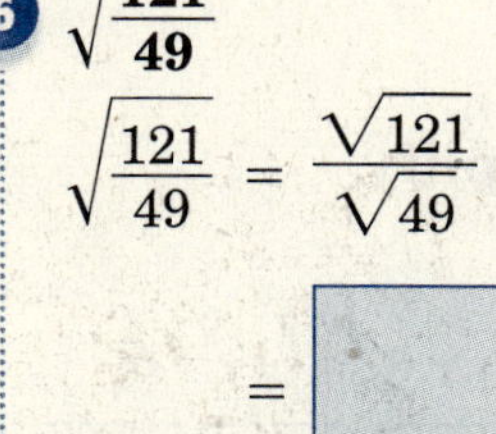

$\sqrt{\dfrac{121}{49}} = \dfrac{\sqrt{121}}{\sqrt{49}}$ Quotient Property

$=$ ______

REMEMBER IT

Simplifying a fraction with a radical in the denominator is called *rationalizing the denominator.*

Your Turn **Simplify each expression.**

a. $\dfrac{\sqrt{20}}{\sqrt{4}}$

b. $\dfrac{\sqrt{144}}{\sqrt{25}}$

EXAMPLES

7 **Simplify $\frac{\sqrt{10}}{\sqrt{7}}$.**

$$\frac{\sqrt{10}}{\sqrt{7}} = \frac{\sqrt{10}}{\sqrt{7}} \cdot \frac{\sqrt{7}}{\sqrt{7}}$$ $\frac{\sqrt{7}}{\sqrt{7}} = 1$

$$= \frac{\sqrt{10 \cdot 7}}{\sqrt{7 \cdot 7}}$$ Product Property of Square Roots

$$= \frac{\sqrt{70}}{7}$$ $\sqrt{7 \cdot 7} = 7$

We used the Identity Property and the Product Property of Square Roots to simplify the above radical expression. The denominator does not have a radical sign.

KEY CONCEPT

Rules for Simplifying Radical Expressions

1. There are no perfect square factors other than 1 in the radicand.
2. The radicand is not a fraction.
3. The denominator does not contain a radical expression.

8 **Simplify $\frac{16}{\sqrt{6}}$.**

$$\frac{16}{\sqrt{6}} = \frac{16}{\sqrt{6}} \cdot \square$$ $\square = 1$

$$= \frac{16 \cdot \sqrt{6}}{\sqrt{6} \cdot \sqrt{6}}$$ Product Property of Square Roots

$$= \frac{16\sqrt{6}}{\square}$$ $\sqrt{6 \cdot 6} = 6$

$$= \frac{16\sqrt{6}}{6} \text{ or } \square$$

We used the Identity Property and the Product Property of Square Roots to simplify the above expression and eliminate the radical in the denominator.

Your Turn **Simplify.**

a. $\frac{\sqrt{7}}{\sqrt{2}}$

HOMEWORK ASSIGNMENT

Page(s):

Exercises:

b. $\frac{4}{\sqrt{5}}$

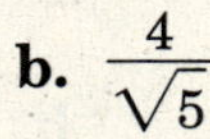

13–2 45°-45°-90° Triangles

What You'll Learn

- Use the properties of 45°-45°-90° triangles.

Build Your Vocabulary (page 252)

The ______ of a square separates the square into two **45°-45°-90° triangles**.

Theorem 13-1 45°-45°-90° Triangle Theorem
In a 45°-45°-90° triangle, the hypotenuse is $\sqrt{2}$ times the length of a leg.

Write It

Describe two different ways to find the length of the hypotenuse of a 45°-45°-90° triangle.

______

______

______

______

EXAMPLE

1 In a scale model of a town, a baseball diamond has sides 36 inches long. What is the distance from first base to third base on the model? Round to the nearest tenth.

$h = s\sqrt{2}$ Theorem 13-1

$= \square\sqrt{2}$ Substitution

$\approx \square$

The distance from first to third base on the scale model is about ______ inches.

Your Turn Find the length of the diagonal of a square whose side measures 22 inches.

EXAMPLE

Remember It

A 45°-45°-90° triangle is isosceles, so the legs are always congruent.

2 **If $\triangle DJT$ is an isosceles right triangle and the measure of the hypotenuse is $\sqrt{200}$, find the measure of either leg.**

$h = s\sqrt{2}$ Theorem 13-1

$\_\_\_\_ = s\sqrt{2}$ Substitution

$\_\_\_\_ = s$

The length of each leg measures ____.

Foldables™ Organize It

On the tab for Lesson 13-2, draw a 45°-45°-90° triangle and label its parts. Then write your own real-world example that can be solved using this information.

Right Triangles and Trigonometry
13-1 Simplifying Square Roots
13-2 45°-45°-90° Triangles
13-3 30°-60°-90° Triangles
13-4 Tangent Ratios
13-5 Sine and Cosine Ratios

Your Turn If $\triangle XYZ$ is an isosceles right triangle and the measure of the hypotenuse is 25, find the measure of either leg.

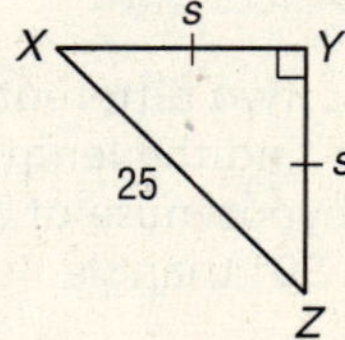

Homework Assignment

Page(s):

Exercises:

13–3 30°-60°-90° Triangles

Build Your Vocabulary (page 252)

The median of an equilateral triangle separates it into two **30°-60°-90°** triangles.

What You'll Learn

- Use the properties of 30°-60°-90° triangles.

Theorem 13-2 30°-60°-90° Triangle Theorem
In a 30°-60°-90° triangle, the hypotenuse is twice the length of the shorter leg, and the longer leg is $\sqrt{3}$ times the length of the shorter leg.

Remember It

The shorter leg is always opposite the 30° angle, and the longer leg is always opposite the 60° angle.

EXAMPLE

1 In $\triangle ABC$, $a = 12$. Find b and c.

$a = b\sqrt{3}$ The longer leg is $\sqrt{3}$ times the length of the shorter leg.

Replace a with ______.

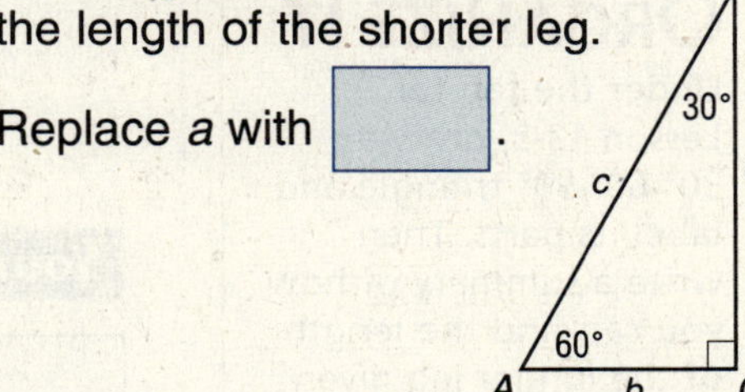

______ $= b\sqrt{3}$

______ $= b$

$c = 2b$ The hypotenuse is twice the shorter leg.

Replace b with ______.

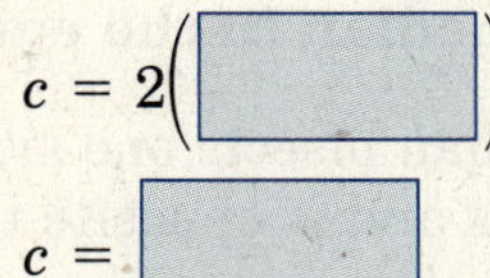

$c = 2($______$)$

$c =$ ______

Your Turn **Refer to Example 1.**

a. If $b = 3.5$, find a and c.

b. If $b = \frac{1}{3}$, find a and c.

EXAMPLE

2 **In $\triangle DEF$, $DE = 18$. Find EF and DF.**

Use Theorem 13–2.

$DE = EF\sqrt{3}$ The longer leg is $\sqrt{3}$ times the shorter leg.

Replace DE with ______.

______ $= EF\sqrt{3}$ Divide each side by $\sqrt{3}$.

______ $= EF$

$DF = 2(EF)$ The hypotenuse is twice the shorter leg.

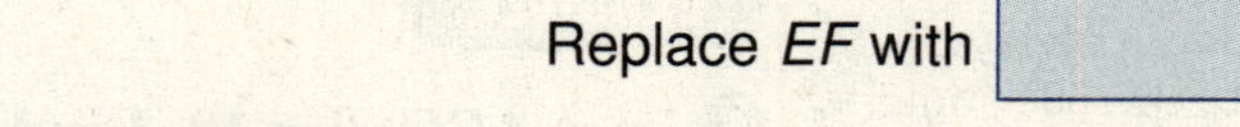

Replace EF with ______.

$DF = 2($______$)$

$DF =$ ______ Associative Property

Your Turn Refer to Example 2. If $DE = 11$, find EF and DF.

EXAMPLE

3 **Find the length, to the nearest tenth, of the median in the equilateral triangle.**

The median bisects one side into two 5-meter segments and is opposite the 60° angle.

$x = 5\sqrt{3}$ Theorem 13-2

$\approx$ ______ meters

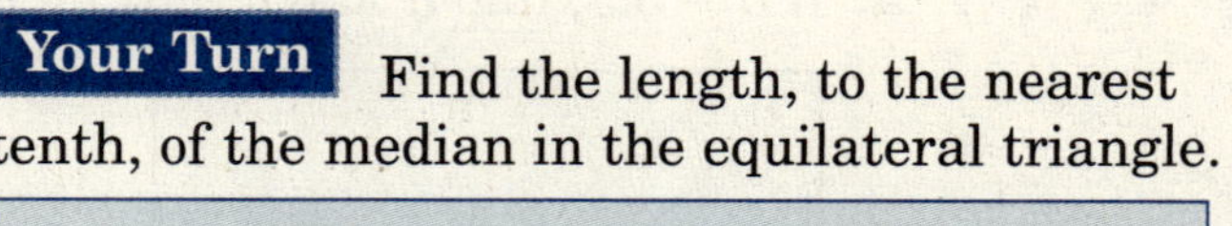

Your Turn Find the length, to the nearest tenth, of the median in the equilateral triangle.

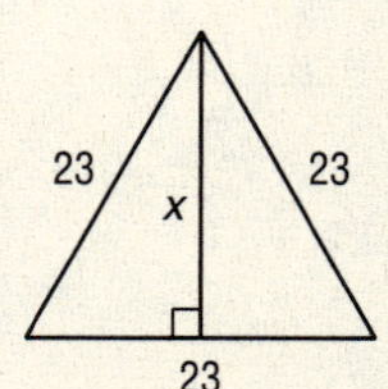

FOLDABLES™

ORGANIZE IT

Under the tab for Lesson 13-3, draw a 30°-60°-90° triangle and label its parts. Then write a summary of how you can find the length of the longer leg given the length of the shorter leg.

HOMEWORK ASSIGNMENT

Page(s):

Exercises:

13–4 Tangent Ratio

What You'll Learn

- Use the tangent ratio to solve problems.

Build Your Vocabulary (page 253)

Trigonometry is the study of the properties of ______.

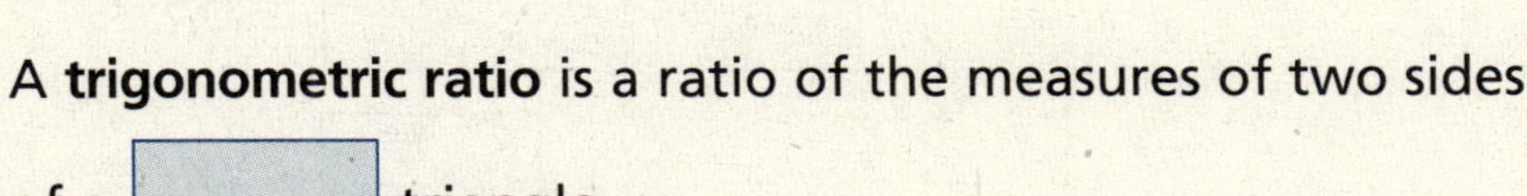

A **trigonometric ratio** is a ratio of the measures of two sides of a ______ triangle.

The **tangent** is the ratio of one ______ to the other.

If A is an acute angle of a right triangle,

$$\tan A = \frac{\text{measure of leg opposite } \angle A}{\text{measure of leg adjacent to } \angle A}.$$

Remember It

The ratio of the measures of the legs of a right triangle can be compared to the ratio of *rise* to *run* in the definition of *slope* of a line.

Example

1 **Find tan *K* and tan *M*.**

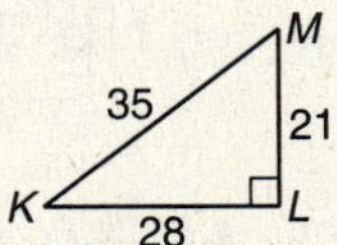

$\tan K = \frac{ML}{KL}$ $\quad \frac{\text{opposite}}{\text{adjacent}}$

$= \frac{21}{28}$ or ______ $\quad$ Substitution

$\tan M = \frac{KL}{ML}$ $\quad \frac{\text{opposite}}{\text{adjacent}}$

$= \frac{28}{21}$ or ______ $\quad$ Substitution

Your Turn Find tan 30°, tan 45°, tan 60°.

EXAMPLES

REMEMBER IT

The symbol tan *B* is read *the tangent of angle B*.

2 **Find QR to the nearest tenth of a meter.**

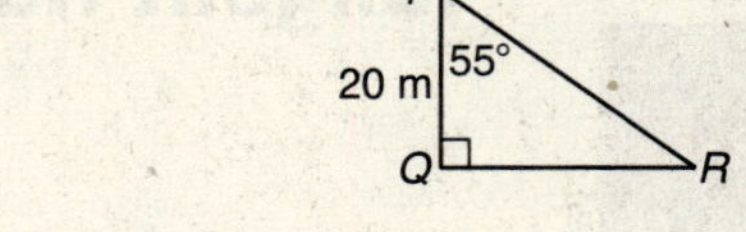

$\tan \boxed{} = \frac{QR}{PQ}$ $\quad$ $\frac{\text{opposite}}{\text{adjacent}}$

$\tan \boxed{} = \frac{QR}{\boxed{}}$ $\quad$ Substitution

$\boxed{} \cdot \boxed{} = QR$ $\quad$ Multiplication Property of Equality

$\boxed{} \approx QR$

3 **A ranger sights the top of a tree at a 40° angle of elevation. Find the height of the tree if it is 80 feet from where the ranger is standing.**

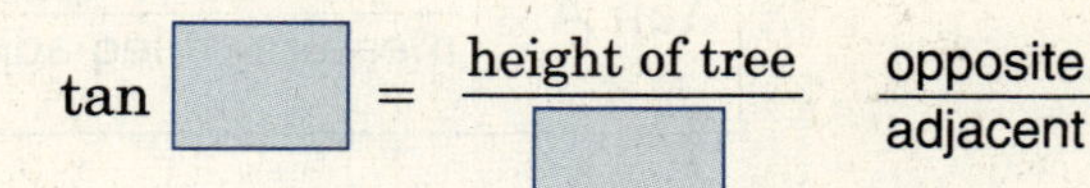

$\tan \boxed{} = \frac{\text{height of tree}}{\boxed{}}$ $\quad$ $\frac{\text{opposite}}{\text{adjacent}}$

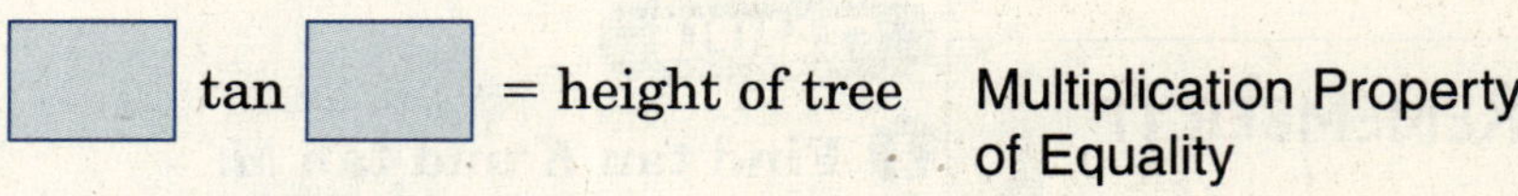

$\boxed{} \tan \boxed{} = \text{height of tree}$ $\quad$ Multiplication Property of Equality

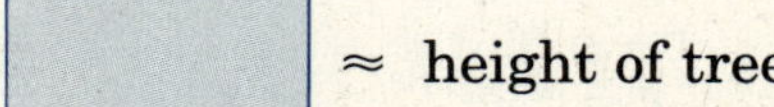

$\boxed{} \approx \text{height of tree}$

The height of the tree is about ______ feet.

FOLDABLES™ ORGANIZE IT

Write the definition of tangent on the tab for Lesson 13-4. Under the tab, sketch and label a triangle. Then express the tangent of one of the angles.

BUILD YOUR VOCABULARY (page 252)

The line of sight and a horizontal line when looking up form the **angle of elevation**.

Angles of elevation can be measured ______ using a **hypsometer**.

The line of sight and a horizontal line when looking

______ form the **angle of depression**.

Remember It

The inverse tangent is also called the *arctangent*.

Your Turn

a. Find YZ to the nearest tenth of a foot.

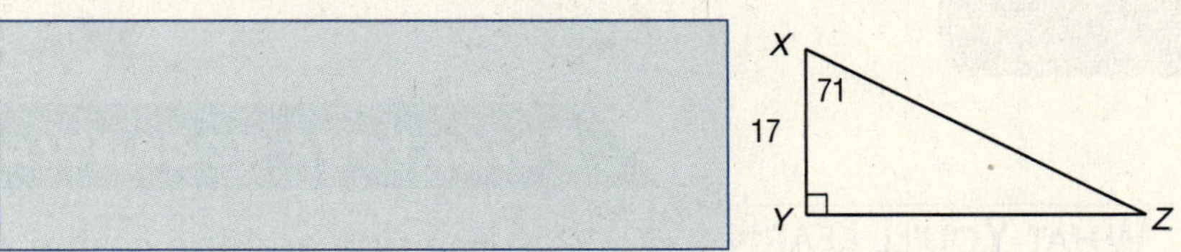

b. The ranger sights the top of another tree at a 52° angle of elevation. Find the height of the tree if it is 20 feet from where he stands.

EXAMPLE

4 Find $m\angle 1$ to the nearest tenth.

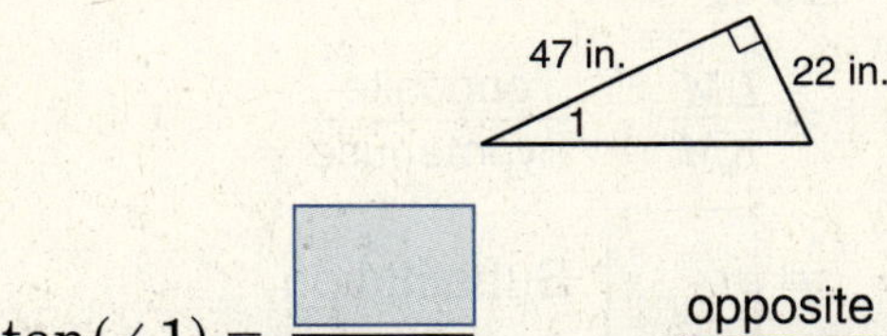

$\tan(\angle 1) = \frac{\square}{\square}$ — opposite/adjacent

$\tan^{-1}\left(\frac{22}{47}\right) \approx \square$ — Definition of arctangent

The measure of $\angle 1$ is about $\square$.

Your Turn

Find $m\angle 2$ to the nearest tenth.

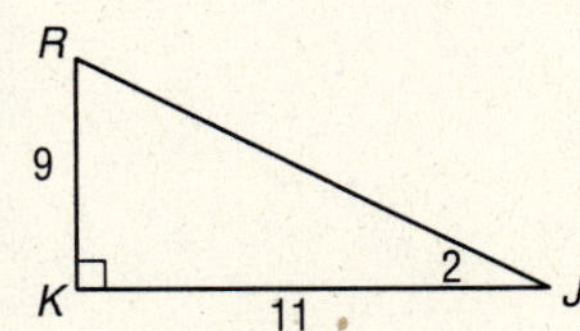

Homework Assignment

Page(s):

Exercises:

13–5 Sine and Cosine Ratios

What You'll Learn

- Use the sine and cosine ratios to solve problems.

Build Your Vocabulary (pages 252–253)

Both the **sine** and the **cosine** ratios relate an angle measure to the ratio of the measures of a triangle's 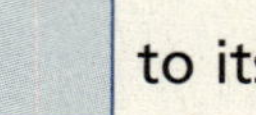 to its ______.

If A is an acute angle of a right triangle,

$\sin A = \dfrac{\text{measure of leg opposite } \angle A}{\text{measure of hypotenuse}}$, and

$\cos A = \dfrac{\text{measure of leg adjacent } \angle A}{\text{measure of hypotenuse}}$.

Foldables™

Organize It

Write the definitions of sine and cosine on the tab for Lesson 13-5. On the back of the tab, describe a similarity and a difference between sine and cosine.

Right Triangles and Trigonometry
13-1 Simplifying Square Roots
13-2 45°-45°-90° Triangles
13-3 30°-60°-90° Triangles
13-4 Tangent Ratios
13-5 Sine and Cosine Ratios

Example

1 Find sin *K*, cos *K*, sin *M*, and cos *M*.

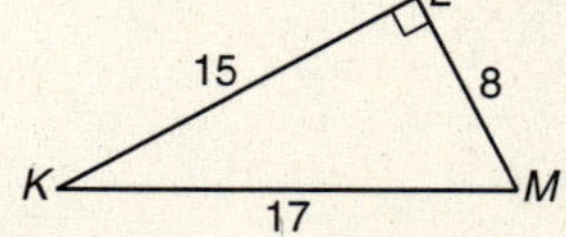

sin K

$= \dfrac{LM}{KM}$ $\quad \dfrac{\text{opposite}}{\text{hypotenuse}}$

= ______ Substitution

≈ ______

sin M

$= \dfrac{KL}{KM}$ $\quad \dfrac{\text{opposite}}{\text{hypotenuse}}$

= ______ Substitution

≈ 

cos K

$= \dfrac{KL}{KM}$ $\quad \dfrac{\text{adjacent}}{\text{hypotenuse}}$

= ______ Substitution

≈ ______

cos M

$= \dfrac{LM}{KM}$ $\quad \dfrac{\text{adjacent}}{\text{hypotenuse}}$

=  Substitution

≈ ______

Your Turn Find sin 30°, cos 30°, sin 45°, cos 45°, sin 60°, cos 60°.

EXAMPLE

2 **Find the value of x to the nearest tenth.**

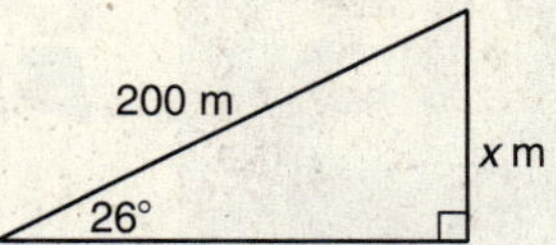

$\sin 26 = \dfrac{x}{200}$ $\quad$ $\dfrac{\text{opposite}}{\text{hypotenuse}}$

$200 \sin 26 = x$ $\quad$ Multiplication Property of Equality

 $\approx x$

EXAMPLE

3 **Find the measure of $\angle K$ to the nearest degree.**

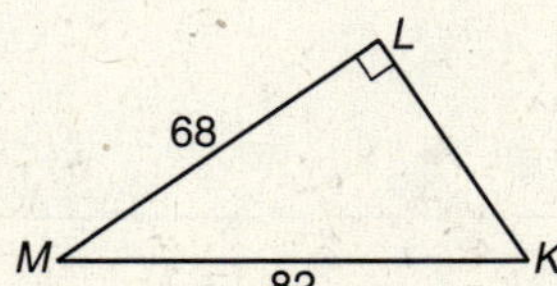

REMEMBER IT

Sin^{-1} and $\cos^{-1}$ are also known as *arcsin* and *arccos*.

$\sin K = \dfrac{LM}{KM}$ $\quad$ $\dfrac{\text{opposite}}{\text{hypotenuse}}$

$\sin K = \dfrac{68}{82}$ $\quad$ Substitution

$m\angle K = \sin^{-1}\left(\dfrac{68}{82}\right)$ $\quad$ Inverse sine

$m\angle K \approx$ ______

Your Turn

a. Find the value of x to the nearest tenth.

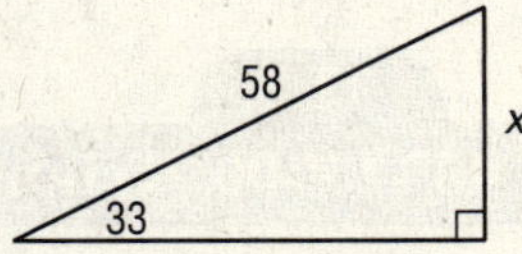

b. Find the measure of $\angle A$ to the nearest degree.

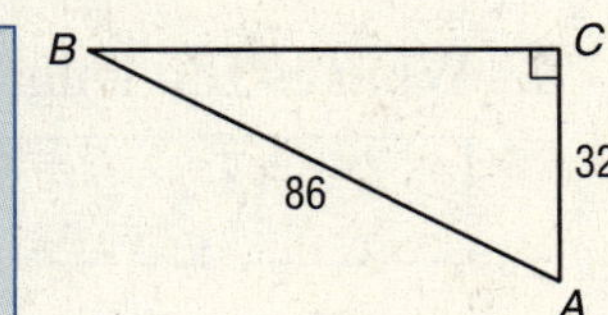

HOMEWORK ASSIGNMENT

Page(s):

Exercises:

Theorem 13-3

If x is a measure of an acute angle of a right triangle, then $\dfrac{\sin x}{\cos x} = \tan x$.

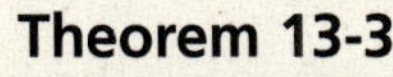

CHAPTER 13

BRINGING IT ALL TOGETHER

STUDY GUIDE

FOLDABLES™	VOCABULARY PUZZLEMAKER	BUILD YOUR VOCABULARY
Use your **Chapter 13 Foldable** to help you study for your chapter test.	To make a crossword puzzle, word search, or jumble puzzle of the vocabulary words in Chapter 13, go to: www.glencoe.com/sec/math/t_resources/free/index.php	You can use your completed **Vocabulary Builder** (pages 252–253) to help you solve the puzzle.

13-1

Simplifying Square Roots

Simplify.

1. $\sqrt{63}$

2. $\frac{1}{\sqrt{3}}$

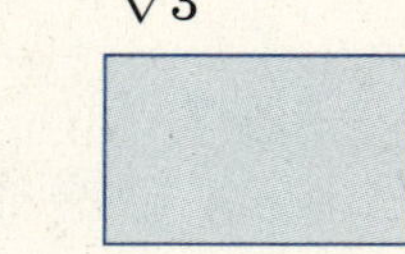

3. $\sqrt{10} \cdot \sqrt{8}$

4. Find the value of x if $\frac{2}{\sqrt{x}} = \frac{2\sqrt{x}}{3}$.

13-2

45°-45°-90° Triangles

A fabric square is cut on the diagonal for a quilt. The perimeter of the square is 116 in.

5. What is the length of each leg/side?

6. What is the length of the hypotenuse/diagonal?

7. What is the measure of each leg of an isosceles right triangle if its hypotenuse measures 10?

13-3

30°-60°-90° Triangles

8. The Gothic arch, similar to the figure, is based on an equilateral triangle. Find the width of the base of the triangle if the median is 4 ft long.

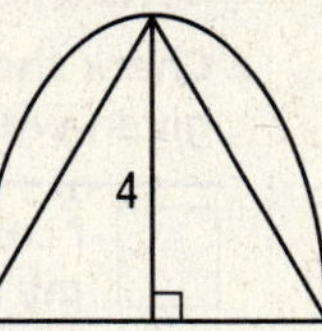

Find the missing measure. Simplify all radicals.

9.

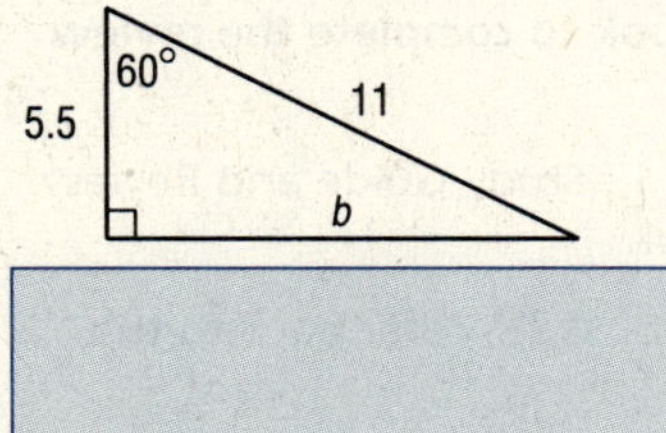

10.

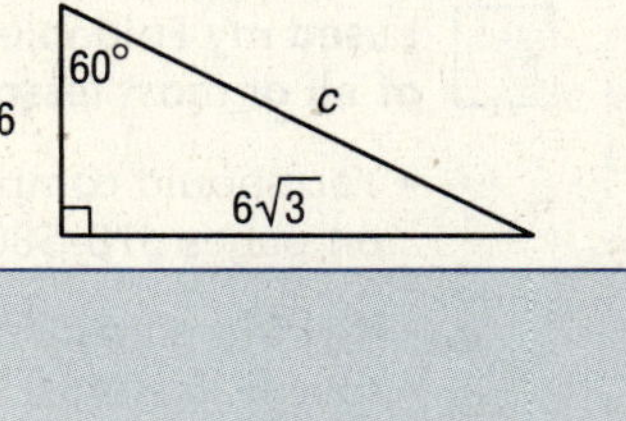

13-4

Tangent Ratio

11. You spot a cat on the roof of a house 80 feet away from where you're standing. Your eye level is 5 feet above ground level, and the angle of elevation from eye level is 33°. How tall is the house?

13-5

Sine and Cosine Ratios

Find the missing measures.

12. If $y = 20$, find x and z.

13. If $z = 2.3$, find x and y.

14. If $x = 9$, find y and z.

ARE YOU READY FOR THE CHAPTER TEST?

Math Online

Visit **geomconcepts.com** to access your textbook, more examples, self-check quizzes, and practice tests to help you study the concepts in Chapter 13.

Check the one that applies. Suggestions to help you study are given with each item.

I completed the review of all or most lessons without using my notes or asking for help.

- You are probably ready for the Chapter Test.
- You may want to take the Chapter 13 Practice Test on page 581 of your textbook as a final check.

I used my Foldable or Study Notebook to complete the review of all or most lessons.

- You should complete the Chapter 13 Study Guide and Review on pages 578–580 of your textbook.
- If you are unsure of any concepts or skills, refer back to the specific lesson(s).
- You may also want to take the Chapter 13 Practice Test on page 581.

I asked for help from someone else to complete the review of all or most lessons.

- You should review the examples and concepts in your Study Notebook and Chapter 13 Foldable.
- Then complete the Chapter 13 Study Guide and Review on pages 578–580 of your textbook.
- If you are unsure of any concepts or skills, refer back to the specific lesson(s).
- You may also want to take the Chapter 13 Practice Test on page 581.

Student Signature

Parent/Guardian Signature

Teacher Signature

Circle Relationships

Use the instructions below to make a Foldable to help you organize your notes as you study the chapter. You will see Foldable reminders in the margin of this Interactive Study Notebook to help you in taking notes.

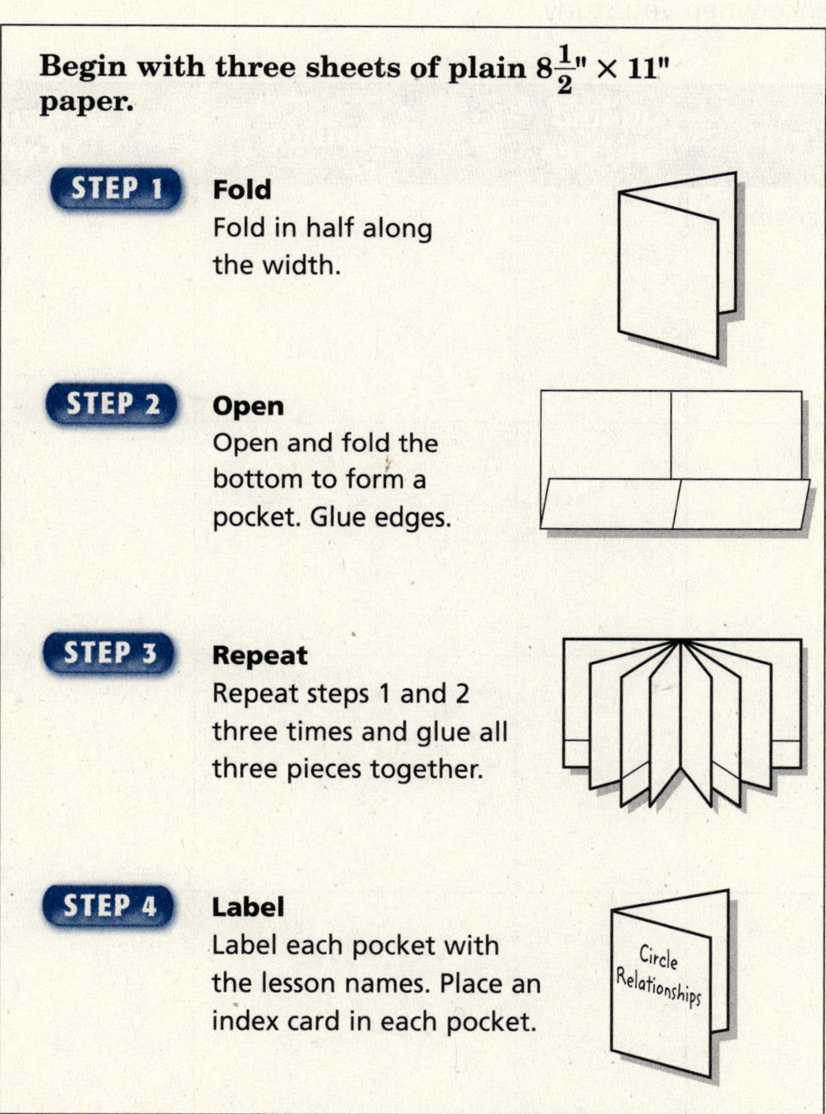

NOTE-TAKING TIP: When taking notes, define new terms and write about the new ideas and concepts you are learning in your own words. Write your own examples that use the new terms and concepts.

Chapter 14

BUILD YOUR VOCABULARY

This is an alphabetical list of new vocabulary terms you will learn in Chapter 14. As you complete the study notes for the chapter, you will see Build Your Vocabulary reminders to complete each term's definition or description on these pages. Remember to add the textbook page number in the second column for reference when you study.

Vocabulary Term	Found on Page	Definition	Description or Example
external secant segment [SEE-kant]			
externally tangent [TAN-junt]			
inscribed angle			
intercepted arc			
internally tangent			

Vocabulary Term	Found on Page	Definition	Description or Example
point of tangency			
secant angle			
secant-tangent angle			
secant segment			
tangent			
tangent-tangent angle			

14–1 Inscribed Angles

WHAT YOU'LL LEARN

- Identify and use properties of inscribed angles.

FOLDABLES™

ORGANIZE IT

Under the tab for *Inscribed Angles*, write the definition of an inscribed angle and draw a picture to illustrate the concept. Record the theorems and other important information from this lesson.

BUILD YOUR VOCABULARY (page 270)

An **inscribed angle** is an angle whose ________ lies on a circle and whose sides contain ________ of the circle.

An **intercepted arc** is an arc of a circle, formed by an angle, such that the ________ of the arc lie on the sides of the angle and all other points of the arc lie on the ________ of the angle.

EXAMPLE

1 **Determine whether $\angle ABC$ is an inscribed angle. Name the intercepted arc for the angle.**

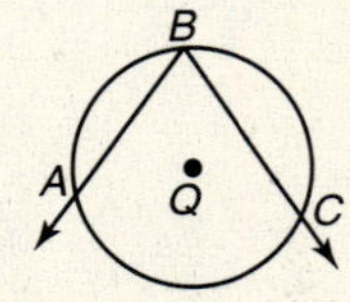

The vertex of $\angle ABC$, point B, is on circle Q. Therefore, $\angle ABC$ is an

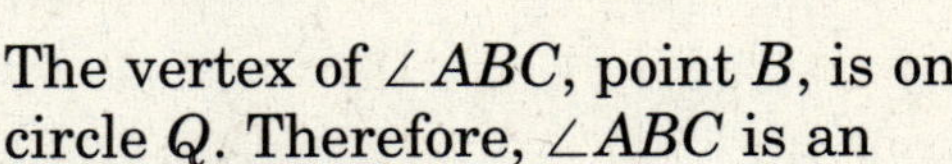

________ angle. The intercepted arc is $\widehat{AC}$.

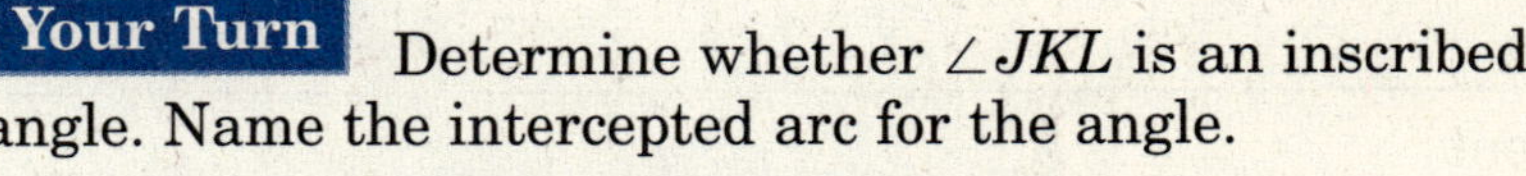

Your Turn Determine whether $\angle JKL$ is an inscribed angle. Name the intercepted arc for the angle.

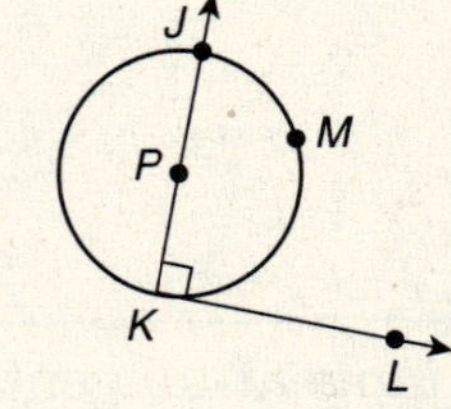

Theorem 14-1
The degree measure of an inscribed angle equals one-half the degree measure of its intercepted arc.

EXAMPLES

Refer to the figure.

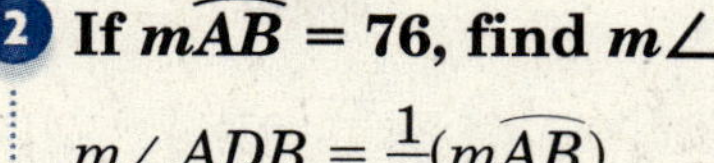

2 **If $m\widehat{AB} = 76$, find $m\angle ADB$.**

$m\angle ADB = \frac{1}{2}(m\widehat{AB})$ Theorem 14-1

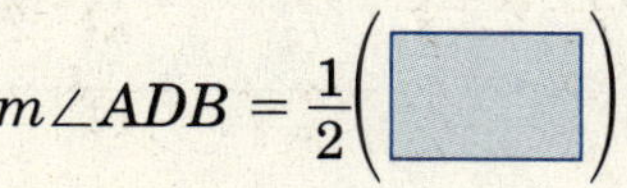

$m\angle ADB = \frac{1}{2}(\square)$ Replace $m\widehat{AB}$.

$m\angle ADB = \square$

3 **If $m\angle BDC = 40$, find $m\widehat{BC}$.**

$m\angle BDC = \frac{1}{2}(m\widehat{BC})$ Theorem 14-1

$40 = \frac{1}{2}(m\widehat{BC})$ Replace $m\angle BDC$.

$2 \cdot 40 = 2 \cdot \frac{1}{2}(m\widehat{BC})$ Multiply each side by 2.

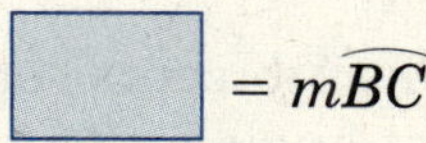

$\square = m\widehat{BC}$

Your Turn **Refer to the figure.**

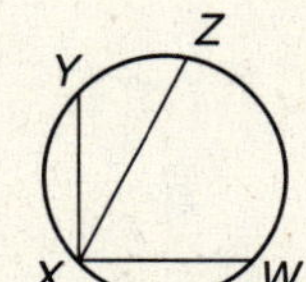

a. If $m\widehat{ZW} = 124$, find $m\angle WXZ$.

b. If $m\angle YXZ = 49$, find $m\widehat{YZ}$.

WRITE IT

What is the difference between a central angle and an inscribed angle?

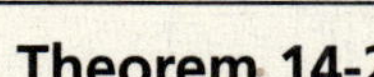

Theorem 14-2
If inscribed angles intercept the same arc or congruent arcs, then the angles are congruent.

EXAMPLE

4 **In circle A, suppose $m\angle TLN = 6y + 7$ and $m\angle TWN = 7y$. Find the value of y.**

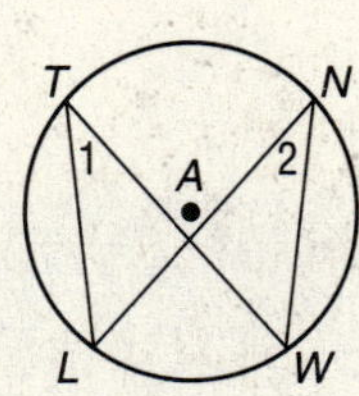

$\angle TLN$ and $\angle TWN$ both intercept $\widehat{TN}$.

$\angle TLN \cong \angle TWN$ Theorem 14-2

$m\angle TLN = m\angle TWN$ Definition of congruent angles

$\square = \square$ Replace $m\angle TLN$ and $m\angle TWN$.

$\square = y$ Subtract $6y$ from each side.

REMEMBER IT

There are 360° in a circle and 180° in a semicircle.

WRITE IT

How does the measure of an inscribed angle relate to the measure of its intercepted arc?

______

______

______

______

Your Turn In the circle, if $m\angle AHM = 10x$ and $m\angle ATM = 20x - 30$, find the value of x.

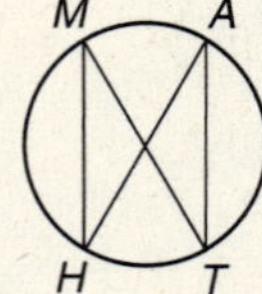

Theorem 14-3
If an inscribed angle of a circle intercepts a semicircle, then the angle is a right angle.

EXAMPLE

5 **In circle G, $m\angle 1 = 6x - 5$ and $m\angle 2 = 3x - 4$. Find the value of x.**

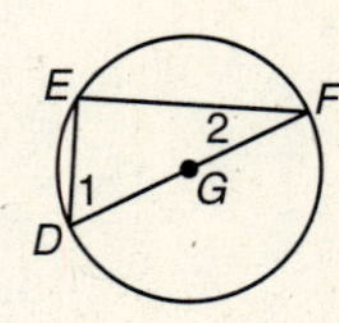

Inscribed angle DEF intercepts semicircle $\overset{\frown}{DF}$. $\angle DEF$ is a right angle by Theorem 14-3. Therefore, $\angle 1$ and $\angle 2$ are complementary.

$m\angle 1 + m\angle 2 = 90$	Complementary angles
$(6x - 5) + (3x - 4) = 90$	Substitution
____ $= 90$	Combine like terms.
$9x - 9 +$ ____ $= 90 +$ ____	Add ____ to each side.
____ $=$ ____	
$\frac{9x}{9} = \frac{99}{9}$	Divide each side by 9.
$x =$ ____ 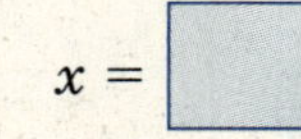	

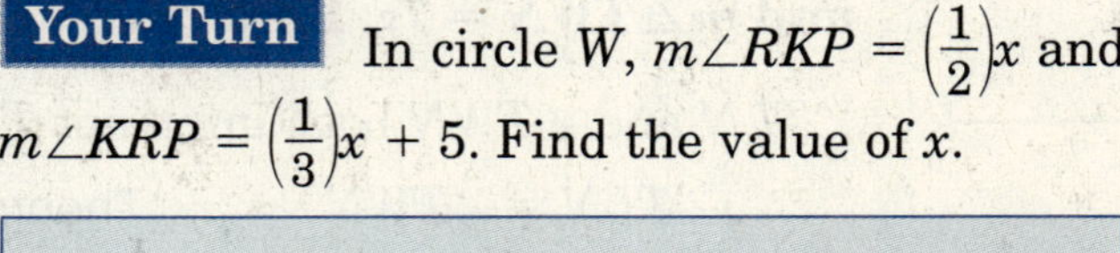
Your Turn In circle W, $m\angle RKP = \left(\frac{1}{2}\right)x$ and $m\angle KRP = \left(\frac{1}{3}\right)x + 5$. Find the value of x.

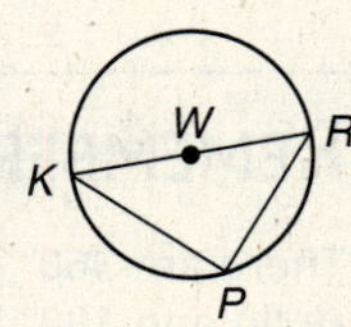

HOMEWORK ASSIGNMENT

Page(s):

Exercises:

14–2 Tangents to a Circle

WHAT YOU'LL LEARN

- Identify and apply properties of tangents to circles.

BUILD YOUR VOCABULARY (page 271)

In a plane, a line is a **tangent** if and only if it intersects a circle in exactly ______ point.

The point of intersection is the **point of tangency**.

Theorem 14-4
In a plane, if a line is tangent to a circle, then it is perpendicular to the radius drawn to the point of tangency.

Theorem 14-5
In a plane, if a line is perpendicular to a radius of a circle at its endpoint on the circle, then the line is a tangent.

EXAMPLE

1 $\overrightarrow{AB}$ **is tangent to circle** C **at** B**. Find** BC**.**

$\overrightarrow{AB} \perp \overline{CB}$ by Theorem 14-4, making $\angle CBA$ a right angle by definition. Therefore, $\triangle ABC$ is a right triangle.

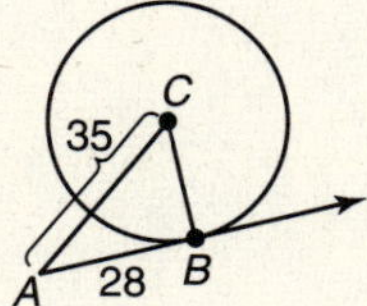

$(BC)^2 + (AB)^2 = (AC)^2$	Pythagorean Theorem
$(BC)^2 + \underline{\quad}^2 = \underline{\quad}^2$	Replace *AB* and *AC*.
$(BC)^2 + \underline{\quad} = \underline{\quad}$	Square ______ and ______.
$(BC)^2 + 784 - 784 = 1225 - 784$	Subtract 784 from each side.
$(BC)^2 = \underline{\quad}$	
$\sqrt{(BC)^2} = \sqrt{441}$	Take the square root of each side.
$BC = \underline{\quad}$	

FOLDABLES™

ORGANIZE IT

Under the tab for *Tangents to a Circle*, write the definition of tangent and draw a picture to illustrate the concept. Record the theorems and other important information from this lesson.

Your Turn $\overline{AE}$ is tangent to circle C at E. Find AE.

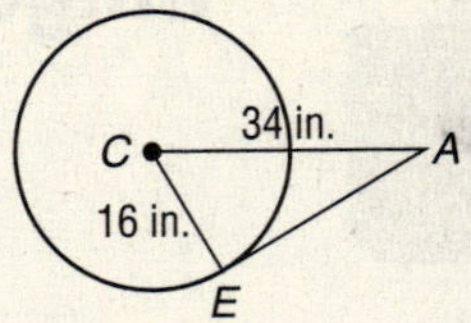

Theorem 14-6
If two segments from the same exterior point are tangent to a circle, then they are congruent.

EXAMPLE

2 **$\overline{EF}$ and $\overline{EG}$ are tangent to circle H. Find the value of x.**

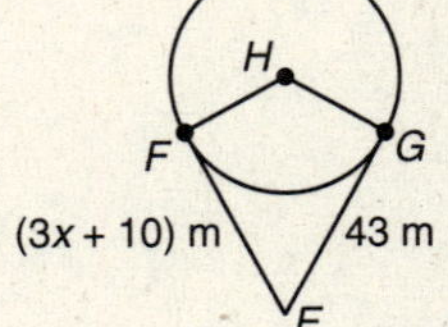

$\overline{EF} \cong \overline{EG}$	Theorem 14-6
____ = ____	Replace $\overline{EF}$ and $\overline{EG}$.
$3x + 10 -$ ____ $= 43 -$ ____	Subtract 10 from each side.
$3x = 33$	
$\frac{3x}{3} = \frac{33}{3}$	Divide each side by 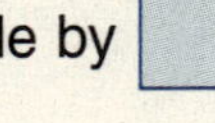 .
$x =$ ____	

Your Turn $\overline{AD}$, $\overline{AC}$, and $\overline{AB}$ are tangents to circles Q and R, respectively. Find the value of x.

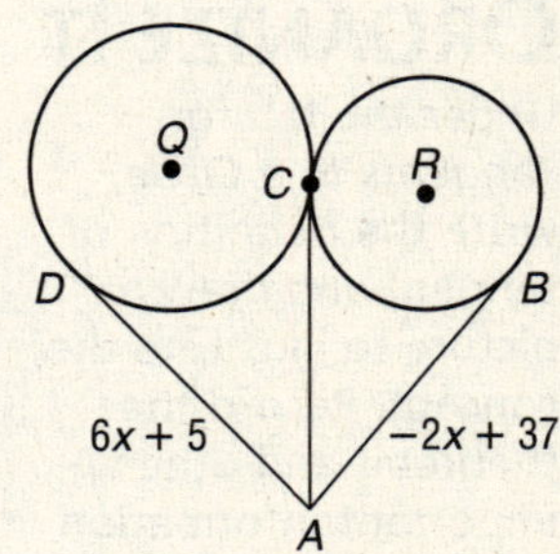

HOMEWORK ASSIGNMENT

Page(s):

Exercises:

BUILD YOUR VOCABULARY (pages 270–271)

If two circles are tangent and one circle is ____ the other, the circles are **internally tangent**.

If two circles are tangent and ____ circle is inside the other, the circles are **externally tangent**.

14–3 Secant Angles

BUILD YOUR VOCABULARY (page 271)

A **secant segment** is a segment that contains a ______ of a circle.

A **secant angle** is the angle formed when two ______ segments intersect.

WHAT YOU'LL LEARN

- Find measures of arcs and angles formed by secants.

FOLDABLES™ ORGANIZE IT

Under the tab for *Secant Angles*, write the definition of a secant segment. Draw a picture of secant angles to illustrate the concept. Record the theorems and other important information from this lesson.

Theorem 14-7
A line or line segment is a secant to a circle if and only if it intersects the circle in two points.

Theorem 14-8
If a secant angle has its vertex inside a circle, then its degree measure is one-half the sum of the degree measures of the arcs intercepted by the angle and its vertical angle.

Theorem 14-9
If a secant angle has its vertex outside a circle, then its degree measure is one-half the difference of the degree measures of the intercepted arcs.

EXAMPLE

1 Find $m\angle 1$.

The vertex of $\angle 1$ is inside circle P.

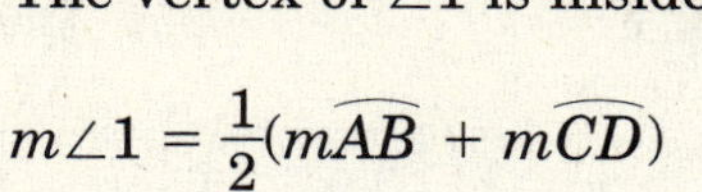

$m\angle 1 = \frac{1}{2}(m\widehat{AB} + m\widehat{CD})$ Theorem 14-8

$m\angle 1 = \frac{1}{2}(\text{______} + \text{______})$ Replace $m\widehat{AB}$ and $m\widehat{CD}$.

$m\angle 1 = \frac{1}{2}(\text{______})$ or ______

Your Turn If $m\widehat{MA} = 40$ and $m\widehat{HT} = 50$, find $m\angle 1$.

EXAMPLE

REMEMBER IT 

The diameter of a circle is also a secant.

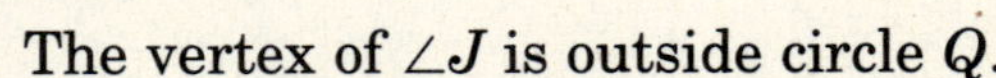

2 Find $m\angle J$.

The vertex of $\angle J$ is outside circle Q.

$m\angle J = \frac{1}{2}(m\overset{\frown}{MN} - m\overset{\frown}{KL})$ Theorem 14-9

$m\angle J = \frac{1}{2}(\square - \square)$

$m\angle J = \frac{1}{2}(\square)$ or $\square$

EXAMPLE

3 Find the value of x. Then find $m\overset{\frown}{CD}$.

The vertex lies inside circle P.

$57 = \frac{1}{2}(m\overset{\frown}{AB} + m\overset{\frown}{CD})$

$57 = \frac{1}{2}[(\square) + (\square)]$

$57 = \frac{1}{2}(9x + 6)$ Combine like terms.

$2 \cdot 57 = 2 \cdot \frac{1}{2}(9x + 6)$ Multiply each side by 2.

$\square = \square$

$114 - 6 = 9x + 6 - 6$ Subtract 6 from each side.

$\square = \square$ Subtraction Property

$\square = x$ Division Property

$m\overset{\frown}{CD} = 6x + 7 = 6(\square) + 7 = \square + 7 = \square$

Your Turn

a. If $m\overset{\frown}{CE} = 85$ and $m\overset{\frown}{BD} = 40$, find $m\angle A$.

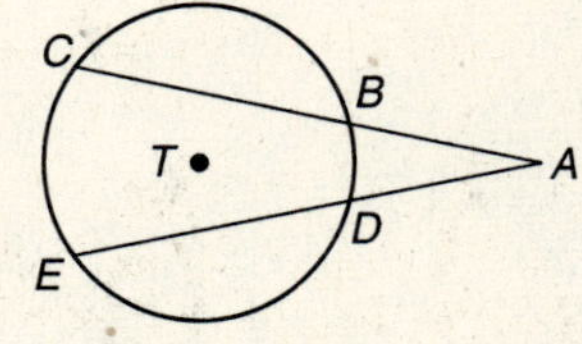

b. Find the value of x. Then find $m\overset{\frown}{TH}$.

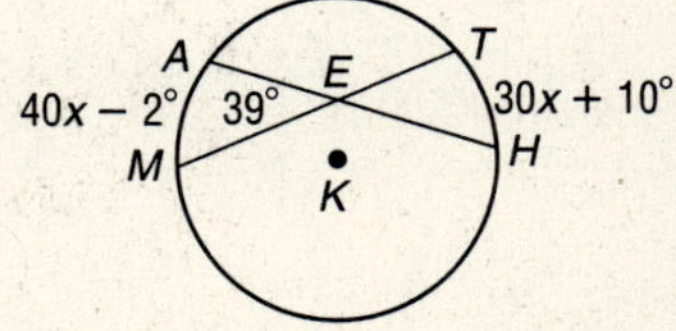

HOMEWORK ASSIGNMENT

Page(s):

Exercises:

14–4 Secant-Tangent Angles

What You'll Learn

- Find measures of arcs and angles formed by secants and tangents.

Theorem 14-10
If a secant-tangent angle has its vertex outside the circle, then its degree measure is one-half the difference of the degree measures of the intercepted arcs.

Theorem 14-11
If a secant-tangent angle has its vertex on the circle, then its degree measure is one-half the degree measure of the intercepted arc.

Key Concept

Secant – Tangent Angles
Vertex Outside the Circle
Secant – tangent angle PQR intercepts $\overset{\frown}{PR}$ and $\overset{\frown}{PS}$.

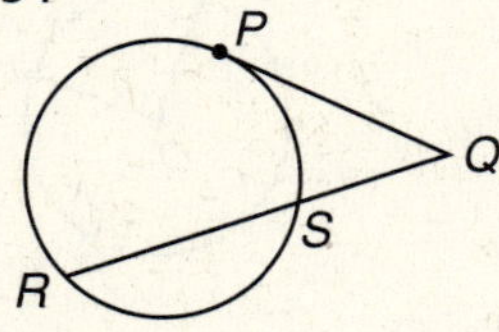

Vertext on the Circle
Secant - tangent angle ABC intercepts $\overset{\frown}{AB}$.

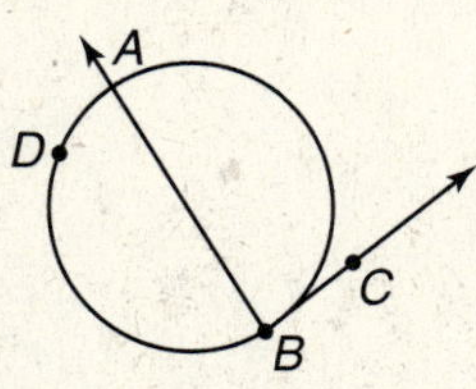

FOLDABLES Under the tab for *Secant-Tangent Angles*, write the definitions of secant-tangent angles and tangent-tangent angles.

EXAMPLES

In the figure, $\overline{AD}$ is tangent to circle K at A.

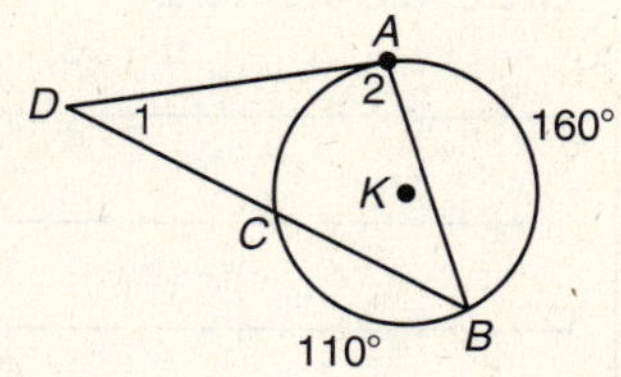

1 Find $m\angle 1$.

Vertex D of the secant-tangent angle is outside circle K. Apply Theorem 14-10.

The degree measure of the whole circle is 360°. So, the measure of $\overset{\frown}{AC}$ is $360° - 160° - 110° = 90°$.

$m\angle 1 = \frac{1}{2}(m\overset{\frown}{AB} - m\overset{\frown}{AC})$ Theorem 14-10

$m\angle 1 = \frac{1}{2}(\square - \square)$ Substitution

$m\angle 1 = \frac{1}{2}(\square)$ or $\square$

2 Find $m\angle 2$.

Vertex A of the secant-tangent angle is on circle K.

$m\angle 2 = \frac{1}{2}(m\overset{\frown}{ACB})$ Theorem 14-11

$m\angle 2 = \frac{1}{2}(\square + \square)$ Substitution

$m\angle 2 = \frac{1}{2}(\square)$ or $\square$

REMEMBER IT

The vertex of a secant-tangent angle cannot be located inside the circle.

REVIEW IT

Explain the difference between a minor arc and a major arc of a circle. (*Lesson 11-2*)

______

______

______

Your Turn

a. $\overline{AZ}$ is tangent to circle D at A. If $m\widehat{AB} = 150$, find $m\angle Z$.

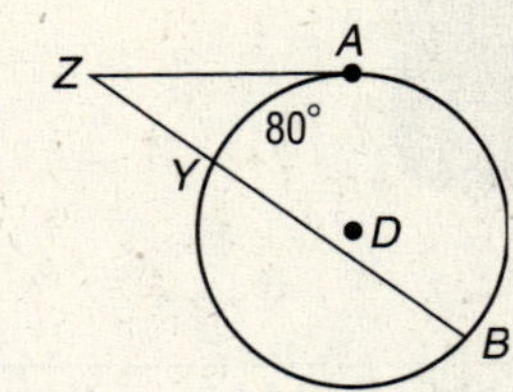

b. $\overrightarrow{EF}$ is tangent to circle D at E. If $m\widehat{EGC} = 230$, find $m\angle FEC$.

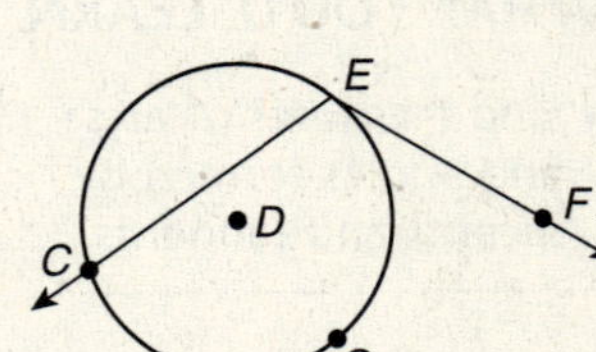

BUILD YOUR VOCABULARY (page 271)

A **tangent-tangent angle** is formed by two ______. Its vertex is always outside the circle.

Theorem 14-12
The degree measure of a tangent-tangent angle is one-half the difference of the degree measures of the intercepted arcs.

EXAMPLE

3 Find $m\angle G$.

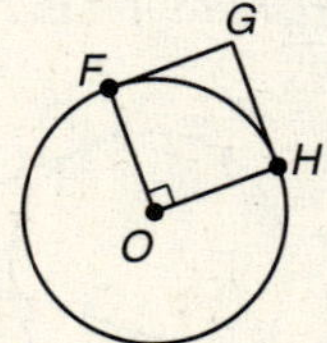

$\angle G$ is a tangent-tangent angle. Apply Theorem 14-12.

By definition of a right angle, $m\angle FOH = 90$. So, $m\widehat{FH} = 90$, because a minor arc is congruent to its central angle.

Since the sum of the measures of a minor arc and its major arc is 360°, major arc $\widehat{FJH}$ is $360° - 90° = 270°$.

$m\angle G = \frac{1}{2}(\text{major arc } \widehat{FJH} - \text{minor arc } \widehat{FH})$

$m\angle G = \frac{1}{2}(270 - 90)$

$m\angle G = \frac{1}{2}\left(\quad\right)$ or ______

HOMEWORK ASSIGNMENT

Page(s):

Exercises:

Your Turn Find $m\angle B$.

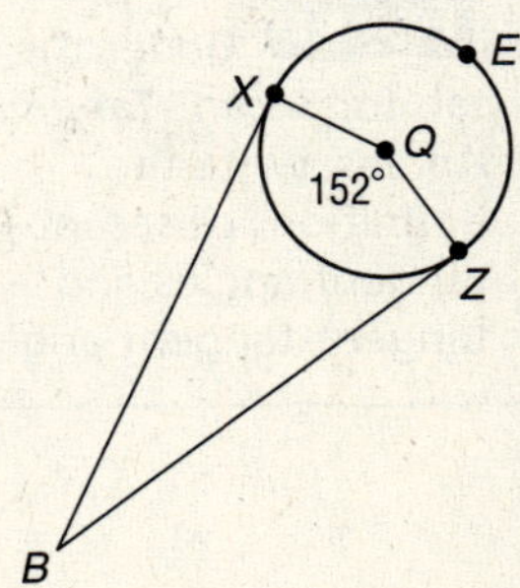

14–5 Segment Measures

WHAT YOU'LL LEARN

- Find measures of chords, secants, and tangents.

BUILD YOUR VOCABULARY (page 270)

A segment is an **external secant segment** if and only if it is the part of a secant segment that is ______ a circle.

FOLDABLES™

ORGANIZE IT

Under the tab for *Segment Measures*, write the definition of an external secant segment. Record the theorems and other main ideas from this lesson.

Theorem 14-13
If two chords of a circle intersect, then the product of the measures of the segments of one chord equals the product of the measures of the segments of the other chord.

Theorem 14-14
If two secant segments are drawn to a circle from an exterior point, then the product of the measures of one secant segment and its external secant segment equals the product of the measures of the other secant segment and its external secant segment.

Theorem 14-15
If a tangent segment and a secant segment are drawn to a circle from an exterior point, then the square of the measure of the tangent segment equals the product of the measures of the secant segment and its external secant segment.

EXAMPLE

1 In circle *A*, find the value of *x*.

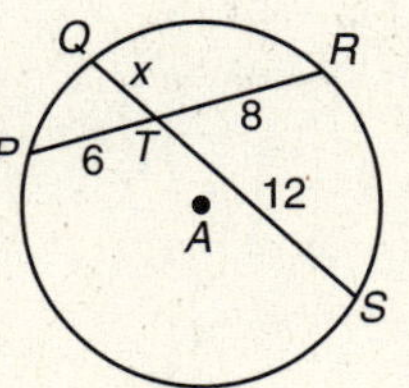

$PT \cdot TR = QT \cdot TS$ Theorem 14-13

$6 \cdot$ ______ $=$ ______ $\cdot 12$ Substitution

$48 = 12x$

$\frac{48}{4} = \frac{12x}{4}$ Divide each side by ______.

______ $= x$ Division Property

Your Turn Find the value of x in the circle.

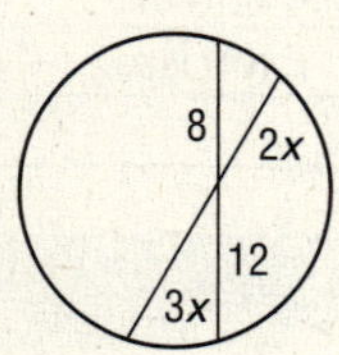

WRITE IT

Explain the difference between Theorem 14-13 and Theorem 14-14 in your own words.

____

____

____

____

____

EXAMPLES

2 Find the value of x to the nearest tenth.

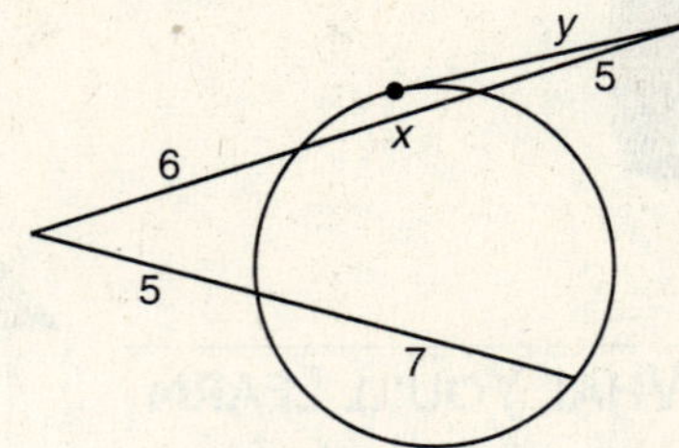

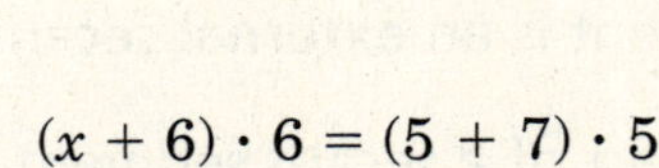

$(x + 6) \cdot 6 = (5 + 7) \cdot 5$ — Theorem 14-14

$\square = 60$ — Distributive Property

$6x + 36 - 36 = 60 - 36$ — Subtract $\square$ from each side.

$6x = \square$

$\frac{6x}{6} = \frac{24}{6}$ — Divide each side by $\square$.

$x = \square$

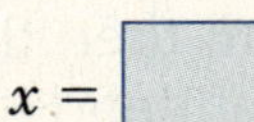

3 Use the value of x to find the value of y.

$y^2 = (x + 5) \cdot 5$ — Theorem 14-15

$y^2 = (4 + 5) \cdot 5$ — Substitution

$y^2 = \square$

$\sqrt{y^2} = \sqrt{45}$ — Take the square root.

$y = \square \approx \square$

Your Turn

a. Find the value of x.

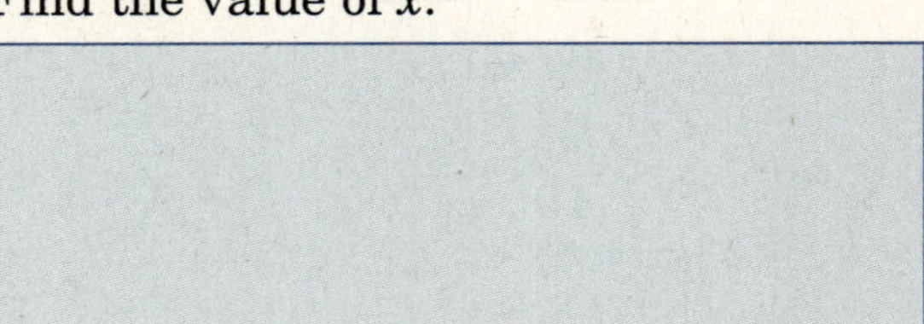

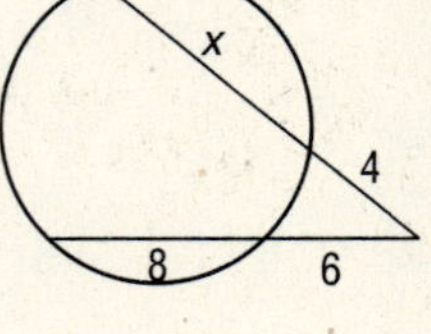

b. Find the value of x.

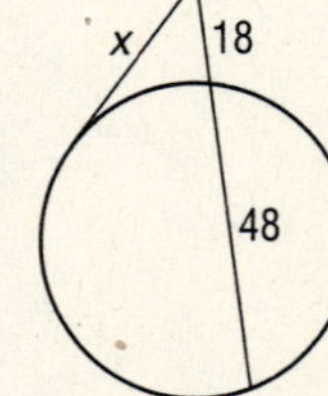

HOMEWORK ASSIGNMENT

Page(s): ______________

Exercises: ______________

____

14–6 Equations of Circles

WHAT YOU'LL LEARN

- Write equations of circles using the center and the radius.

Theorem 14-16 General Equation of a Circle
The equation of a circle with center at (h, k) and a radius of r units is $(x - h)^2 + (y - k)^2 = r^2$.

FOLDABLES™

ORGANIZE IT

Under the tab for *Equations of Circles*, write the General Equation of a Circle, and draw a picture, labeling the center and radius. Record several examples to help you remember the main idea.

EXAMPLE

1 Write the equation of a circle with center at (−4, 0) and a radius of 5 units.

$(x - h)^2 + (y - k)^2 = r^2$ Equation of a Circle

$[x - (\square)]^2 + (y - \square)^2 = \square$ $(h, k) = (-4, 0)$, $r = 5$

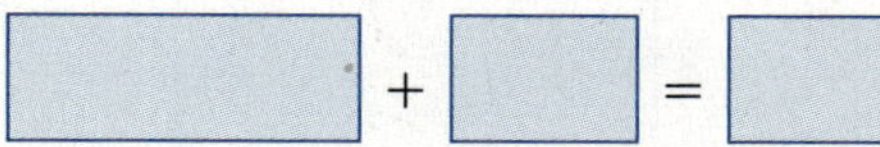

$\square + \square = \square$

The equation for the circle is $\square$.

EXAMPLE

2 Find the coordinates of the center and the measure of the radius of a circle whose equation is

$$\left(x + \frac{3}{2}\right)^2 + \left(y - \frac{1}{2}\right)^2 = \frac{1}{4}.$$

Rewrite the equation.

$(x - h)^2 + (y - k)^2 = r^2$

$[x - (\square)]^2 + (y - \square)^2 = (\square)^2$

Since $h = \square$, $k = \square$, and $r = \square$, the center of the circle is at $\square$. Its radius is $\square$.

Your Turn

a. Write the equation of a circle with center $C(5, -3)$ and a radius of 6 units.

b. Find the coordinates of the center and the measure of the radius of a circle whose equation is $(x + 2)^2 + (y + 7)^2 = 81$.

HOMEWORK ASSIGNMENT

Page(s):

Exercises:

CHAPTER 14

BRINGING IT ALL TOGETHER

STUDY GUIDE

FOLDABLES™	VOCABULARY PUZZLEMAKER	BUILD YOUR VOCABULARY
Use your **Chapter 14 Foldable** to help you study for your chapter test.	To make a crossword puzzle, word search, or jumble puzzle of the vocabulary words in Chapter 14, go to: www.glencoe.com/sec/math/t_resources/free/index.php	You can use your completed **Vocabulary Builder** (pages 270–271) to help you solve the puzzle.

14-1 Inscribed Angles

In circle P, $\overline{AC}$ is a diameter; $m\widehat{CD} = 68$ and $m\widehat{BE} = 96$. Find each of the following.

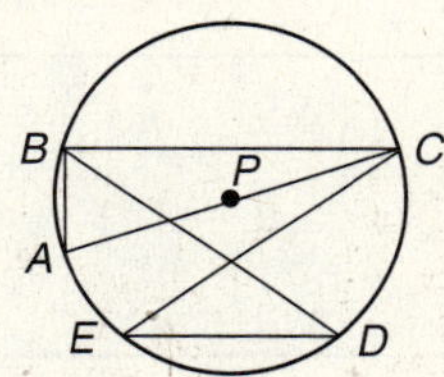

1. $m\angle ABC$

2. $m\angle CED$

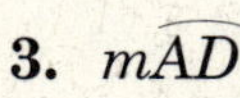

3. $m\widehat{AD}$

In circle A, $\overline{HE}$ is a diameter.

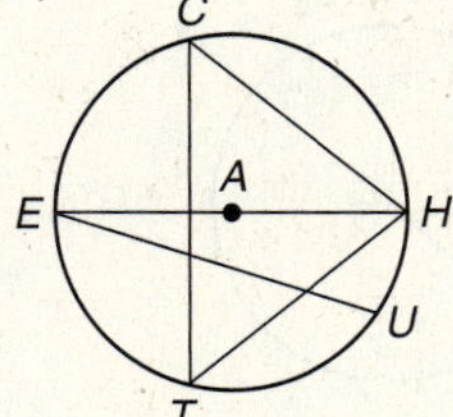

4. If $m\angle HTC = 52$, find $m\widehat{CH}$.

5. Find $m\widehat{HCE}$.

6. If $m\angle HTC = 52$, find $m\widehat{CEH}$.

14-2 Tangents to a Circle

Underline the best term to complete the statement.

7. If a line is tangent to a circle, then it is perpendicular to the radius drawn to the [point of tangency/vertex].

8. $\overline{AB}$ is tangent to circle C. Find the value of x.

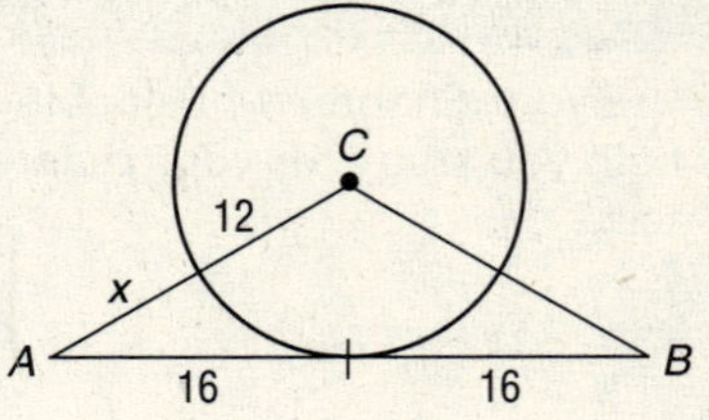

9. Circle P is inscribed in right $\triangle CTA$. Find the perimeter of $\triangle CTA$ if the radius of circle P is 5, $CT = 18$, and $JT = 11$.

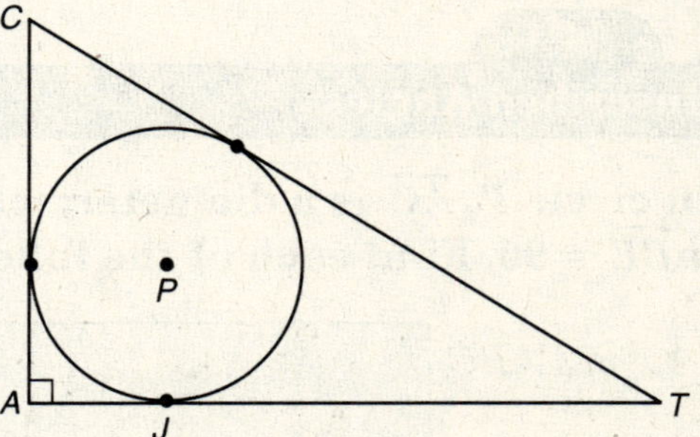

14-3 Secant Angles

Underline the best term to complete the statement.

10. A [radius/secant segment] is a line segment that intersects a circle in exactly two points.

Find the value of x.

11.

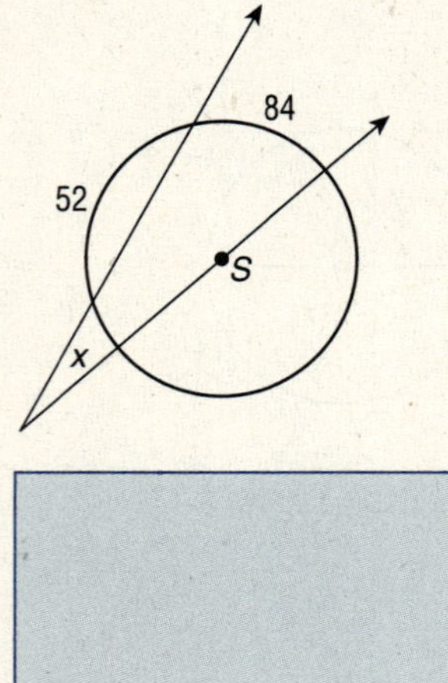

12.

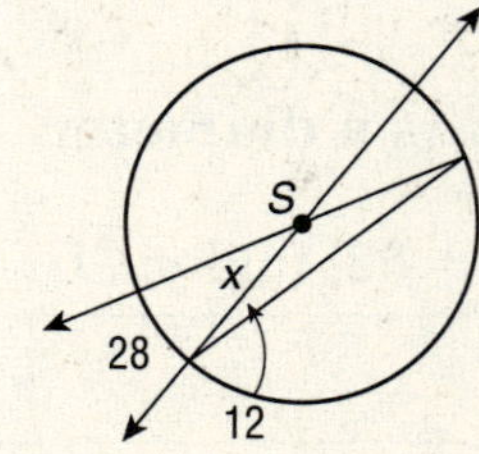

14-4
Secant–Tangent Angles

Underline the best term to complete the statement.

13. The measure of a(n) [tangent-tangent/inscribed] angle is always one-half the difference of the measures of the intercepted arcs.

Find the value of x. Assume that segments that appear to be tangent are tangent.

14.

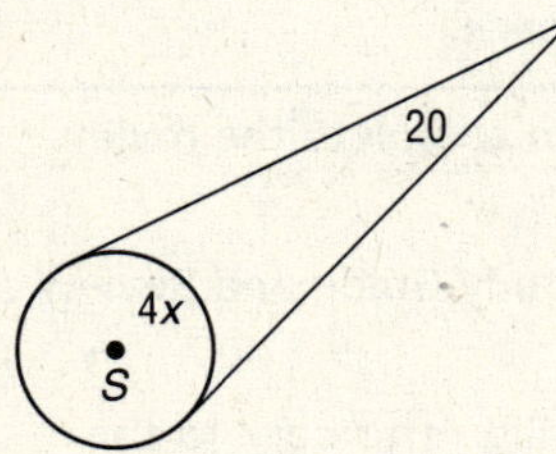

15.

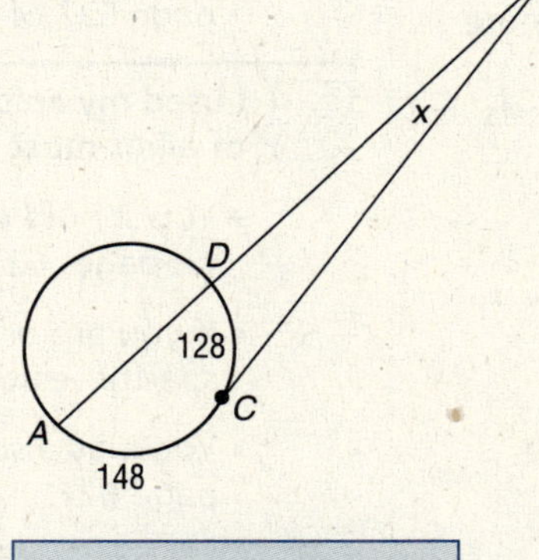

14-5
Segment Measures

Find the value of x.

16.

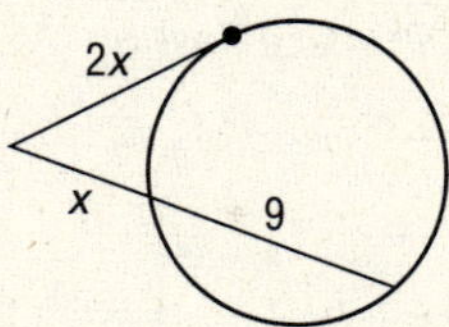

17.

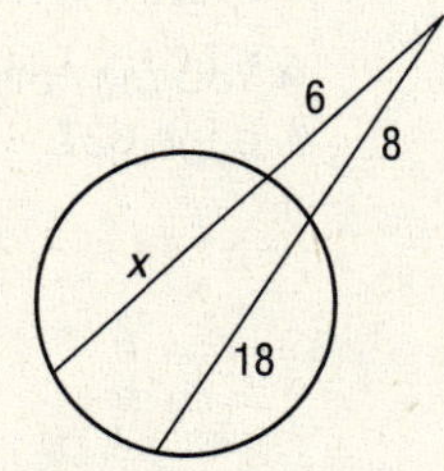

14-6
Equations of Circles

18. Write the equation of the circle with center $(-5, 9)$ and radius $2\sqrt{5}$.

19. What are the coordinates of the center and length of the radius for the circle $(x + 4)^2 + y^2 = 121$.

ARE YOU READY FOR THE CHAPTER TEST?

Visit **geomconcepts.com** to access your textbook, more examples, self-check quizzes, and practice tests to help you study the concepts in Chapter 14.

Check the one that applies. Suggestions to help you study are given with each item.

☐ **I completed the review of all or most lessons without using my notes or asking for help.**

- You are probably ready for the Chapter Test.
- You may want to take the Chapter 14 Practice Test on page 627 of your textbook as a final check.

☐ **I used my Foldable or Study Notebook to complete the review of all or most lessons.**

- You should complete the Chapter 14 Study Guide and Review on pages 624–626 of your textbook.
- If you are unsure of any concepts or skills, refer back to the specific lesson(s).
- You may also want to take the Chapter 14 Practice Test on page 627.

☐ **I asked for help from someone else to complete the review of all or most lessons.**

- You should review the examples and concepts in your Study Notebook and Chapter 14 Foldable.
- Then complete the Chapter 14 Study Guide and Review on pages 624–626 of your textbook.
- If you are unsure of any concepts or skills, refer back to the specific lesson(s).
- You may also want to take the Chapter 14 Practice Test on page 627.

Student Signature

Parent/Guardian Signature

Teacher Signature

Formalizing Proof

Use the instructions below to make a Foldable to help you organize your notes as you study the chapter. You will see Foldable reminders in the margin of this Interactive Study Notebook to help you in taking notes.

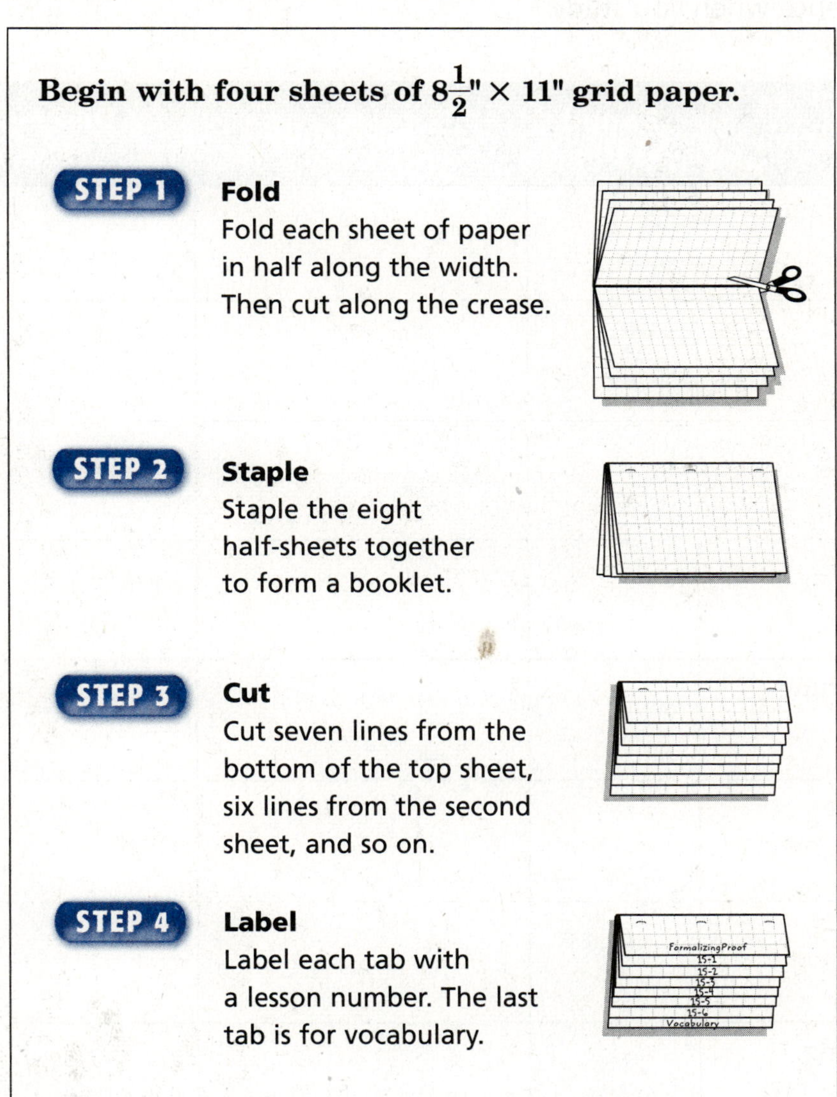

Begin with four sheets of $8\frac{1}{2}$" × 11" grid paper.

STEP 1 **Fold**
Fold each sheet of paper in half along the width. Then cut along the crease.

STEP 2 **Staple**
Staple the eight half-sheets together to form a booklet.

STEP 3 **Cut**
Cut seven lines from the bottom of the top sheet, six lines from the second sheet, and so on.

STEP 4 **Label**
Label each tab with a lesson number. The last tab is for vocabulary.

NOTE-TAKING TIP: To help you organize data, create a study guide or study cards when taking notes, solving equations, defining vocabulary words and explaining concepts.

Chapter 15

BUILD YOUR VOCABULARY

This is an alphabetical list of new vocabulary terms you will learn in Chapter 15. As you complete the study notes for the chapter, you will see Build Your Vocabulary reminders to complete each term's definition or description on these pages. Remember to add the textbook page number in the second column for reference when you study.

Vocabulary Term	Found on Page	Definition	Description or Example
compound statement			
conjunction			
contrapositive			
coordinate proof			
deductive reasoning [dee-DUK-tiv]			
disjunction			
indirect proof			
indirect reasoning			
inverse			
Law of Detachment			

Vocabulary Term	Found on Page	Definition	Description or Example
Law of Syllogism [SIL-oh-jiz-um]			
logically equivalent			
negation			
paragraph proof			
proof			
proof by contradiction			
statement			
truth table			
truth value			
two-column proof			

15–1 Logic and Truth Tables

BUILD YOUR VOCABULARY (pages 290–291)

A **statement** is any sentence that is either true or false, but not both.

Every ______ has a **truth value**, true (T) or false (F).

If a statement is represented by *p*, then ______ *p* is the **negation** of the statement.

The relationship between the ______ of a statement are organized on a **truth table**.

When two statements are ______, they form a **compound statement**.

A **conjunction** is a ______ statement formed by joining two statements with the word ______.

A **disjunction** is a ______ statement formed by joining two statements with the word ______.

WHAT YOU'LL LEARN

- Find the truth values of simple and compound statements.

FOLDABLES™

ORGANIZE IT

Under the tab for Lesson 15-1, list and define the following symbols used in the lesson: ~ , ∧, ∨, and →. Under the last tab, list the vocabulary words and their definitions from Lesson 15-1.

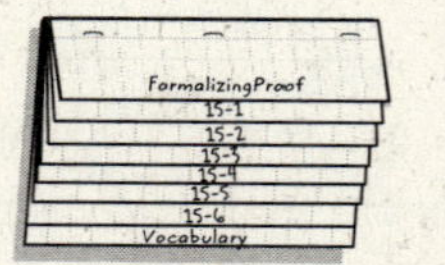

EXAMPLES

Let *p* represent "An octagon has eight sides" and *q* represent "Water does not boil at 90°C."

1 **Write the negation of statement *p*.**

$\sim p$: An octagon ______ have eight sides.

2 **Write the negation of statement *q*.**

$\sim q$: Water ______ boil at 90°C.

Write in *if-then* form: All natural numbers are whole numbers. (*Lesson 1-4*)

______

______

______

______

Your Turn **Let p represent "Tofu is a protein source" and q represent "π is not a rational number."**

a. Write the negation of statement p.

b. Write the negation of statement q.

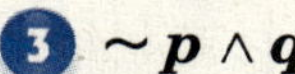

Let p represent "$9^2 = 99$", q represent "An equilateral triangle is equiangular", and r represent "A rectangular prism has six faces." Write the statement for each conjunction or disjunction. Then find the truth value.

REMEMBER IT In the Negation truth table, p does not have to be a true statement and $\sim p$ is not necessarily a false statement.

3 $\sim p \wedge q$

$9^2 \neq 99$ and an equilateral triangle is equiangular. Because p is ______, $\sim p$ is ______. Therefore, $\sim p \wedge q$ is ______ because both $\sim p$ and q are ______.

4 $p \vee \sim r$

$9^2 = 99$ or a rectangular prism does not have six faces. Because r is ______, $\sim r$ is ______. Therefore, $p \vee \sim r$ is ______ because both p and $\sim r$ are ______.

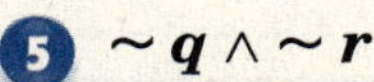

5 $\sim q \wedge \sim r$

An equilateral triangle is not equiangular and a rectangular prism does not have six faces. Because q is ______, $\sim q$ is ______; and because r is ______, $\sim r$ is ______. Therefore, $\sim q \wedge \sim r$ is ______ because both $\sim q$ and $\sim r$ are ______.

Your Turn **Let p represent "0.5 is an integer", q represent "A rhombus has four congruent sides", and r represent "A parallelogram has congruent diagonals." Write the statement for each conjunction or disjunction. Then find the truth value.**

a. $\sim p \wedge q$ **b.** $\sim p \vee r$ **c.** $\sim q \wedge \sim r$

EXAMPLE

6 Construct a truth table for the conjunction $\sim (p \wedge q)$.

p	q	$p \wedge q$	$\sim (p \wedge q)$

Make columns with the headings p, q, ______, and $\sim (p \wedge q)$. Then, list all possible combinations of truth values for p and q. Use these truth values to complete the last two columns of the ______ and its ______.

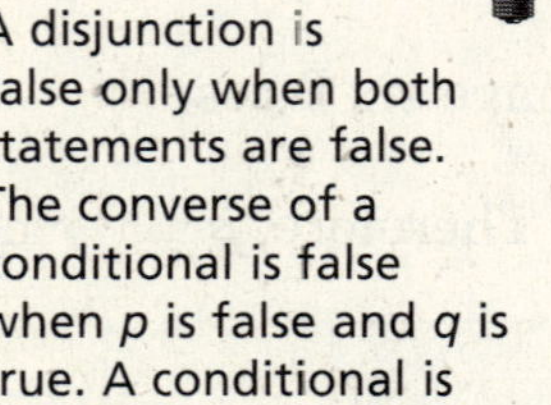

REMEMBER IT

A disjunction is false only when both statements are false. The converse of a conditional is false when p is false and q is true. A conditional is false only when p is true and q is false.

Your Turn Construct a truth table for the disjunction $\sim (p \vee q)$.

HOMEWORK ASSIGNMENT

Page(s):

Exercises:

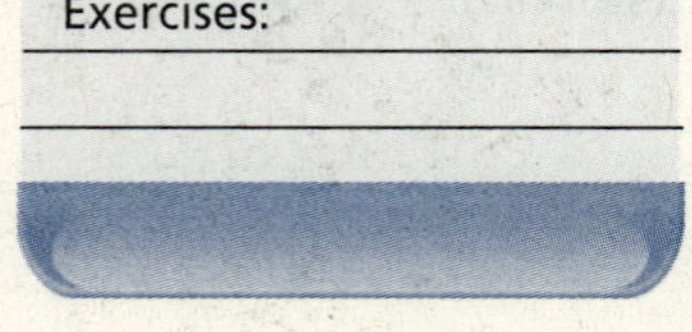

BUILD YOUR VOCABULARY (pages 290–291)

The **inverse** of a conditional is formed by ______ both p and q.

The **contrapositive** of a conditional statement is formed by negating the ______ of the statement.

Two statements are **logically equivalent** if their truth tables are the ______.

15–2 Deductive Reasoning

What You'll Learn

- Use the Law of Detachment and the Law of Syllogism in deductive reasoning.

Deductive reasoning is the process of using facts, rules, definitions, and properties in a logical order.

The **Law of Detachment** allows us to reach logical ______ from ______ statements.

The **Law of Syllogism** is similar to the Transitive Property of Equality.

Key Concept

Law of Detachment
If $p \to q$ is a true conditional and p is true, then q is true.

FOLDABLES™ Under the tab for Lesson 15-2, summarize the Law of Detachment and the Law of Syllogism in your own words.

EXAMPLES

Use the Law of Detachment to determine a conclusion that follows from statements (1) and (2). If a valid conclusion does not follow, then write *no valid conclusion*.

1 **(1) In a plane, if a line is perpendicular to one of two parallel lines, then it is perpendicular to the other line.**

(2) $\overleftrightarrow{AB} \parallel \overleftrightarrow{CD}$ and $\overleftrightarrow{EF} \perp \overleftrightarrow{AB}$.

p: $\overleftrightarrow{AB} \parallel \overleftrightarrow{CD}$ and ______.

q: $\overleftrightarrow{EF} \perp$ ______

Statement (1) indicates that $p \to q$ is ______, and statement (2) indicates that p is ______. So, ______ is true. Therefore, $\overleftrightarrow{EF} \perp \overleftrightarrow{CD}$.

2 **(1) Two nonvertical lines have the same slope if and only if they are parallel.**

(2) $\overleftrightarrow{AB}$ is a vertical line.

p: Two lines are nonvertical and ______.

q: Two lines have the same ______.

Statement (2) indicates that p is ______. Therefore, there is no valid conclusion.

Your Turn **Use the Law of Detachment to determine a conclusion that follows from statements (1) and (2). If a valid conclusion does not follow, then write *no valid conclusion.***

a. (1) If a figure is an isosceles triangle, then it has two congruent angles.
(2) A figure is an isosceles triangle.

b. (1) If a hexagon is regular, each interior angle measures 120°.
(2) The hexagon is regular.

Remember It

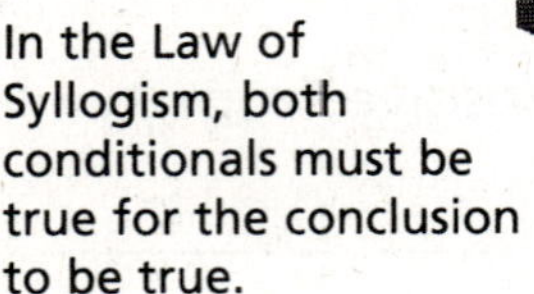

In the Law of Syllogism, both conditionals must be true for the conclusion to be true.

EXAMPLE

3 **Use the Law of Syllogism to determine a conclusion that follows from statements (1) and (2).**

(1) If $m\angle K = 90$, then $\angle K$ is a right angle.
(2) If $\angle K$ is a right angle, then $\triangle JKL$ is a right triangle.

p: $m\angle K =$ ______

q: $\angle K$ is a right angle.

r: $\triangle JKL$ is a ______ triangle.

Use the Law of Syllogism to conclude $p \rightarrow r$.

Therefore, if ______, then $\triangle JKL$ is a ______ triangle.

Your Turn Use the Law of Syllogism to determine a conclusion that follows from statements (1) and (2).

(1) If it is rainy tomorrow, then Alan cannot play golf.
(2) If Alan cannot play golf, then he will watch television.

Homework Assignment

Page(s):
Exercises:

15–3 Paragraph Proofs

WHAT YOU'LL LEARN

- Use paragraph proofs to prove theorems.

BUILD YOUR VOCABULARY (page 291)

A **proof** is a logical argument in which each statement is backed up by a ________ that is accepted as ________.

Statements and reasons are written in ________ form in a **paragraph proof**.

EXAMPLES

Write a paragraph proof for the conjecture.

1 **In $\triangle RST$, if $\overline{TX} \perp \overline{RS}$ and $\overline{TX}$ bisects $\angle RTS$, then $\overline{RX} \cong \overline{XS}$.**

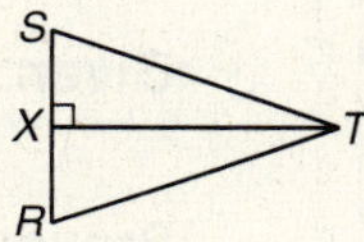

Given: $\overline{TX} \perp \overline{RS}$; $\overline{TX}$ bisects $\angle RTS$.
Prove: $\overline{RX} \cong \overline{XS}$

Proof: If $\overline{TX} \perp \overline{RS}$, then $\angle RXT$ and $\angle TXS$ are ________ angles and $\triangle RXT$ and ________ are right triangles. If $\overline{TX}$ bisects $\angle RTS$, then $\angle RTX \cong \angle STX$ by the definition of angle ________. Also, $\overline{TX} \cong$ ________ since congruence is ________. So, $\triangle RTX \cong \triangle STX$ by the ________ Theorem. Therefore, $\overline{RX} \cong \overline{XS}$ because ________ parts of congruent triangles are congruent (CPCTC).

FOLDABLES™

ORGANIZE IT

Under the tab for Lesson 15-3, summarize what information is listed as "Given" and "Prove" in a paragraph proof.

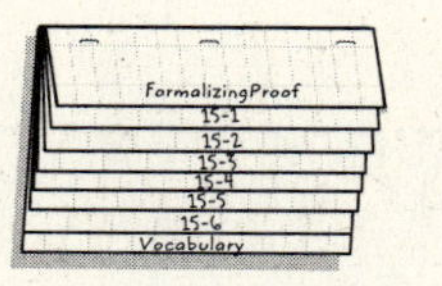

REVIEW IT

What can you say about corresponding angles formed when parallel lines are cut by a transversal? (*Lesson 4-3*)

____

____

2 **If ∠1 and ∠2 are congruent, then ℓ is parallel to *m*.**

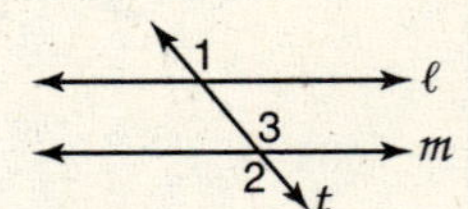

Given: ∠1 ≅ ∠2

Prove: ℓ ∥ ______

Proof: Vertical angles are congruent so ∠2 ≅ ∠3. Since ∠1 ≅ ∠2, ∠1 ≅ ______ by substitution. If two lines in a ______ are cut by a ______ so that corresponding angles are ______, then the lines are ______. Therefore, ______.

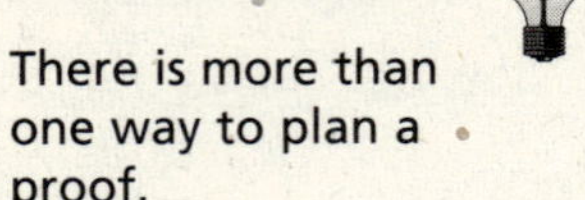

REMEMBER IT

There is more than one way to plan a proof.

Your Turn **Write a paragraph proof for each conjecture.**

a. If *A* is the midpoint of $\overline{DC}$ and $\overline{EB}$, then △*DAE* ≅ △*CAB*.

Given: *A* is the midpoint of $\overline{DC}$ and $\overline{EB}$

Prove: △*DAE* ≅ △*CAB*

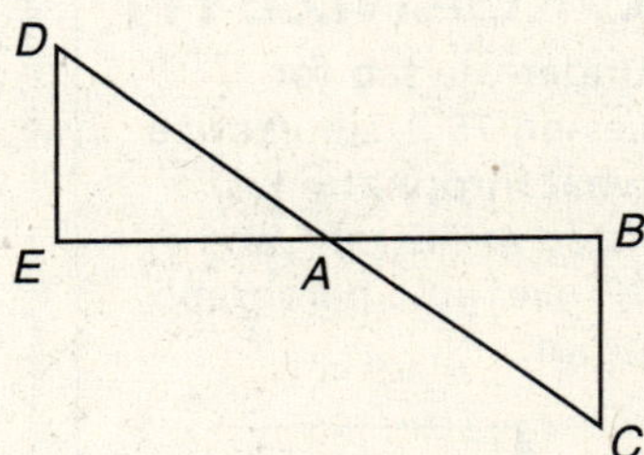

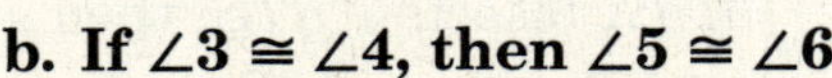

b. If ∠3 ≅ ∠4, then ∠5 ≅ ∠6.

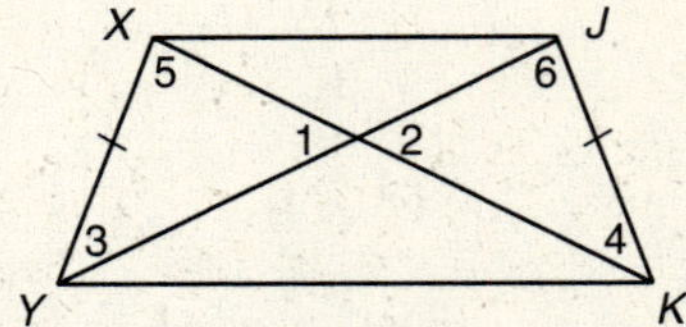

HOMEWORK ASSIGNMENT

Page(s): _______________

Exercises: _______________

15–4 Preparing for Two-Column Proofs

What You'll Learn

- Use properties of equality in algebraic and geometric proofs.

Build Your Vocabulary (page 291)

A **two-column proof** is a deductive argument with ________ and ________ organized in two columns.

Example

1 **Justify the steps for the proof of the conditional. If $\angle XWY \cong \angle XYW$, then $\angle AWX \cong \angle BYX$.**

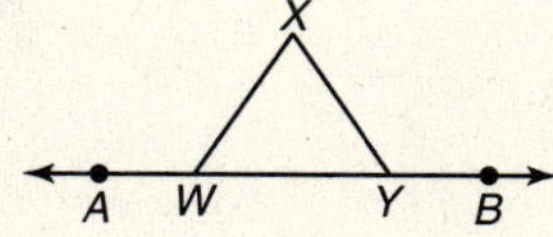

Statements	Reasons
1. ________ $\cong$ ________	1. Given
2. $m\angle XWY = m\angle XYW$	2. ________
3. $m\angle AWX + m\angle XWY = 180$; $m\angle BYX + m\angle XYW = 180$	3. ________
4. $m\angle AWX + m\angle XWY =$ ________ $+$ ________	4. Substitution
5. $m\angle AWX =$ ________	5. Subtraction property
6. ________ $\cong$ ________	6. Definition of congruent angles

Foldables™

Organize It

Under the tab for Lesson 15-4, summarize the Properties of Equality from Lesson 2-2.

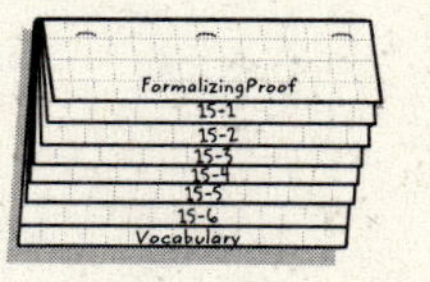

Your Turn Justify the steps for the proof of the conditional. If $m\angle AOC = m\angle BOD$, then $m\angle AOB = m\angle COD$.

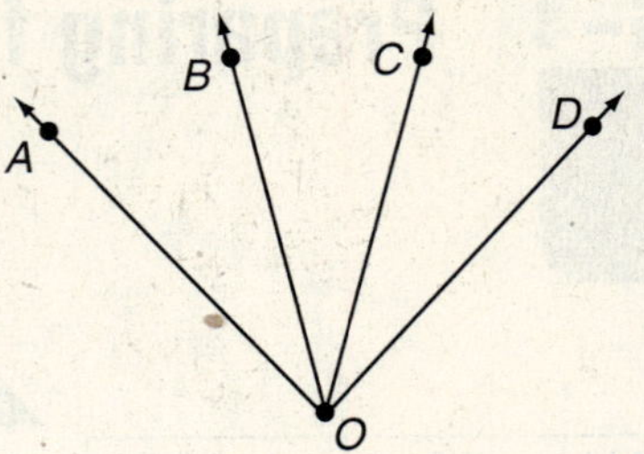

Given: $m\angle AOC = m\angle BOD$
Prove: $m\angle AOB = m\angle COD$

Proof:

Statements	Reasons
1.	1.
2.	2.
3.	3.
4.	4.
5.	5.

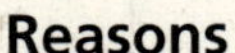

REMEMBER IT

You cannot write a statement unless you give a reason to justify it.

WRITE IT

What information is always in the first statement of a proof? What information can always be found in the last satement?

______

______

______

______

______

EXAMPLE

2 Show that if $A = \frac{1}{2}bh$, then $b = \frac{2A}{h}$.

Given: $A = \frac{1}{2}bh$

Prove: $b = \frac{2A}{h}$

Proof:

Statements	Reasons
1. $A = \frac{1}{2}bh$	1. Given
2. ______ $= bh$	2. Multiplication property
3. $\frac{2A}{h} = b$	3. ______
4. $b =$ ______	4. Symmetric property

Your Turn Show that if $PV = nRT$, then $R = \frac{PV}{nT}$.

Given:

Prove:

Proof:

Statements	Reasons
1.	1.
2.	2.
3.	3.

HOMEWORK ASSIGNMENT

Page(s):

Exercises:

15–5 Two-Column Proofs

What You'll Learn

- Use two-column proofs to prove theorems.

Foldables™

Organize It

Under the tab for Lesson 15-5, summarize the process to write a two-column proof.

Example

1 Write a two-column proof for the conjecture.

If $\angle 1 = \angle 2$, then quadrilateral $ABCD$ is a trapezoid.

Given: $\angle 1 = \angle 2$

Prove: $ABCD$ **is a trapezoid**

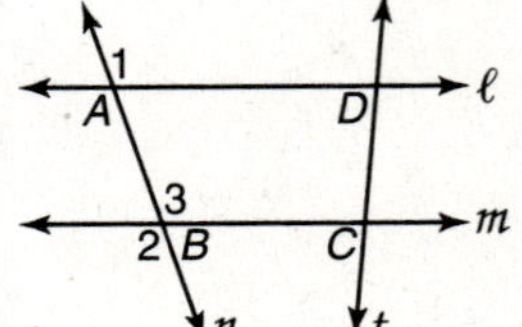

Proof:

Statements	Reasons
1. ____ ≅ ____	1. Given
2. $\angle 2 \cong$ ____	2. ____
3. ____ ≅ ____	3. Substitution
4. ____	4. If two lines in a plane are cut by a transversal so that corresponding angles are congruent, then the lines are parallel.
5. Quadrilateral $ABCD$ is a trapezoid.	5. ____

Your Turn Write a two-column proof. If $\triangle XYZ$ is isosceles with $\overline{XZ} \cong \overline{XY}$ and $\overline{OZ} \cong \overline{NY}$, then $\overline{OY} \cong \overline{NZ}$.

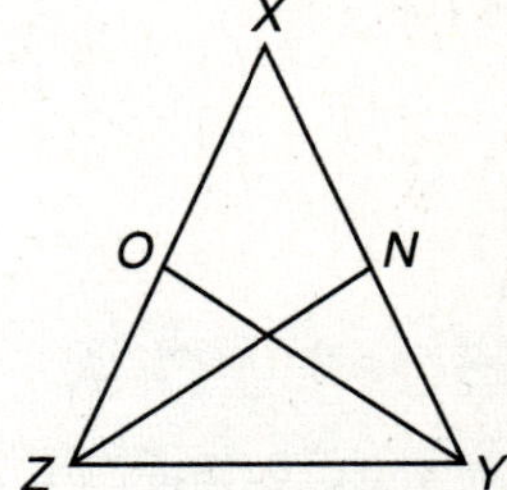

Given: $\triangle XYZ$ is isosceles with $\overline{XZ} \cong \overline{XY}$ and $\overline{OZ} \cong \overline{NY}$

Prove: $\overline{OY} \cong \overline{NZ}$

Proof:

Statements	Reasons
1.	1.
2.	2.
3.	3.
4.	4.
5.	5.
6.	6.

EXAMPLE

2 Write a two-column proof.

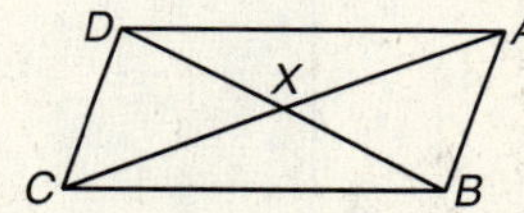

Given: X is the midpoint of both $\overline{BD}$ and $\overline{AC}$.

Prove: $\triangle DXC \cong \triangle BXA$

Proof:

Statements	Reasons
1. X is the midpoint of both $\overline{BD}$ and $\overline{AC}$.	1. Given
2. $\overline{DX} \cong \overline{BX}$; ____ $\cong$ ____	2.
3. ____ $\cong$ ____	3. Vertical angles are congruent.
4. ____ $\cong$ ____	4.

Your Turn Write a two-column proof.

Given: AD and CE bisect each other.

Prove: $AE \parallel CD$

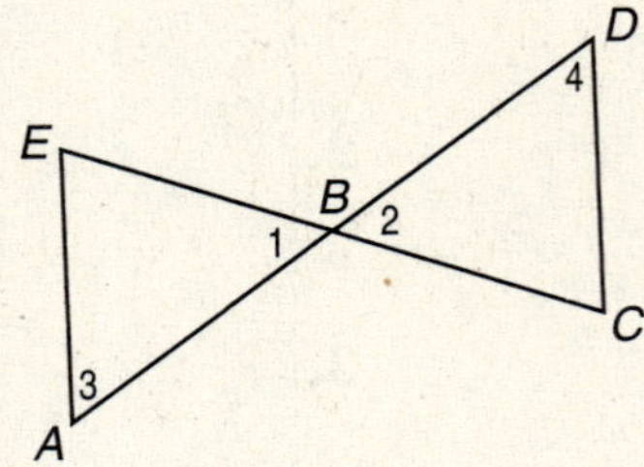

Proof:

Statements	Reasons
1.	1.
2.	2.
3.	3.
4.	4.
5.	5.
6.	6.

HOMEWORK ASSIGNMENT

Page(s):

Exercises:

15–6 Coordinate Proofs

BUILD YOUR VOCABULARY (page 290)

A proof that uses _____ on a coordinate plane is a **coordinate proof**.

WHAT YOU'LL LEARN

- Use coordinate proofs to prove theorems.

EXAMPLE

1 Position and label a rectangle with length b and height d on a coordinate plane.

- Use the origin as a _____.
- Place one side on the x-axis and one side on the _____.
- Label the _____ A, B, C and D.
- Label the coordinates D(_____), C(_____, 0), B(_____), and A(0, _____).

KEY CONCEPT

Guidelines for Placing Figures on a Coordinate Plane

1. Use the origin as a vertex or center.
2. Place at least one side of a polygon on an axis.
3. Keep the figure within the first quadrant, if possible.
4. Use coordinates that make computations as simple as possible.

FOLDABLES Under the tab for Lesson 15-6, summarize the Guidelines for Placing Figures on a Coordinate Plane.

Your Turn Position and label an isosceles triangle with base m units long and height n units on a coordinate plane.

EXAMPLE

2 Write a coordinate proof to prove that the opposite sides of a parallelogram are congruent.

Given: parallelogram $ABDC$

Prove: $\overline{AB} \cong \overline{CD}$ and $\overline{AC} \cong \overline{BD}$

REVIEW IT

What is slope and how would you determine the slope of a line? (*Lesson 4-6*)

______

______

______

______

Proof:

Label the vertices $A(0, 0)$, $B(a, 0)$, $D(a + b, c)$, and $C(b, c)$. Use the Distance Formula to find AB, CD, AC, and BD.

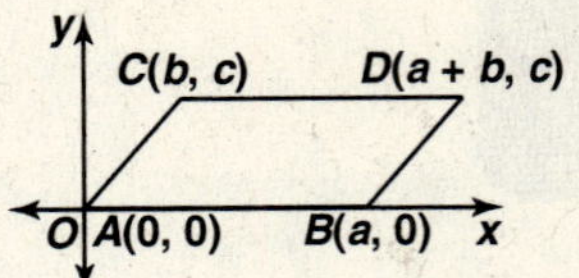

$$AB = \sqrt{(a - 0)^2 + (0 - 0)^2}$$

$$= \sqrt{a^2} \text{ or } a$$

$$CD = \sqrt{[(a + b) - b]^2 + (c - c)^2} = \square \text{ or } a$$

$$AC = \sqrt{\left(b - \square\right)^2 + \left(c - \square\right)^2} = \sqrt{b^2 + c^2}$$

$$BD = \sqrt{\left[(a + b) - \square\right]^2 + \left(c - \square\right)^2}$$

$$= \square$$

So, $AB = CD$ and $AC = BD$.

Therefore, ______ and ______; opposite sides of a parallelogram are ______.

REVIEW IT

What is the Distance Formula? (*Lesson 6-7*)

______

______

______

Your Turn Write a coordinate proof to prove that parallelogram $WXYZ$ is a rectangle by proving the diagonals are congruent.

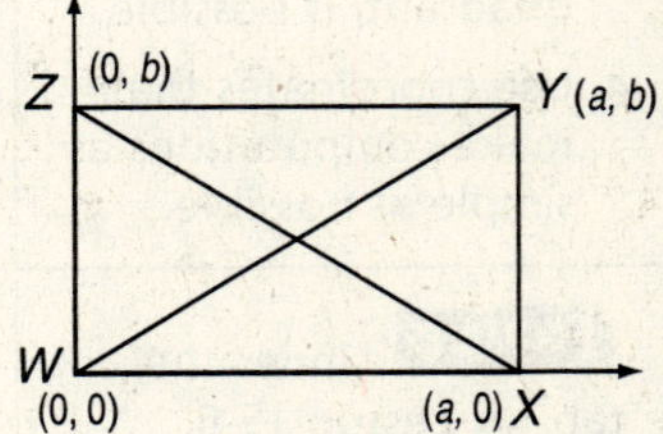

EXAMPLE

3 **Write a coordinate proof to prove that the length of the segment joining the midpoints of two sides of a triangle is one-half the length of the third side.**

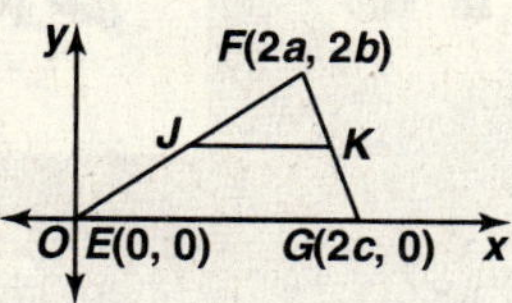

REVIEW IT

What is the Midpoint Formula? (*Lesson 2-5*)

______

______

______

Given: $\triangle EFG$ with midpoints J and K, of $\overline{EF}$ and $\overline{FG}$

Prove: $JK = \frac{1}{2}EG$

Label the vertices E ______, F ______, and $G(2c, 0)$.

Use the Midpoint Formula to find the coordinates of J and K, and the Distance Formula to find JK and EG.

Coordinates of J: $\left(\frac{0 + 2a}{2}, \frac{0 + 2b}{2}\right) =$ ______

Coordinates of K: $\left(\frac{2a + 2c}{2}, \frac{2b + 0}{2}\right) = (a + c, b)$

$JK = \sqrt{[(a + c) - a]^2 + (b - b)^2} = \sqrt{c^2}$ or c

$EG = \sqrt{(2c - 0)^2 + (0 - 0)^2} = \sqrt{(2c)^2}$ or ______

$\frac{1}{2}EG = \frac{1}{2}$ ______ $=$ ______

Therefore, ______.

Your Turn Write a coordinate proof to prove that the length of a median segment joining the midpoints of two legs of a trapezoid is one-half the sum of the length of the bases.

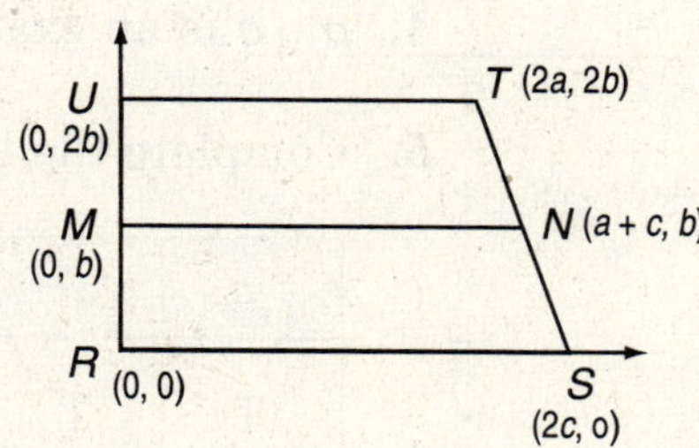

______

HOMEWORK ASSIGNMENT

Page(s): ______

Exercises: ______

______

CHAPTER 15

BRINGING IT ALL TOGETHER

STUDY GUIDE

FOLDABLES™	VOCABULARY PUZZLEMAKER	BUILD YOUR VOCABULARY
Use your **Chapter 15 Foldable** to help you study for your chapter test.	To make a crossword puzzle, word search, or jumble puzzle of the vocabulary Chapter 15, go to: www.glencoe.com/sec/math/t_resources/free/index.php	You can use your completed **Vocabulary Builder** (pages 290–291) to help you solve the puzzle.

15-1 Logic and Truth Tables

Indicate whether the statement is *true* or *false*.

1. A table that lists all truth values of a statement is a truth table. ______

2. $p \to q$ is an example of a disjunction. ______

3. $\sim p \to \sim q$ is the inverse of a conditional statement. ______

4. $p \vee q$ is an example of a conjunction. ______

5. Complete the truth table.

p	q	$\sim p$	$\sim q$	$p \wedge q$	$p \vee q$	$\sim p \to \sim q$
T	T					
T	F					
F	T					
F	F					

15-2

Deductive Reasoning

Draw a conclusion from statements (1) and (2).

6. (1) All functions are relations.

(2) $x = y^2$ is a relation.

7. (1) Integers are rational numbers.

(2) (-6) is an integer.

8. (1) If it is Saturday, I see my friends.

(2) If I see my friends, we laugh.

15-3

Paragraph Proofs

Indicate whether the statement is *true* or *false*.

9. A proof is a logical argument where each statement is backed up by a reason accepted as true.

Write a paragraph proof.

10. **Given:** $m\angle 1 = m\angle 2$; $m\angle 3 = m\angle 4$

Prove: $m\angle 1 + m\angle 4 = 90$

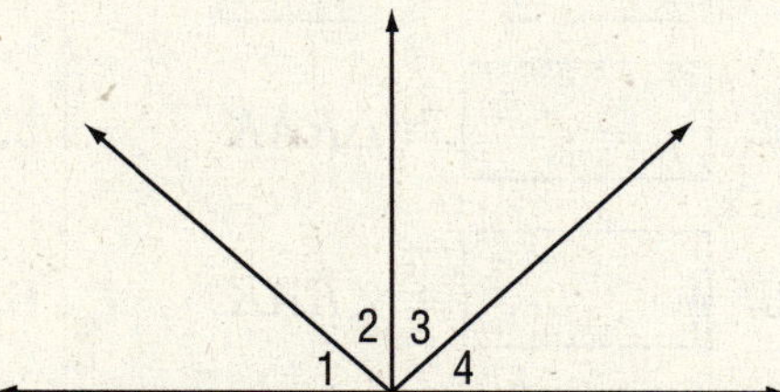

15-4

Preparing for Two-Column Proofs

Complete the statement.

11. A proof containing statements and reasons and is organized by steps is a ________ proof.

Complete the proof.

12. Given: $\overline{AB}$ and $\overline{AR}$ are tangent to circle K.
Prove: $\angle BAK \cong \angle RAK$

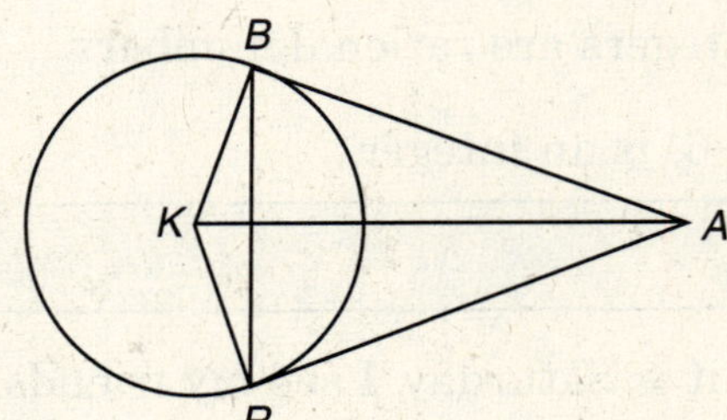

Proof:

Statements	Reasons
1. $\overline{AB}$ and $\overline{AR}$ are tangent to circle K	**1.**
2. ________ $\cong$ ________	**2.** If 2 segments from the same exterior point are tangent to a circle, then they are $\cong$.
3. $\overline{BK} \cong \overline{RK}$	**3.**
4. ________ $\cong$ ________	**4.**
5. ________ $\cong \triangle RAK$	**5.**
6. ________ $\cong \angle RAK$	**6.**

15-5

Two Column Proofs

13. Write a two-column proof.

Given: $\overline{AB}$ is tangent to circle X at B.
$\overline{AC}$ is tangent to circle X at C.
Prove: $\overline{AB} \cong \overline{AC}$

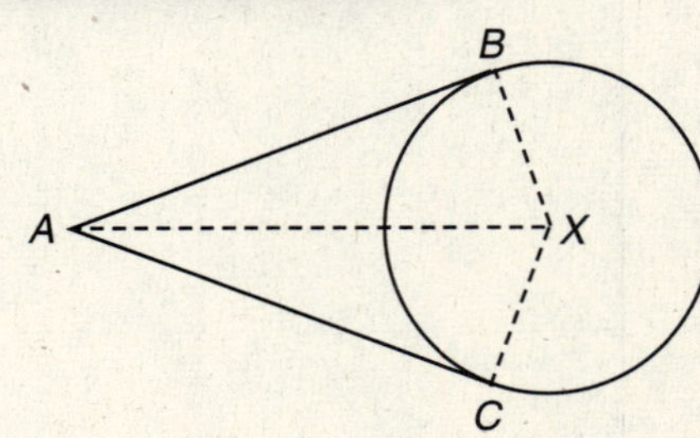

Proof:

Statements	Reasons
1. $\overline{AB}$ is tangent to circle X at B. $\overline{AC}$ is tangent to circle X at C.	1. ______
2. Draw $\overline{BX}$, $\overline{CX}$, and $\overline{AX}$.	2. Through any 2 ______ there is 1 ______.
3. $\angle ABX$ and $\angle ACX$ are ______.	3. If a line is tangent to a circle, then it is $\perp$ to the radius drawn to the point of tangency.
4. $\overline{BX} \cong \overline{CX}$	4. ______
5. ______ $\cong$ ______	5. Reflexive Property
6. $\triangle AXB \cong$ ______	6. HL
7. ______	7. CPCTC

15-6 Coordinate Proofs

Complete the statement.

14. The vertex or center of the figure should be placed on the ______.

15. Position and label a rhombus on a coordinate plane with base r and height t.

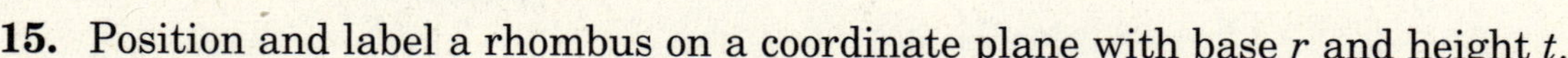

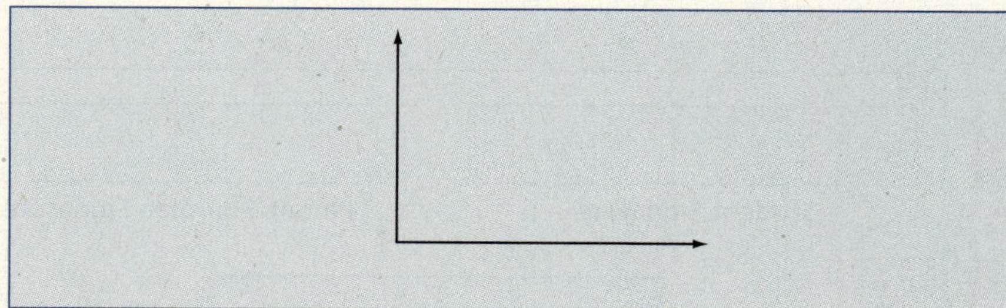

ARE YOU READY FOR THE CHAPTER TEST?

Visit **geomconcepts.com** to access your textbook, more examples, self-check quizzes, and practice tests to help you study the concepts in Chapter 15.

Check the one that applies. Suggestions to help you study are given with each item.

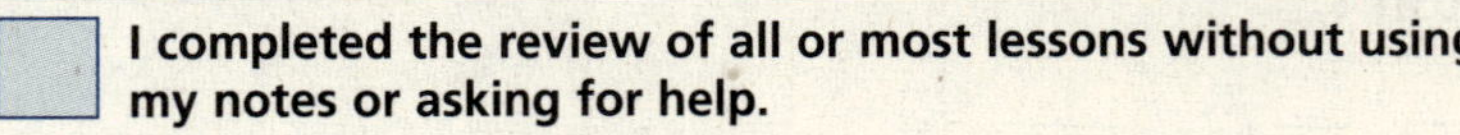

☐ **I completed the review of all or most lessons without using my notes or asking for help.**

- You are probably ready for the Chapter Test.
- You may want to take the Chapter 15 Practice Test on page 671 of your textbook as a final check.

☐ **I used my Foldable or Study Notebook to complete the review of all or most lessons.**

- You should complete the Chapter 15 Study Guide and Review on pages 668–670 of your textbook.
- If you are unsure of any concepts or skills, refer back to the specific lesson(s).
- You may also want to take the Chapter 15 Practice Test on page 671.

☐ **I asked for help from someone else to complete the review of all or most lessons.**

- You should review the examples and concepts in your Study Notebook and Chapter 15 Foldable.
- Then complete the Chapter 15 Study Guide and Review on pages 668–670 of your textbook.
- If you are unsure of any concepts or skills, refer back to the specific lesson(s).
- You may also want to take the Chapter 15 Practice Test on page 671.

Student Signature

Parent/Guardian Signature

Teacher Signature

More Coordinate Graphing and Transformations

Use the instructions below to make a Foldable to help you organize your notes as you study the chapter. You will see Foldable reminders in the margin of this Interactive Study Notebook to help you in taking notes.

Begin with six sheets of graph paper and an $8\frac{1}{2}$" × 11" poster board.

STEP 1 **Staple**
Staple the six sheets of graph paper onto the poster board.

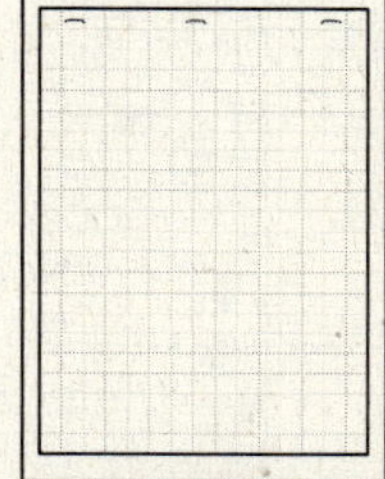

STEP 2 **Label**
Label the six pages with the lesson titles.

NOTE-TAKING TIP: When taking notes, mark anything you do not understand with a question mark. Be sure to ask your instructor to explain the concepts or sections before your next quiz or exam.

BUILD YOUR VOCABULARY

This is an alphabetical list of new vocabulary terms you will learn in Chapter 16. As you complete the study notes for the chapter, you will see Build Your Vocabulary reminders to complete each term's definition or description on these pages. Remember to add the textbook page number in the second column for reference when you study.

Vocabulary Term	Found on Page	Definition	Description or Example
center of rotation			
composition of transformations			
dilation [dye-LAY-shun]			
elimination [ee-LIM-in-AY-shun]			
reflection			
rotation			
substitution [SUB-sti-TOO-shun]			
system of equations			
translation			
turn			

16–1 Solving Systems of Equations by Graphing

BUILD YOUR VOCABULARY (page 314)

A set of two or more equations is called a **system of equations.**

WHAT YOU'LL LEARN

- Solve systems of equations by graphing.

FOLDABLES™

ORGANIZE IT

On the page labeled *Solving Systems of Equations by Graphing,* sketch graphs of systems of equations. Explain why each graph produces the result that it does.

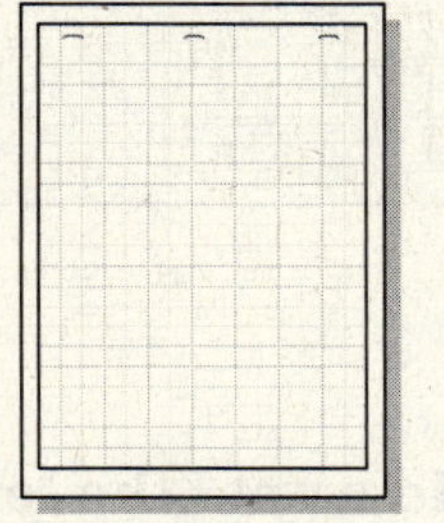

EXAMPLES

Solve each system of equations by graphing.

1 $y = x - 1$
$y = -x + 3$

Find ordered pairs by choosing values for x and finding the corresponding y-values.

$y = x - 1$			
x	$x - 1$	y	(x, y)
3	2	2	(3, 2)
2	1	1	(2, 1)
1	0	0	(1, 0)

$y = -x + 3$			
x	$-x + 3$	y	(x, y)
3	0	0	(3, 0)
2	1	1	(2, 1)
1	2	2	(1, 2)

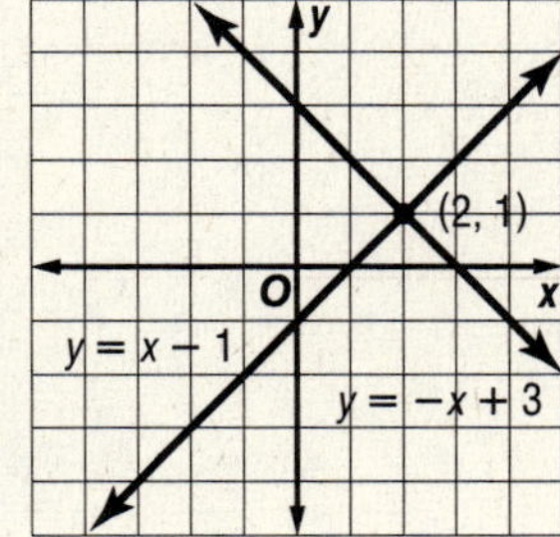

Graph the ordered pairs and draw the graphs of the equations. The graphs intersect at the point whose coordinates are ______. Therefore, the solution of the system of equations is ______.

2 $y = -2x$
$y = -2x + 3$

Use the slope and y-intercept to graph each equation.

Equation	Slope	y-intercept
$y = -2x$	-2	0
$y = -2x + 3$	-2	3

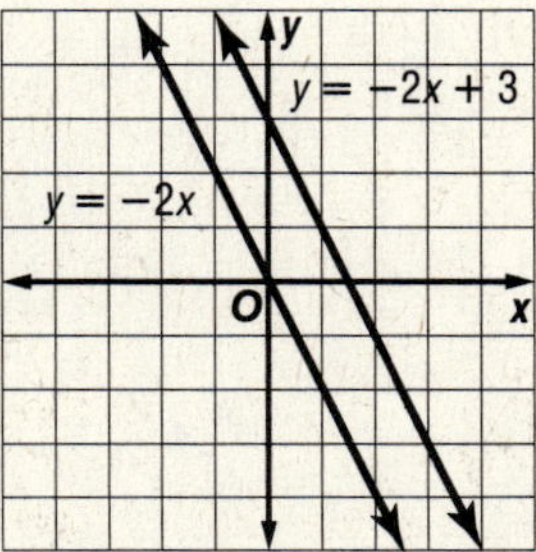

The slope of each line is  so the graphs are  and do not intersect. Therefore, there is ______.

REVIEW IT

Explain how to graph $6x - 2y = 8$ using the slope-intercept method. *(Lesson 4-6)*

____

____

WRITE IT

Explain how to solve a system of equations by graphing.

____

____

____

Your Turn **Solve each system of equations by graphing.**

a. $x - 2y = 2$
$3x + y = 6$

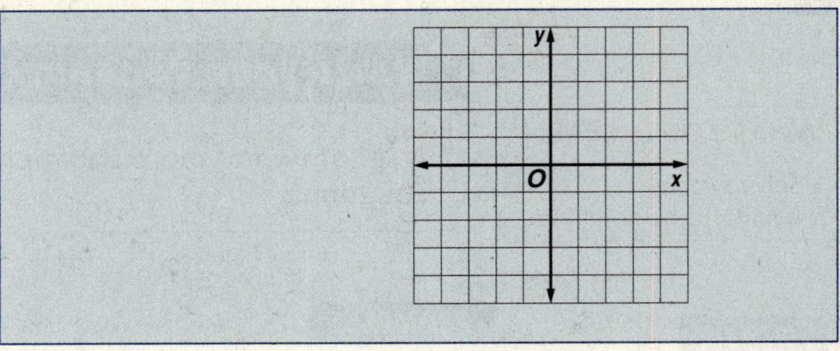

b. $3x + 2y = 12$
$3x + 2y = 6$

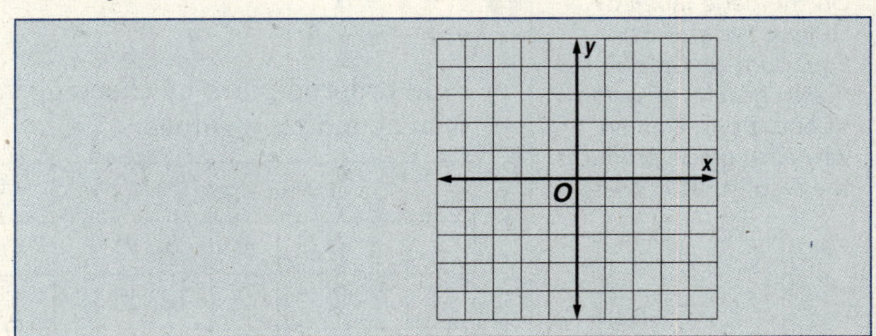

EXAMPLE

3 **Toshiro wants a wildflower garden. He wants the length to be 1.5 times the width and he has 100 meters of fencing to put around the garden. If w represents the width of the garden and ℓ represents the length, solve the system of equations below to find the dimensions of the wildflower garden.**

$\ell = 1.5w$
$2w + 2\ell = 100$

Solve the second equation for ℓ.

$2w + 2\ell = 100$ — The perimeter is 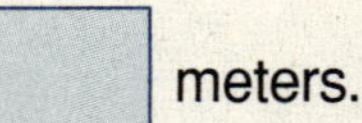 meters.

$2w + 2\ell - 2w = 100 - 2w$ — Subtract  from each side.

[   ] $= 100 - 2w$

$\frac{2\ell}{2} = \frac{100 - 2w}{2}$ — Divide.

$\ell =$ [   ]

Use a graphing calculator to graph the equations ______

and ______ to find the coordinates of the intersection point. *Note that these equations can be written as $y = 1.5x$ and $y = 50 - x$ and then graphed.*

Enter: [◆] [y =] 1.5 [×] [ENTER] 50 [–] [×] [◆] [GRAPH]

Next, use the intersection tool on [F5] to find the coordinates of the point of intersection.

The solution is ______. Since $w =$ ______ and $\ell =$ ______, the width of the garden is ______ meters and the length is ______ meters.

Check your answer by examining the original problem.
Is the length of the garden 1.5 times the width? ✔
Does the garden have a perimeter of 100 meters? ✔
The solution checks.

REMEMBER IT

Check the solution to a system of equations by substituting it into each equation.

Your Turn Ruth wants to enclose an area of her yard for her children to play. She has 72 meters of fence. The length of the play area is 4 meters greater than 3 times the width. What are the dimensions of the play area?

HOMEWORK ASSIGNMENT

Page(s):

Exercises:

16–2 Solving Systems of Equations by Using Algebra

WHAT YOU'LL LEARN

- Solve systems of equations by using the substitution or elimination method.

BUILD YOUR VOCABULARY (page 314)

One algebraic method for solving a system of equations is called **substitution**.

Another algebraic method for solving systems of equations is called **elimination**.

FOLDABLES™

ORGANIZE IT

On the page labeled *Solving Systems of Equations by Using Algebra*, write a system of equations and solve it using substitution and elimination. Explain the process you used with each method.

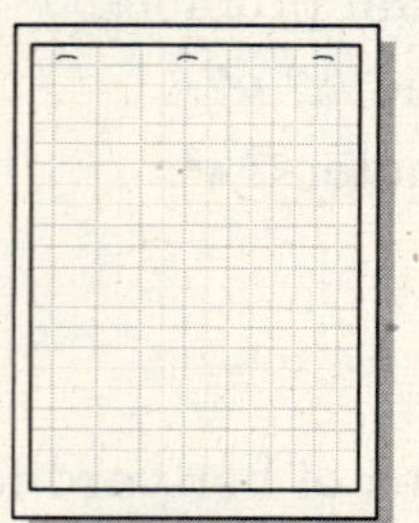

EXAMPLE

1 Use substitution to solve the system of equations.

$\boldsymbol{y = x + 4}$
$\boldsymbol{2x + y = 1}$

Substitute $x + 4$ for y in the second equation.

$2x + y = 1$

$2x + \boxed{} = 1$

$3x + 4 = 1$ Combine like terms.

$3x + 4 - \boxed{} = 1 - \boxed{}$ Subtract $\boxed{}$ from each side.

$3x = -3$

$\frac{3x}{3} = \frac{-3}{3}$ Divide each side by $\boxed{}$.

$x = \boxed{}$ Division Property

Substitute -1 for x in the first equation and solve for y.

$y = (-1) + 4 = \boxed{}$

The solution to this system of equations is $\boxed{}$.

Your Turn Use substitution to solve $2x - y = 4$ and $x = y + 5$.

EXAMPLE

2 Use elimination to solve the system of equations.

$3x - 2y = 4$
$4x + 2y = 10$

$$\begin{array}{r} 3x - 2y = 4 \\ (+)\ 4x + 2y = 10 \\ \hline 7x + 0 = 14 \end{array}$$

Add the equations to eliminate the *y* terms.

$$\frac{7x}{7} = \frac{14}{7}$$

Divide each side by 7.

$x = \square$

The value of x in the solution is $\square$.

Now substitute in either equation to find the value of y.

$3x - 2y = 4$

$3(\square) - 2y = 4$

$\square - 2y = 4$

$6 - 2y - 6 = 4 - 6$ Subtract $\square$ from each side.

$\square = \square$ Subtraction Property

$$\frac{-2y}{-2} = \frac{-2}{-2}$$

Divide each side by $\square$.

$y = \square$

The value of y in the solution is $\square$.

The solution to the system is $\square$.

Your Turn Use elimination to solve $x + y = 7$ and $2x - y = -1$.

EXAMPLE

WRITE IT

Explain the difference between solving a system of equations by substitution or by the elimination method.

____

____

____

____

____

3 **Use elimination to solve the system of equations.**

$$3x + y = 6$$
$$x - 2y = 9$$

$$3x + y = 6 \xrightarrow{(\times 2)} 6x + 2y = 12$$
$$x - 2y = 9 \longrightarrow +\ x - 2y = 9$$

$7x + 0 = 21$ Combine like terms.

$\frac{7x}{7} = \frac{21}{7}$ Divide.

$x =$ ▢

Substitute 3 into either equation to solve for y.

$3x + y = 6$

$3(\ ▢\ ) + y = 6$ Replace x with ▢.

$9 + y = 6$

$9 + y - 9 = 6 - 9$ Subtract ▢ from each side.

$y =$ ▢ Subtraction Property

The solution of this system is ▢.

Your Turn Use elimination to solve $7x + 3y = -1$ and $4x + y = 3$.

▢

HOMEWORK ASSIGNMENT

Page(s):

Exercises:

16–3 Translations

WHAT YOU'LL LEARN

- Investigate and draw translations on a coordinate plane.

BUILD YOUR VOCABULARY (page 314)

A **translation** is a slide of a figure from one position to another.

FOLDABLES

ORGANIZE IT

On the page labeled *Translations*, sketch graphs of several different translations. Explain why each translation produces the result it does.

EXAMPLE

1 **Graph $\triangle LMN$ with vertices $L(0, 3)$, $M(4, 2)$, and $N(-3, -1)$. Then find the coordinates of its vertices if it is translated by (5, 0). Graph the translation image.**

To find the coordinates of the vertices of $\triangle L'M'N'$, add 5 to each x-coordinate and add 0 to each y-coordinate of $\triangle LMN$: $(x + 5, y + 0)$.

$L(0, 3) + (5, 0) \rightarrow L'(0 + 5, 3 + 0) = L'$ ______

$M(4, 2) + (5, 0) \rightarrow M'(4 + 5, 2 + 0) = M'$ ______

$N(-3, -1) + (5, 0) \rightarrow N'(-3 + 5, -1 + 0) = N'$ ______

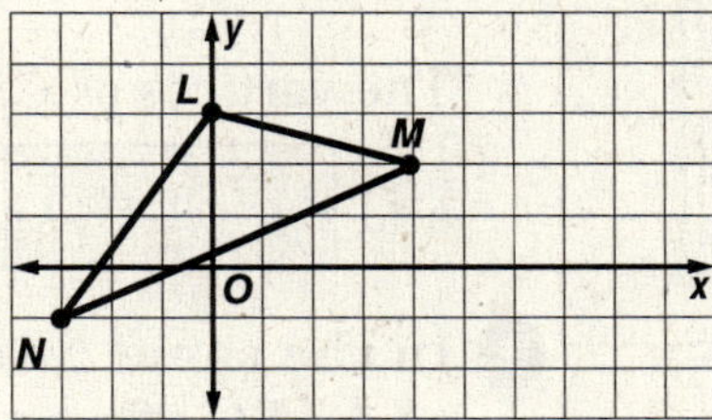

Your Turn Graph $\triangle ABC$ with vertices $A(1, 2)$, $B(-3, -1)$, and $C(2, 1)$. Then find the coordinates of its vertices if it is translated by $(3, -2)$. Graph the translation image.

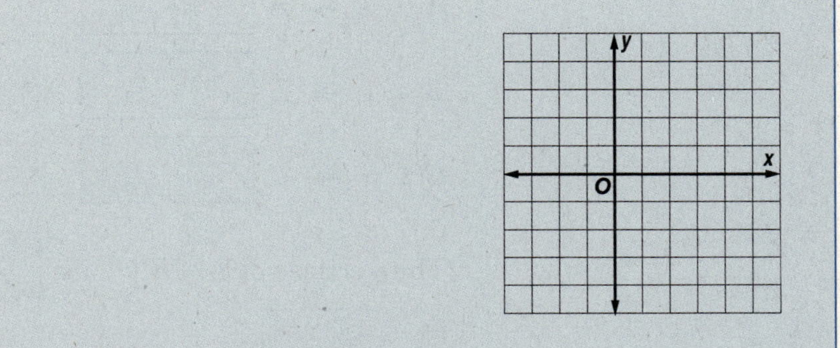

HOMEWORK ASSIGNMENT

Page(s):

Exercises:

16–4 Reflections

WHAT YOU'LL LEARN

- Investigate and draw reflections on a coordinate plane.

BUILD YOUR VOCABULARY (page 314)

A **reflection** is the flip of a figure over a line to produce a mirror image.

FOLDABLES

ORGANIZE IT

On the pages labeled *Reflections*, sketch graphs of several different reflections. Explain why each reflection produces the result it does.

EXAMPLES

1 **Graph $\triangle ABC$ with vertices $A(0, 0)$, $B(4, 1)$, and $C(1, 5)$. Then find the coordinates of its vertices if it is reflected over the x-axis and graph its reflection image.**

To find the coordinates of the vertices of $\triangle A'B'C'$, use the definition of reflection over the x-axis: $(x, y) \rightarrow (x, -y)$.

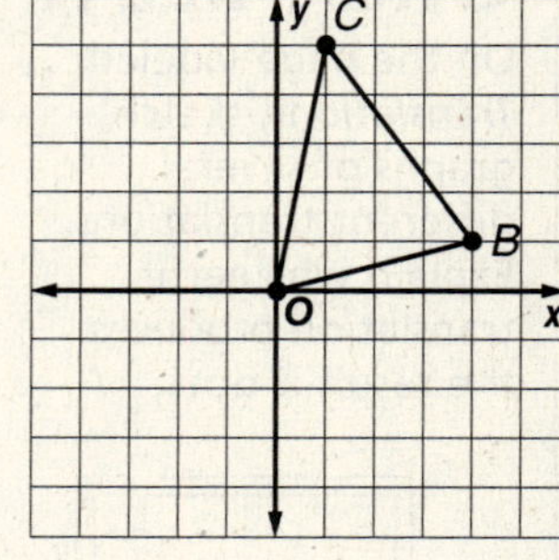

$A(0, 0) \rightarrow A'$ ______

$B(4, 1) \rightarrow B'$ ______

$C(1, 5) \rightarrow C'$ ______

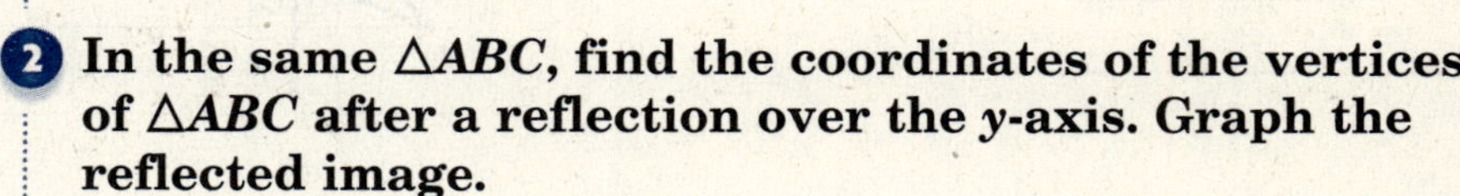

The vertices of $\triangle A'B'C'$ are ______, ______, and ______.

2 **In the same $\triangle ABC$, find the coordinates of the vertices of $\triangle ABC$ after a reflection over the y-axis. Graph the reflected image.**

To find the coordinates of A'', B'', and C'', use the definition of reflection over the y-axis: $(x, y) \rightarrow (-x, y)$.

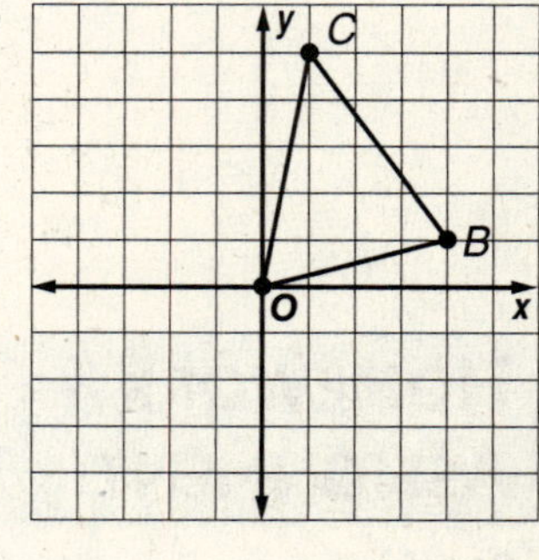

$A(0, 0) \rightarrow A''$ ______

$B(4, 1) \rightarrow B''$ ______

$C(1, 5) \rightarrow C''$ ______

The vertices of $\triangle A''B''C''$ are , , and ______.

Your Turn

a. Graph quadrilateral $QUAD$ with vertices $Q(-3, 3)$, $U(3, 2)$, $A(4, -4)$, and $D(-4, -1)$. Then find the coordinates of its vertices if it is reflected over the y-axis. Graph its reflection image.

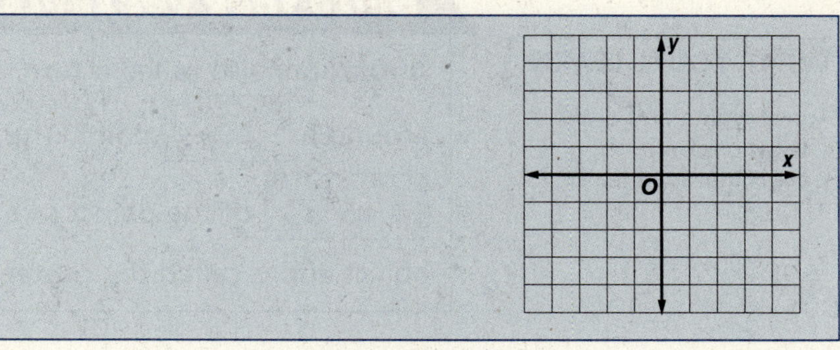

b. Graph $\triangle STU$ with vertices $S(1, 2)$, $T(4, 4)$, and $U(3, -3)$. Then find the coordinates of its vertices if it is reflected over the y-axis and graph its reflection image.

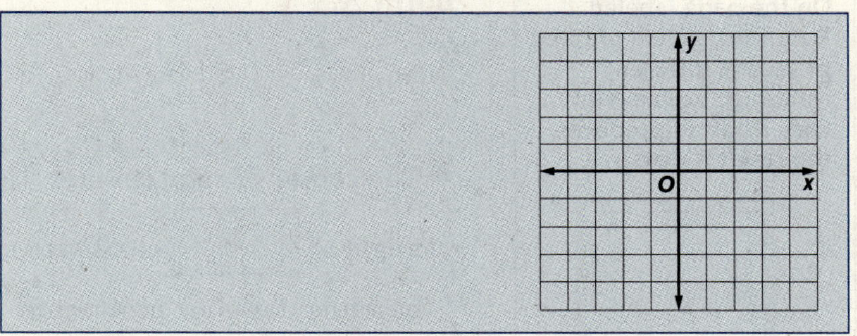

WRITE IT

Reflect a figure over the x-axis and then reflect its image over the y-axis. Is this double reflection the same as a translation? Explain.

____

____

____

____

HOMEWORK ASSIGNMENT

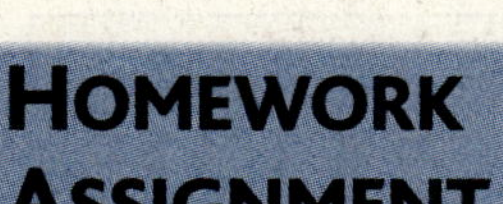

Page(s): ______________________

Exercises: ______________________

16–5 Rotations

What You'll Learn

- Investigate and draw rotations on a coordinate plane.

Build Your Vocabulary (page 314)

A **rotation**, also called **a turn**, is a movement of a figure around a ______ point. The fixed point may be in the ______ of the object or a point ______ the object and is called the **center of rotation**.

Foldables™

Organize It

On the page labeled *Rotations*, sketch graphs of several different rotations. Explain why each rotation produces the result it does.

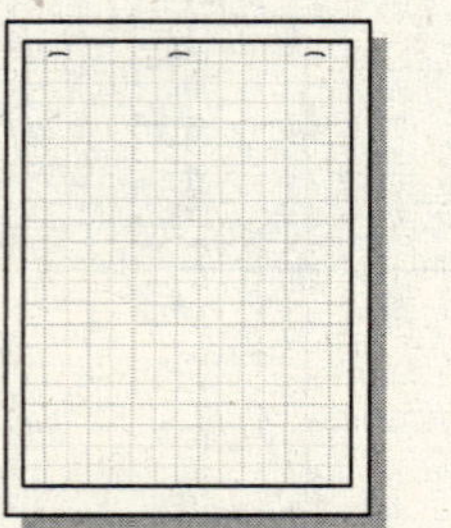

Example

1 Rotate $\triangle ABC$ 270° clockwise about point A.

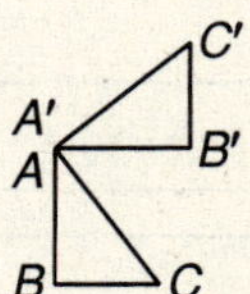

- The center of rotation is A. Use a protractor to draw an angle of ______ clockwise about point A, using $\overline{AB}$ as a baseline for your protractor.
- Draw segment $\overline{A'B'}$ ______ to $\overline{AB}$.
- Trace the figure on a piece of paper and rotate the top paper clockwise, until the figure is rotated ______ clockwise.
- Draw $\triangle A'B'C'$ congruent to $\triangle ABC$.

Your Turn Rotate $\triangle XYZ$ 60° counterclockwise about point Y.

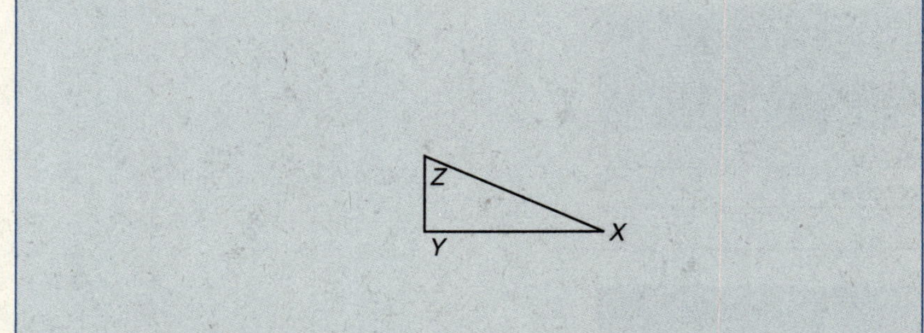

EXAMPLE

2 **Graph $\triangle XYZ$ with vertices $X(-2, 1)$, $Y(2, -3)$, and $Z(3, 5)$. Then find the coordinates of the vertices after the triangle is rotated 180° clockwise about the origin. Graph the rotation image.**

- Draw a segment from the origin to point X.
- Use a protractor to reproduce $\overline{OX}$ at a 180° angle so that $OX = OX'$.
- Repeat this procedure with points Y and Z.

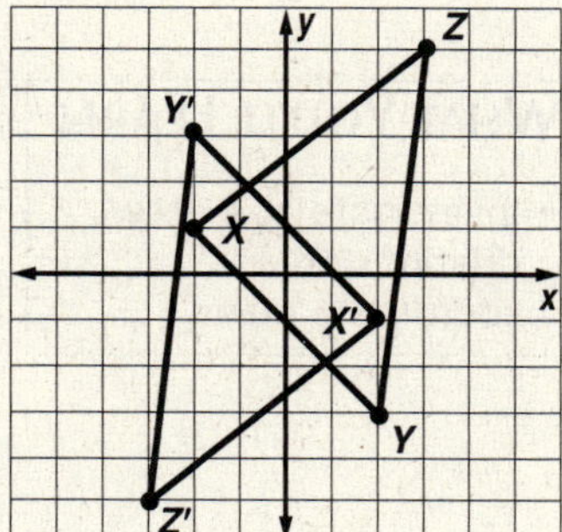

The rotation image $\triangle X'Y'Z'$ has vertices X' ______, Y' ______, and Z' ______.

Your Turn Rotate $\triangle ABC$ 90° counterclockwise around the origin. The vertices are $A(0, 4)$, $B(3, 1)$, and $C(4, 3)$.

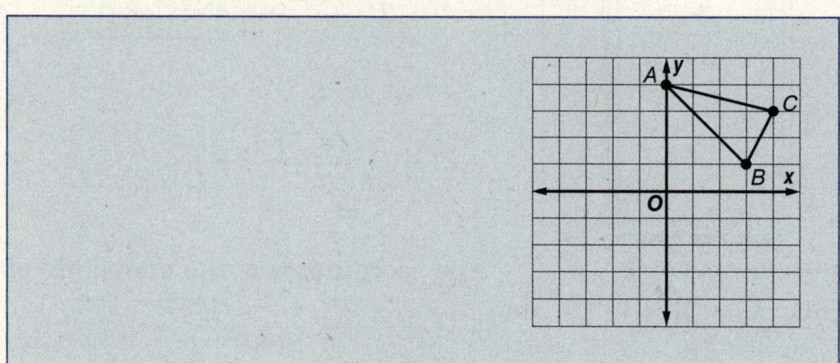

HOMEWORK ASSIGNMENT

Page(s):

Exercises:

16–6 Dilations

WHAT YOU'LL LEARN

- Investigate and draw dilations on a coordinate plane.

BUILD YOUR VOCABULARY (page 314)

A **dilation** is a transformation that alters the size of a figure, but not its shape. It enlarges or reduces a figure by a ______ k.

FOLDABLES

ORGANIZE IT

On the page labeled *Dilations*, sketch graphs of several different dilations. Explain why each dilation produces the result it does.

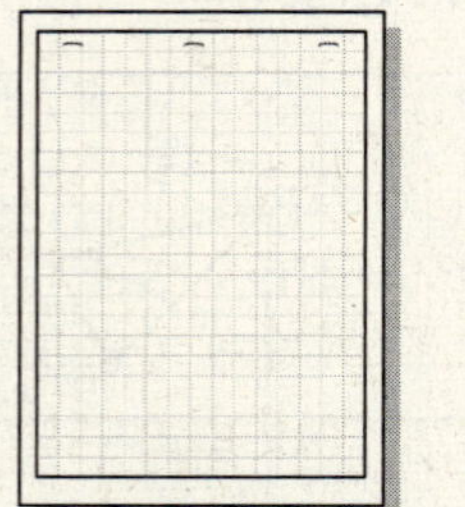

EXAMPLE

1 **Graph $\overline{AB}$ with vertices $A(0, 2)$ and $B(2, 1)$. Then find the coordinates of the dilation image of $\overline{AB}$ with a scale factor of 3, and graph its dilation image.**

Since $k > 1$, this is an enlargement. To find the dilation image, multiply each coordinate in the ordered pairs by 3.

preimage ⟶ **image**

$A(0, 2) \xrightarrow{(\times 3)} A'$ ______

$B(2, 1) \xrightarrow{(\times 3)} B'$ ______

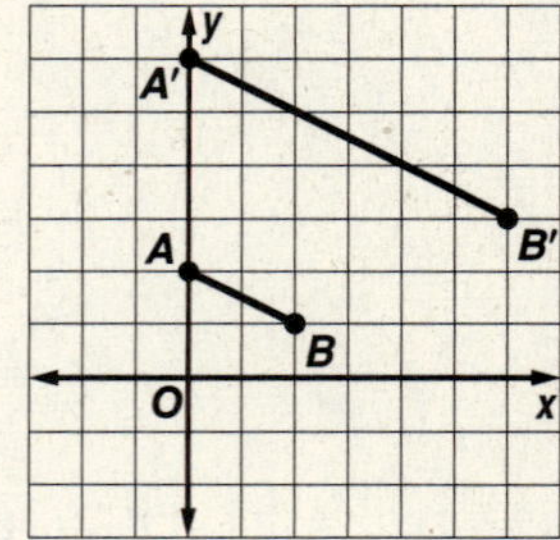

The coordinates of the endpoints of the dilation image are

A' ______ and B' ______.

Your Turn Graph $\triangle JKL$ with vertices $J(1, -2)$, $K(4, -3)$, and $L(6, -1)$. Then find the coordinates of the dilation image of $\triangle JKL$ with a scale factor of 2, and graph its dilation.

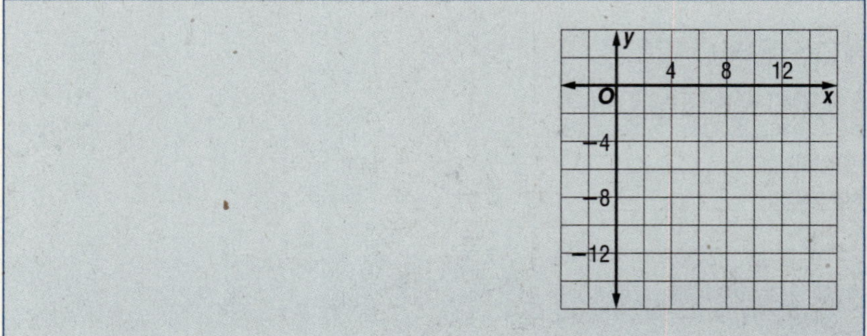

EXAMPLE

WRITE IT

How can you determine whether a dilation is a reduction or an enlargement?

____

____

____

____

2 **Graph $\triangle DEF$ with vertices $D(3, 3)$, $E(0, -3)$, and $F(-6, 3)$. Then find the coordinates of the dilation image with a scale factor of $\frac{1}{3}$ and graph its dilation image.**

Since $k < 1$, this is a reduction.

preimage ⟶ image

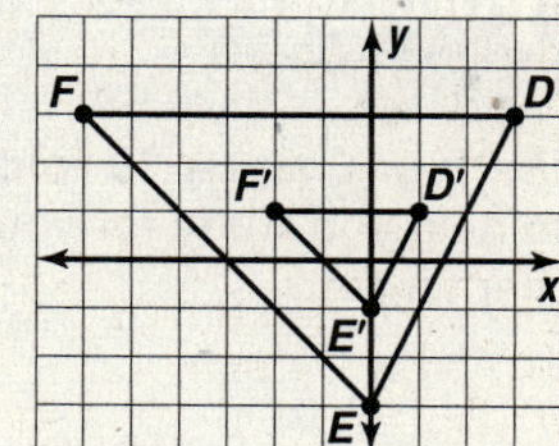

$D(3, 3) \xrightarrow{\times \frac{1}{3}} D'$ [ ]

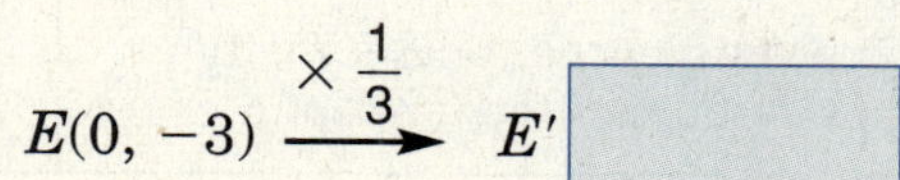

$E(0, -3) \xrightarrow{\times \frac{1}{3}} E'$ [ ]

$F(-6, 3) \xrightarrow{\times \frac{1}{3}} F'$ [ ]

The coordinates of the vertices of the dilation image are

D' [ ], E' [ ], and F' [ ].

Your Turn Graph quadrilateral $MNOP$ with vertices $M(1, 2)$, $N(3, 3)$, $O(3, 5)$, and $P(1, 4)$. Then find the coordinates of the dilation image with a scale factor of $\frac{2}{3}$ and graph its dilation image.

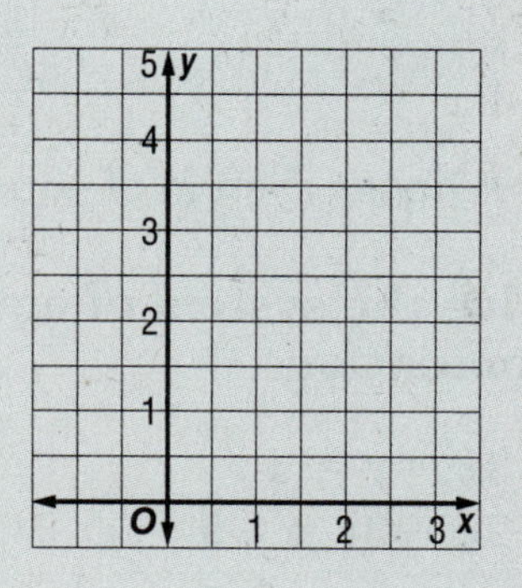

HOMEWORK ASSIGNMENT

Page(s):

Exercises:

CHAPTER 16

BRINGING IT ALL TOGETHER

STUDY GUIDE

FOLDABLES™	VOCABULARY PUZZLEMAKER	BUILD YOUR VOCABULARY
Use your **Chapter 16 Foldable** to help you study for your chapter test.	To make a crossword puzzle, word search, or jumble puzzle of the vocabulary words in Chapter 16, go to: www.glencoe.com/sec/math/t_resources/free/index.php	You can use your completed **Vocabulary Builder** (page 314) to help you solve the puzzle.

16-1 Solving Systems of Equations by Graphing

Solve each system of equations by graphing.

1. $x - y = 6$
$y = 9$

2. $x + y = 27$
$3x - y = 41$

3. $y = 4x + 2$
$12x - 3y = 9$

16-2 Solving Systems of Equations by Using Algebra

Complete each statement.

4. Substitution and elimination are methods for solving

.

5. A linear system of equations can have at most  solution.

Solve the system of equations using substitution or elimination.

6. $3x - y = 4$
$2x - 3y = -9$

7. $y = 3x - 8$
$y = 4 - x$

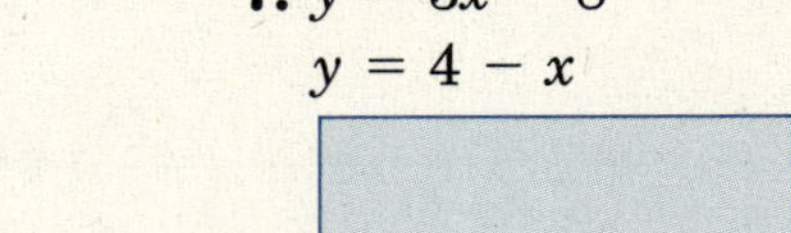

8. $2x + 7y = 3$
$x = 1 - 4y$

9. $3x - 5y = 11$
$x - 3y = 1$

16-3

Translations

Complete the statement.

10. When a figure is moved from one position to another without turning, it is called a ____________.

Find the coordinates of the vertices after the translation. Graph each preimage and image.

11. rectangle $WXZY$ with vertices $W(-2, -2)$, $X(-2, -10)$, $Z(-7, -10)$, and $Y(-7, -2)$ translated $(6, 9)$

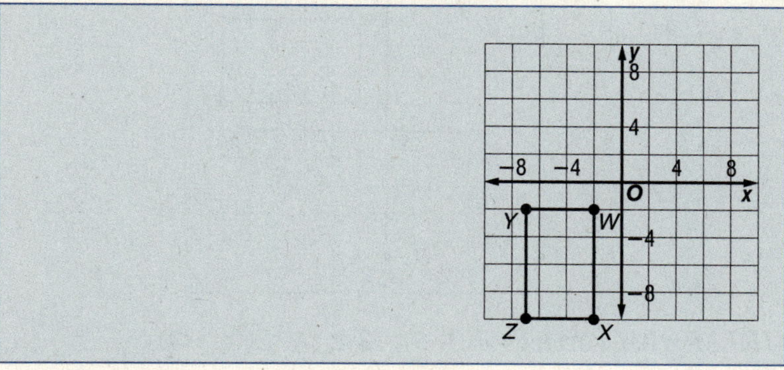

12. $\triangle ABC$ with vertices $A(4, 0)$, $B(2, -1)$, and $C(0, 1)$ translated $(0, -4)$

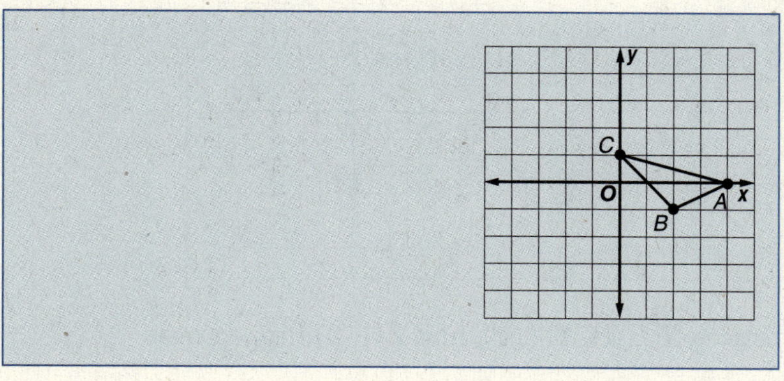

13. $\triangle JKL$ with vertices $J(-5, -2)$, $K(-2, 7)$, and $L(1, -6)$ translated $(6, 2)$

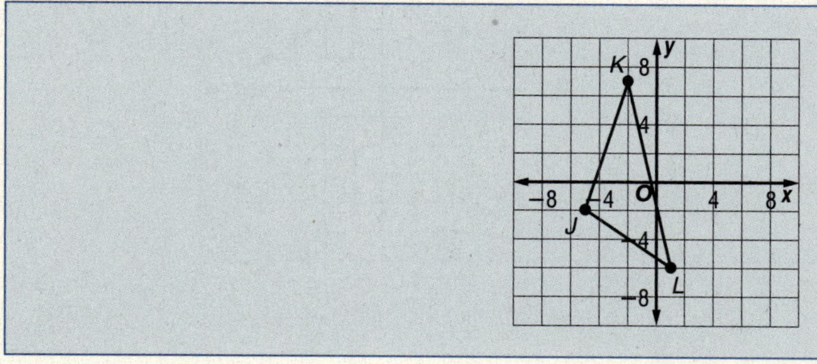

16-4 Reflections

Complete the statement.

14. A ______ is a flip of a figure over a line.

Find the coordinates of the vertices after the reflection. Graph each preimage and image.

15. quadrilateral $ABCD$ with vertices $A(1, 1)$, $B(1, 4)$, $C(6, 4)$, and $D(6, 1)$ flipped over the x-axis

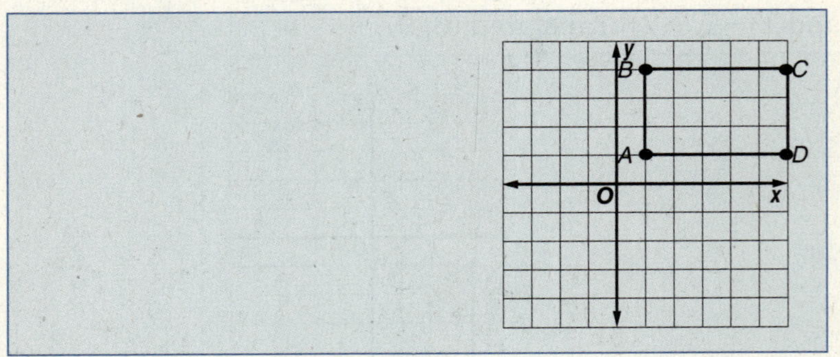

16. quadrilateral $JKLM$ with vertices $J(3, 5)$, $K(4, 0)$, $L(0, -3)$, and $M(-1, 2)$ flipped over the y-axis

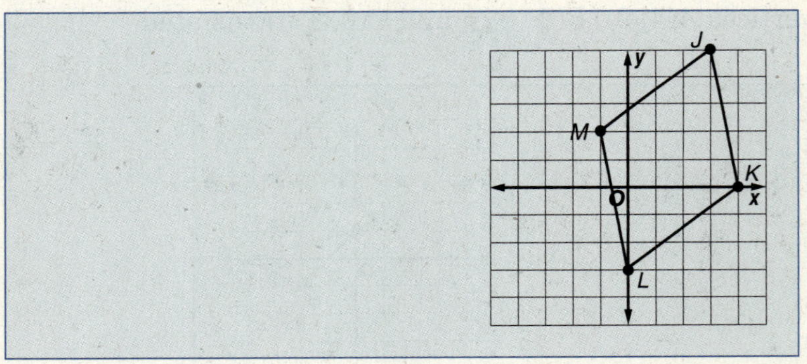

17. $\triangle XYZ$ with vertices $X(1, 1)$, $Y(4, 1)$, and $Z(1, 3)$ flipped over the x-axis

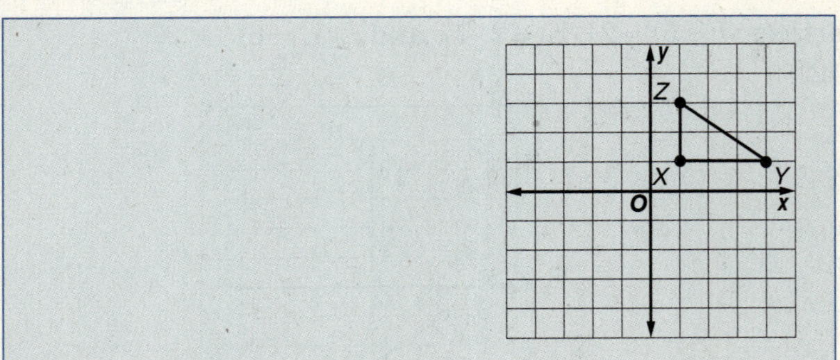

16-5

Rotations

Find the coordinates of the vertices after a rotation about the origin. Graph the preimage and image.

18. $\triangle RST$ with vertices $R(-4, 1)$, $S(-1, 5)$, and $T(-6, 9)$ rotated 90° counterclockwise

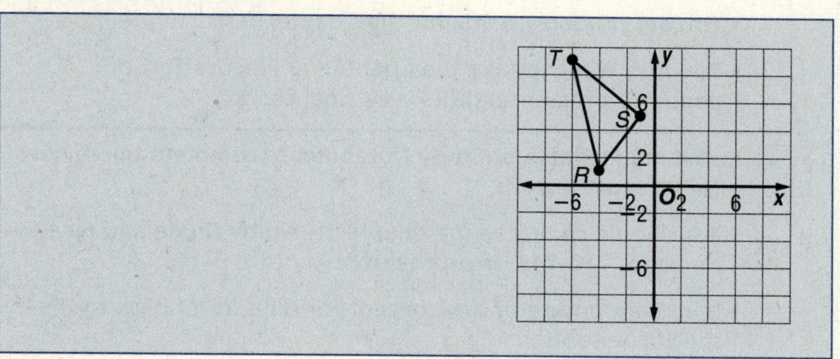

16-6

Dilations

Underline the best term to complete the statement.

19. A [dilation/rotation] alters the size of a figure but does not change its shape.

20. A figure is [reduced/enlarged] in a dilation if the scale factor is between 0 and 1.

Find the coordinates of the dilation image for the given scale factor. Graph the preimage and image.

21. quadrilateral $STUV$ with vertices $S(2, 1)$, $T(0, 2)$, $U(-2, 0)$, and $V(0, 0)$ and scale factor 3

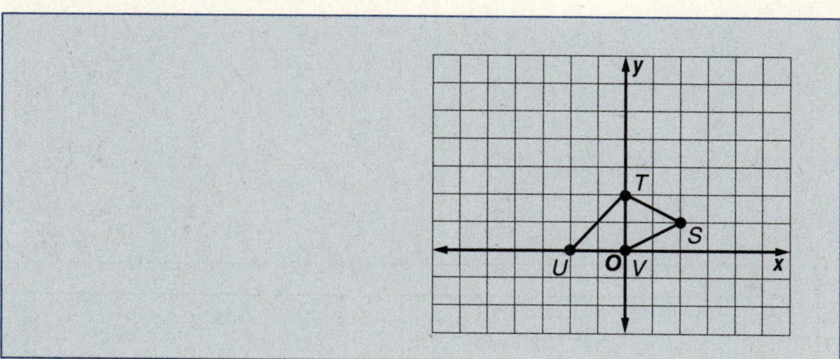

ARE YOU READY FOR THE CHAPTER TEST?

Checklist

Visit **geomconcepts.net** to access your textbook, more examples, self-check quizzes, and practice tests to help you study the concepts in Chapter 16.

Check the one that applies. Suggestions to help you study are given with each item.

☐ **I completed the review of all or most lessons without using my notes or asking for help.**

- You are probably ready for the Chapter Test.
- You may want to take the Chapter 16 Practice Test on page 713 of your textbook as a final check.

☐ **I used my Foldable or Study Notebook to complete the review of all or most lessons.**

- You should complete the Chapter 16 Study Guide and Review on pages 710–712 of your textbook.
- If you are unsure of any concepts or skills, refer back to the specific lesson(s).
- You may also want to take the Chapter 16 Practice Test on page 713 of your textbook.

☐ **I asked for help from someone else to complete the review of all or most lessons.**

- You should review the examples and concepts in your Study Notebook and Chapter 16 Foldable.
- Then complete the Chapter 16 Study Guide and Review on pages 710–712 of your textbook.
- If you are unsure of any concepts or skills, refer back to the specific lesson(s).
- You may also want to take the Chapter 16 Practice Test on page 713 of your textbook.

Student Signature

Parent/Guardian Signature

Teacher Signature